GREEN'S FUNCTIONS *and* CONDENSED MATTER

Techniques of Physics

Editors

N. H. MARCH

Department of Theoretical Chemistry, University of Oxford, Oxford, England

H. N. DAGLISH

Post Offices Research Centre, Martlesham Heath, Ipswich, England

Techniques of physics find wide application in biology, medicine, engineering and technology generally. This series is devoted to techniques which have found and are finding application. The aim is to clarify the principles of each technique, to emphasize and illustrate the applications and to draw attention to new fields of possible employment.

1. D. C. Champeney: Fourier Transforms and their Physical Applications.
2. J. B. Pendry: Low Energy Electron Diffraction.
3. K. G. Beauchamp: Walsh Functions and their Applications.
4. V. Cappellini, A. G. Constantinides and P. Emiliani: Digital Filters and their Applications.
5. G. Rickayzen: Green's Functions and Condensed Matter.

GREEN'S FUNCTIONS *and* CONDENSED MATTER

G. RICKAYZEN

The University of Kent at Canterbury

1980

ACADEMIC PRESS
A Subsidiary of Harcourt Brace Jovanovich, Publishers
London New York Toronto Sydney San Francisco

ACADEMIC PRESS INC. (LONDON) LTD.
24/28 Oval Road
London NW1

United States Edition published by
ACADEMIC PRESS INC.
111 Fifth Avenue
New York, New York 10003

British Library Cataloguing in Publication Data
Rickayzen, G
Green's functions and condensed matter. –
(Techniques of physics; 5 ISSN 0309-5392).
1. Matter – Properties
2. Green's functions
3. Mathematical physics
I. Title II. Series
530.4 QC 171.2 80-49951

ISBN 0-12-587950-4

Printed in Great Britain by
Alden Press, Oxford, London and Northampton

Preface

Green's functions are now one of the well established tools for dealing with problems in condensed matter. Like wave functions they contain the information necessary to explain observations but compared with wave functions they carry little surplus information. They therefore provide an economical description of matter, especially condensed matter. However although the description is economical it can be quite complex and sophisticated. Green's functions can easily be used to describe systems in which quasi-particles with long lifetimes exist but they can also be used where this is not the case. Furthermore their equations of motion suggest methods of approximation which are based not solely on the smallness of an interaction parameter. They are therefore useful both for finding solutions and embodying them and they have stood the test of time.

This book has developed from lectures given at various places and times but particularly from a course given to research students and staff, both theoretical and experimental at the University of Kent at Canterbury. I am very grateful to my colleagues and former students for their comments and criticisms which I believe have improved the presentation. The readership at which the book is aimed is the same. It is assumed that the reader has a good working knowledge of quantum mechanics, including statistical quantum mechanics and some familiarity with the occupation number representation, although the main formulae and the correspondence with wave mechanics are provided in an Appendix.

Although Green's functions have undergone an enormous development since Green introduced them into the theory of electromagnetism in 1828 they are conceptually unchanged and serve the same purpose. We therefore begin the discussion with a description of Green's function in classical physics from a modern point of view. From there we progress to the definition and properties of Green's functions in quantum physics and show how they arise naturally in the response of quantum systems to external fields. Most of the book is taken up with the application of Green's functions. This could not possibly be comprehensive and we have chosen examples which show the variety of their uses. Most of the applications are to fermions but the application to superfluid ^{4}He is concerned with bosons and that to magnetism with spin operators. The discussion of Landau's theory of Fermi liquids illustrates their usefulness in a description of quasi-particles. The discussion of the Coulomb interaction displays their usefulness in describing collective modes. The chapters on superconductivity, superfluidity and magnetism show how Green's functions are used to describe phase transitions. There are a number of illustrations of the use of Green's functions in describing irreversible processes and in obtaining transport coefficients. There are also examples of the use of Green's functions in describing disordered systems where at the present time they provide the only satisfactory general analytic technique. An example of the failure of the quasi-particle picture is provided by the theory of the strong-coupling superconductors. In the final chapter it is shown, albeit briefly, how Green's functions play a part in the renormalization group method of studying critical phenomena. It is to be hoped that these examples will whet the reader's appetite.

An enterprise of this kind cannot be accomplished without the help of others to whom I am grateful. In particular I should like to thank Dr Alan Evans, Mr John Ashby and Mr Giancarlo Jug for reading and criticizing the manuscript, Mrs Betty Jones, Mrs Joy Rampe, Mrs Gill Rickayzen and Miss Mary Watts for the typing of the manuscript and Mr Dave Murray and Mr Asher Rickayzen for help with the illustrations.

Contents

Chapter 1

Introduction or "Why Green's Functions?"

§1.1 Classical Green's functions

Anyone who has used the Coulomb potential due to a point charge has used a Green's function. If the charge is e, situated at the point with position vector $\boldsymbol{r}'$, the electric potential at a second point $\boldsymbol{r}$ is

$$\phi(\boldsymbol{r}) = \frac{e}{4\pi\epsilon_0|\boldsymbol{r}-\boldsymbol{r}'|}.$$

From this potential it is possible to write down the potential due to a number of charges or, in the limit of a continuous distribution, that due to a charge distribution $\rho(\boldsymbol{r})$. The resulting potential is

$$\phi(\boldsymbol{r}) = \int d^3r' \frac{\rho(\boldsymbol{r}')}{4\pi\epsilon_0 R}, \qquad R = |\boldsymbol{r}-\boldsymbol{r}'|, \tag{1.1}$$

and satisfies Poisson's equation

$$\nabla^2\phi(\boldsymbol{r}) = -\rho(\boldsymbol{r})/\epsilon_0. \tag{1.2}$$

Consequently, the Green's function,

$$G(R) = (4\pi\epsilon_0 R)^{-1}, \tag{1.3}$$

enables the solution of Poisson's equation for an arbitrary charge density to be written as a quadrature. This illustrates the great power and usefulness of the Green's function. In general, Green's functions provide explicit solutions of differential equations.

To exemplify some of the methods that are used in this book, let us prove that (1.1) is a solution of equation (1.2). First, we note that the Green's function is the Coulomb potential due to a unit charge situated at $\boldsymbol{r}'$. Now the charge density $\rho(\boldsymbol{r})$ of a unit point charge at $\boldsymbol{r}'$ is zero if $\boldsymbol{r} \neq \boldsymbol{r}'$ and has an integral over all space which is the total charge present, namely unity. Hence,

$$\rho(\boldsymbol{r}) = 0, \qquad \boldsymbol{r} = \boldsymbol{r}', \qquad \int d^3 r \rho(\boldsymbol{r}) = 1.$$

The function which satisfies these equations is Dirac's three-dimensional delta function, $\delta(\boldsymbol{r}-\boldsymbol{r}')$. Hence

$$\rho(\boldsymbol{r}) = \delta(\boldsymbol{r}-\boldsymbol{r}') = \delta(x-x')\delta(y-y')\delta(z-z').$$

The Green's function (1.3) must therefore satisfy Poisson's equation,

$$\nabla^2 G(R) = -\delta(\boldsymbol{r}-\boldsymbol{r}')/\epsilon_0 = -\delta(\boldsymbol{R})/\epsilon_0, \qquad \boldsymbol{R} = \boldsymbol{r}-\boldsymbol{r}'. \tag{1.4}$$

We now show that the Green's function (1.3) is indeed a solution of equation (1.4). We usually have to satisfy the boundary condition that the potential tends to zero at infinity. We can, therefore, introduce the Fourier transforms,

$$\hat{G}(\boldsymbol{k}) = \int d^3 R \exp(-i\boldsymbol{k}.\boldsymbol{R}) G(R)$$

and

$$\int d^3 R \exp(-i\boldsymbol{k}.\boldsymbol{R})\delta(\boldsymbol{R}) = 1. \tag{1.5}$$

Equation (1.5) is transformed to

$$-k^2 \hat{G}(\boldsymbol{k}) = -\epsilon_0^{-1},$$

whence

$$\hat{G}(\boldsymbol{k}) = (\epsilon_0 k^2)^{-1},$$

$$G(\boldsymbol{R}) = \frac{1}{(2\pi)^3}\int d^3 k \frac{\exp(i\boldsymbol{k}.\boldsymbol{R})}{\epsilon_0 k^2}$$

$$= (4\pi\epsilon_0 R)^{-1}. \tag{1.6}$$

We confirm that $G(R)$ satisfies Poisson's equation for a point charge.

Now the potential given by equation (1.3) can be written

$$\phi(\boldsymbol{r}) = \int d^3 r' G(R)\rho(\boldsymbol{r}'). \tag{1.7}$$

Hence

$$\nabla^2\phi(\boldsymbol{r}) = \int d^3r'\nabla^2 G(R)\rho(\boldsymbol{r}')$$

and we can use equation (1.4) to obtain

$$\begin{aligned}\nabla^2\phi(\boldsymbol{r}) &= -\int d^3r'[\delta(\boldsymbol{r}-\boldsymbol{r}')/\epsilon_0]\,\rho(\boldsymbol{r}')\\ &= -\rho(\boldsymbol{r})/\epsilon_0.\end{aligned}$$

This confirms that $\phi(\boldsymbol{r})$ satisfies Poisson's equation.

The solution is unique because in using the Fourier transform we have assumed that $\phi(\boldsymbol{r})$ tends to zero sufficiently rapidly as $|r|\to\infty$. This ensures that $\phi(\boldsymbol{r})$ automatically satisfies the spatial boundary conditions, an advantage of the use of a Green's function.

A second familiar example of the use of Green's functions comes from the solution of Maxwell's equations. As these involve time as well as space we consider them in some detail, too. We look for the solution of the equations for the potentials $\boldsymbol{A}$, ϕ,

$$\begin{aligned}\Box\boldsymbol{A} &\equiv \left[\nabla^2 - \frac{1}{c^2}\frac{\partial^2}{\partial t^2}\right]\boldsymbol{A}(\boldsymbol{r},t) = -\mu\boldsymbol{j}(\boldsymbol{r},t),\\ \Box\phi &\equiv \left[\nabla^2 - \frac{1}{c^2}\frac{\partial^2}{\partial t^2}\right]\phi(\boldsymbol{r},t) = -\rho(\boldsymbol{r},t)/\epsilon_0.\end{aligned} \tag{1.8}$$

Both equations can be solved using the Green's function $G(\boldsymbol{R},s)$ which satisfies

$$\Box G(\boldsymbol{r}-\boldsymbol{r}',t-t') = -\delta(\boldsymbol{r}-\boldsymbol{r}')\delta(t-t') \equiv -\delta(\boldsymbol{R})\delta(s) \tag{1.9}$$

We again use the spatial Fourier transform $G(\boldsymbol{k},s)$ which now satisfies

$$\frac{1}{c^2}\frac{\partial^2 G(\boldsymbol{k},s)}{\partial s^2} + k^2 G(\boldsymbol{k},s) = \delta(s). \tag{1.10}$$

Here $\boldsymbol{k}$ is a parameter. This equation has the same form as for a simple harmonic oscillator with a unit impulsive force at time $s=0$. G plays the role of displacement, c^{-2} is the inertia and k^2 the "spring" constant. The solution depends on the "displacement" G and "velocity" $\partial G/\partial s$, at the initial time $s=0$. It is a general feature of problems involving time that we need to define boundary conditions which, in turn, are determined by the physical nature of the problem. Thus a given set of differential equations is satisfied by many different Green's functions, and the physics of the problem is required to determine the appropriate one to use. This means that one deals with the boundary conditions at the stage of determining the Green's function (independently of ρ and j in this case), and once we have found G we can find $\boldsymbol{A}$ and ϕ for a large number of different problems. This is another advantage of the use of Green's functions.

For the moment let us now solve equation (1.10) with $G = 0 = \partial G/\partial s$ at $s = 0$. Since the impulse is of unit strength, the momentum of the oscillator is unity immediately after $s = 0$; since its "mass" is c^{-2}, the "velocity" is $\partial G/\partial s = c^2$. Formally this result can be obtained by the following argument. Since $\partial G/\partial s$ exists G is continuous at $s = 0$. If equation (1.10) is integrated from $s = -\epsilon(< 0)$ to $s = \epsilon$, one finds

$$k^2 \int_{-\epsilon}^{\epsilon} ds\, G(\boldsymbol{k}, s) + \frac{1}{c^2}\frac{\partial G}{\partial s}\bigg|_{-\epsilon}^{\epsilon} = 1.$$

Now let ϵ tend to zero. Since G is continuous the integral tends to zero. Hence

$$\lim_{\epsilon \to +0} \frac{\partial G}{\partial s}(\boldsymbol{k}, \epsilon) = c^2$$

as we argued before.

The solution of equation (1.10) for $s > 0$ and $G = 0$, $\partial G/\partial s = c^2$ at $s = 0$ is then

$$G(\boldsymbol{k}, s) = (c/k) \sin(kcs).$$

We can also include the fact that $G(\boldsymbol{k}, s)$ is zero for $s < 0$ by writing

$$G(\boldsymbol{k}, s) = (c/k) \sin(kcs)\theta(s),$$

where the theta-function is defined by

$$\theta(s) = \begin{cases} 1, & s \geqslant 0 \\ 0, & s < 0. \end{cases}$$

The spatial Fourier transform of $G(\boldsymbol{k}, s)$, and therefore a solution of equation (1.10), is

$$\begin{aligned} G(\boldsymbol{R}, s) &= c\theta(s) \int \frac{d^3k}{(2\pi)^3} \exp(i\boldsymbol{k}\,.\,\boldsymbol{R}) \frac{\sin(kcs)}{k} \\ &= \frac{c\theta(s)}{2\pi^2 R} \int_0^{\infty} dk \sin(kR) \sin(kcs) \\ &= \frac{c\theta(s)}{8\pi^2 R} \int_{-\infty}^{\infty} dk \{\cos[k(R - cs)] - \cos[k(R + cs)]\} \\ &= \frac{c\theta(s)}{8\pi^2 R} \operatorname{Re} \int_{-\infty}^{\infty} dk \{\exp[ik(R - cs)] - \exp[ik(R + cs)]\} \end{aligned}$$

$$= \frac{c\theta(s)}{4\pi R}\{\delta(R-cs)-\delta(R+cs)\}$$

where we have used the Fourier transform of the δ-function in the last step. Since R and s are positive, the last term is zero, $\theta(s)$ is redundant in the first, and

$$G(\boldsymbol{R}, s) = \frac{c}{4\pi R}\,\delta(R-cs).$$

Hence a pulse at one point produces an effect at another point a distance R away, at a later time R/c. The solution G, therefore, describes the retarded interaction. The corresponding solution of equation (1.4) is

$$\begin{aligned}\phi(\boldsymbol{r}, t) &= \frac{1}{\epsilon_0}\int G(\boldsymbol{r}-\boldsymbol{r}', t-t')\rho(\boldsymbol{r}', t')\,d^3r'dt'\\ &= \frac{c}{4\pi\epsilon_0}\int d^3r'\int \frac{dt'\delta(R-cs)}{R}\,\rho(r', t')\\ &= \frac{c}{4\pi\epsilon_0}\int \frac{d^3r'}{R}\,\rho\left(\boldsymbol{r}', t-\frac{R}{c}\right), \end{aligned} \tag{1.11}$$

the usual form of the retarded potential. We can prove that $\phi(\boldsymbol{r}, t)$ does indeed satisfy (1.8) by acting on it with the D'Alembertian. Thus

$$\begin{aligned}\Box\phi &= \Box\int d^3r'dt'G(\boldsymbol{r}-\boldsymbol{r}', t-t')\rho(\boldsymbol{r}', t')/\epsilon_0\\ &= \int d^3r'dt'\Box G(r-r', t-t')\rho(\boldsymbol{r}', t')\\ &= -\int d^3r'dt'\,\delta(\boldsymbol{r}-\boldsymbol{r}')\delta(t-t')\rho(\boldsymbol{r}', t')\\ &= -\rho(\boldsymbol{r}, t)/\epsilon_0\end{aligned}$$

From our examples we have learnt a number of things. The response of our systems to linear perturbations can be expressed in terms of Green's functions which are independent of the perturbation. The choice of Green's function to be used depends on the boundary conditions for the problem. A complete set of boundary conditions determines the Green's function uniquely and so this form of solution already includes the boundary conditions. The form of the solution as an integral also expresses the superposition principle for a linear system, namely that the effect arising from the sum of two causes is the sum of the effects arising from the causes separately. The use of Green's functions reduces the solution of partial differential equations to a number of quadratures. The power of their use is well known from examples in electromagnetism, where they were originally introduced by Green (1828).

In all of the problems we have to discuss, the spatial boundary condition is that the Green's function tends to zero at infinity. This is usually ensured if one finds the spatial Fourier transform of G. The temporal boundary condition usually has to be dealt with more carefully.

§ 1.2 Linear response of quantum systems

We now show that the linear response of a quantum system can analogously be expressed in terms of functions which we call Green's functions. The response of a quantum mechanical system to an external field or force is to be obtained by solving the appropriate time-dependent Schrödinger equation or, equivalently, for macroscopic systems to be described statistically, by solving the equation for the density matrix:

$$i\hbar \frac{\partial \rho}{\partial t} = [H + H', \rho]. \tag{1.12}$$

Here H is the Hamiltonian of the unperturbed system, which we assume to be independent of time, and H' is linear in the external field. Usually, the system is in equilibrium before the external field acts. Hence, the initial value of ρ is, for a grand canonical ensemble, (see, for example, Kubo 1971)

$$\rho_0 = \exp[-\beta(H - \mu N)]/Z, \tag{1.13}$$

where

$$\beta = (k_B T)^{-1}, \qquad Z = \operatorname{Tr} \exp[-\beta(H - \mu N)],$$

$$\operatorname{Tr} \rho_0 = 1,$$

T being the absolute temperature, and μ the chemical potential. Provided that H conserves particle number and so commutes with N, ρ_0 is a solution of equation (1.12) with H' zero.

To find the linear response of the system, we need to solve equation (1.12) to first order in H'. We begin by making a canonical transformation to remove the term linear in H from the equation. The appropriate transformation is

$$\rho(t) = S(t)^+ \sigma(t) S(t'), \tag{1.14}$$

where S is unitary and satisfies

$$-i\hbar(\partial S/\partial t) = HS, \tag{1.15}$$

$$+i\hbar(\partial S^+/\partial t) = S^+H. \tag{1.16}$$

This takes us to the Heisenberg representation for the Hamiltonian H. If H does not depend explicitly on t, then

$$S(t) = \exp(iHt/\hbar). \tag{1.17}$$

In general

$$\begin{aligned}\frac{\partial\rho}{\partial t} &= \frac{\partial}{\partial t}(S^+\sigma S) \\ &= \frac{\partial S^+}{\partial t}\sigma S + S^+\sigma\frac{\partial S}{\partial t} + S^+\frac{\partial\sigma}{\partial t}S \\ &= -\frac{i}{\hbar}S^+[H,\sigma]S + S^+\frac{\partial\sigma}{\partial t}S.\end{aligned} \tag{1.18}$$

From equations (1.12) and (1.18) it follows that

$$i\hbar\frac{\partial\sigma}{\partial t} = [\hat{H}'(t), \sigma], \tag{1.19}$$

where

$$\hat{H}'(t) = SH'(t)S^+ \tag{1.20}$$

and is still linear in $H'(t)$. At the initial time $t = 0$,

$$\sigma(t) = S(0)\rho(0)S^+(0) = \rho_0. \tag{1.21}$$

Hence the solution of equation (1.19) to first order in H' is, by iteration,

$$\sigma(t) = \rho_0 - \frac{i}{\hbar}\int_0^t [\hat{H}'(t'), \rho_0]\, dt'. \tag{1.22}$$

If H is independent of time, S and ρ_0 commute and

$$\rho(t) = \rho_0 - \frac{i}{\hbar}S^+(t)\int_0^t [\hat{H}(t'), \rho_0]\, dt' S(t). \tag{1.23}$$

Usually one is interested in comparing the value of some macroscopic variable (for example, charge density, magnetic moment) with observation. The variable is represented by a macroscopic operator M, say, and the quantity to be compared with experiment is the quantum and thermal average of M at time t. This is

$$\begin{aligned}\langle M\rangle &= \mathrm{Tr}\,\rho(t)M \\ &= \mathrm{Tr}\,\rho_0 M - \frac{i}{\hbar}\mathrm{Tr}\left\{MS^+(t)\int_0^t [\hat{H}(t'), \rho_0]dt'S(t)\right\}.\end{aligned}$$

The first term is the equilibrium value of the quantity. The second is the change $\delta\langle M\rangle$ induced by the field. If one uses the cyclic property of the trace of a product of operators, namely,

$$\mathrm{Tr}(ABC\ldots) = \mathrm{Tr}(BC\ldots A) = \mathrm{Tr}(C\ldots AB),$$

one finds that

$$\delta\langle M\rangle = \frac{i}{\hbar}\int_0^t \mathrm{Tr}\{\rho_0[\hat{H}(t'), \hat{M}(t)]\}\,dt', \tag{1.24}$$

where, for any operator α, we write

$$\hat{\alpha}(t) = S(t)\alpha S^+(t). \tag{1.25}$$

The result (1.24) is often known as Kubo's formula after its discoverer (Kubo 1957).

It is useful to show the dependence of $\hat{H}(t)$ on the external field explicitly. If at time t this is a classical field $A(t)$, we can write

$$H(t) = A(t)B, \tag{1.26}$$

where B is an operator belonging to the system being perturbed. (A could be the magnetic field at a point and B the magnetic moment density operator.) In general, $H(t)$ would be a sum of terms of the type (1.26) but since we are considering only the linear response the effects of these terms can be considered separately and the total response obtained by adding the individual contributions. Thus the use of (1.26) does not limit the applicability of the result. If (1.26) is used in equation (1.24) one finds

$$\delta\langle M\rangle = (i/\hbar)\int_0^t A(t')\,\mathrm{Tr}\,\rho_0[\hat{B}(t'), \hat{M}(t)]\,dt'$$

$$= \int_0^\infty G(t,t')A(t')\,dt', \tag{1.27}$$

where

$$G(t,t') = -(i/\hbar)\,\mathrm{Tr}\,\{\rho_0[\hat{M}(t), \hat{B}(t')]\}\,\theta(t-t'). \tag{1.28}$$

This result, which yields the dependence of the effect on the cause, is analogous to the classical result (1.11) and therefore $G(t, t')$ is called a Green's function. The θ-function ensures that the effect at time t depends on the cause only at preceding times. Hence, the Green's function in (1.28) is called a retarded Green's function. It is possible to define an advanced Green's function,

$$G^A(t,t') = -(i/\hbar)\,\mathrm{Tr}\,\{\rho_0[\hat{B}(t'), \hat{M}(t)]\}\,\theta(t'-t), \tag{1.29}$$

the use of which is more formal.

The part of $G(t, t')$ which multiplies the θ-function, that is

$$\Phi_{BM}(t, t') = (i/\hbar) \operatorname{Tr} \rho_0 [\hat{M}(t), \hat{B}(t')],$$

is often called the response function or after-effect function.

Any definition of a Green's function to within a constant factor is possible and many different definitions are to be found in the literature. Care is always required in checking the constant of proportionality when comparing equations involving Green's functions.

It follows from this discussion that we know the linear properties of a quantum mechanical system once we know the appropriate Green's functions (1.29). These are sufficient to describe most electromagnetic emission, absorption and transmission experiments, accoustic attenuation, resonance experiments such as electron spin and nuclear spin, and many others. As we shall see in §2.8, they also determine the thermodynamic potential and hence the thermodynamic properties. Hence the Green's functions contain all the information we need about the system and, if we can find the Green's functions without first determining all the eigenstates and energies of the unperturbed system, we may save a considerable amount of work and obtain useful results economically. In any case, even if it were possible to compute the eigenvalues and eigenstates of a macroscopic system it would not be possible to hold the information in a useful store. However, we can hope to hold the information contained in the Green's functions.

We have stated that absorption can be related to Green's functions using the linear response of the system to an appropriate external field. While this is true, it is common to discuss absorption in terms of Fermi's golden rule. We therefore show explicitly how this relates absorption to Green's functions.

Suppose that a classical field $A(\omega)$ frequency ω interacts with a quantum system, through a term in the Hamiltonian

$$A(\omega)H' \exp(i\omega t) + A^*(\omega)H' \exp(-i\omega t).$$

Then, according to Fermi's rule the rate at which the quantum system changes its state from $|m\rangle$ to $|n\rangle$ is

$$w_{mn} = (2\pi/\hbar)\, |\langle n|A(\omega)H'|m\rangle|^2\, \delta(E_n - E_m - \hbar\omega).$$

If we sum over all possible final states $|n\rangle$ and thermally average over initial states $|m\rangle$, the rate at which energy $\hbar\omega$ is absorbed from the field is

$$W(\omega) = \frac{2\pi|A(\omega)|^2}{\hbar Z} \sum_{m,n} \exp(-\beta E_m)\, |\langle n|H'|m\rangle|^2\, \delta(E_n - E_m - \hbar\omega).$$

If we introduce the Fourier transform of the δ-function this can be written

$$W(\omega) = \frac{|A(\omega)|^2}{\hbar Z} \int_{-\infty}^{\infty} dt \sum_{m,n} \exp(-\beta E_m) \langle m|H'|n\rangle$$

$$\times \exp[i(E_m - E_n + \hbar\omega)t/\hbar] \langle n|H'|m\rangle$$

$$= \frac{|A(\omega)|^2}{\hbar Z} \int_{-\infty}^{\infty} dt$$

$$\times \sum_{m,n} \langle m|\exp(-\beta H)\exp(iHt/\hbar)H' \exp(-iHt/\hbar)|n\rangle$$

$$\times \langle n|H'|m\rangle \exp(i\omega t),$$

where H is the Hamiltonian of the unperturbed system so that

$$H|m\rangle = E_m|m\rangle.$$

The sums over m and n can be performed to yield

$$W(\omega) = \frac{|A(\omega)|^2}{\hbar^2 Z} \int_{-\infty}^{\infty} dt \operatorname{Tr}[\exp(-\beta H)H'(t)H'] \exp(i\omega t).$$

The integrand has a similar structure to a Green's function although it is not exactly of the same form. However, the methods of §3.2 can be used to relate the two explicitly. An example using this expression is given in §7.2 where the elastic scattering of neutrons by phonons is discussed.

§1.3 The simple harmonic oscillator

As a simple example of the use of a Green's function in quantum mechanics we consider the problem of a one-dimensional simple harmonic oscillator of unit mass forced to oscillate by a time dependent force $f(t)$. This simple problem is a useful guide to the physical interpretation of the formal properties of Green's functions. If x is the displacement of the oscillator, the interaction term in the Hamiltonian due to the external force is the potential

$$H'(t) = -f(t)x.$$

According to the general theory of the previous section the displacement at time t is (for an ensemble of oscillators)

$$\langle x(t)\rangle = -\frac{i}{\hbar}\int_0^t \operatorname{Tr}\{\rho_0[\hat{x}(t'), \hat{x}(t)]\} f(t')\,dt' \tag{1.30}$$

and the relevant Green's function is

$$G(t, t') = -(i/\hbar)\,\mathrm{Tr}\,\{\rho_0\,[\hat{x}(t), \hat{x}(t')]\}\,\theta(t-t'). \qquad (1.31)$$

Heisenberg's equations of motion for a quantum oscillator are the same as the classical ones, with the same solutions. Hence

$$\hat{x}(t) = x\cos(\omega t) + (p/\omega)\sin(\omega t).$$

Using the commutation relations of x and p this leads to

$$\begin{aligned}[\hat{x}(t), \hat{x}(t')] &= [x\cos(\omega t), (p/\omega)\sin(\omega t')] \\ &\quad + [(p/\omega)\sin(\omega t), x\cos(\omega t')] \\ &= -(i\hbar/\omega)\sin[\omega(t-t')], \\ G(t, t') &= -(1/\omega)\sin[\omega(t-t')]\,\theta(t-t'). \qquad (1.32)\end{aligned}$$

As is often the case in quantum mechanics, the simple harmonic oscillator provides an exactly soluble model. The Green's function oscillates with the angular frequency ω of the oscillator. This is not surprising as it is the response to a δ-function force. This is partly the result of having no dissipation in the system. If there were dissipation the system would eventually return to equilibrium after an external force were removed. One would then have the condition,

$$\lim_{t\to\infty} G(t, t') = 0. \qquad (1.33)$$

This is the case for all macroscopic systems. In this sense the problem of the simple harmonic oscillator is artificial and atypical. Nevertheless, the simple harmonic oscillator provides a useful model of G for some purposes. In particular one would expect that for a system which can oscillate at several frequencies, the Green's functions will contain components which oscillate at these frequencies. In quantum mechanics, these frequencies become the excitation energies divided by $\hbar$. Hence, we expect that an analysis of the Green's functions will yield the excitation energies of the system.

The parallel between the quantum and classical systems can usefully be pushed further. Suppose we look at the classical response of a simple harmonic oscillator to a force $f(t)$ and include a dissipative term. The equation of motion is

$$\ddot{x} + (\dot{x}/\tau) + \omega^2 = f(t).$$

The classical's Green's function is found by solving

$$\ddot{x} + (\dot{x}/\tau) + \omega^2 x = -\delta(t) \qquad (1.34)$$

with

$$x(0) = 0 = \dot{x}(0).$$

The solution can be found in the same way as that of equation (1.10). For the case $\omega\tau \gg 1$ (little dissipation, ordinary damping)

$$G(t-t') = -(1/\omega)\exp[-(t-t')/\tau]\sin[\omega(t-t')]\,\theta(t-t'). \tag{1.35}$$

This agrees with the quantum mechanical result in the limit of no dissipation ($\tau \to \infty$). Also, as expected it satisfies condition (1.33). It is, therefore, a reasonable model Green's function for a system in which energy is dissipated. Unfortunately, just because energy is not conserved, it is not possible to write down a simple Hamiltonian from which to derive equation (1.34) and so we cannot obtain the analogue of (1.35) as easily from quantum mechanics.

As we shall later be led to consider the time Fourier transform of Green's functions, we look now at the Fourier transform of (1.35). We have,

$$\begin{aligned}
G(s) &= \int_{-\infty}^{\infty} dt \exp(ist) G(t) \\
&= -\int_{-\infty}^{\infty} dt \exp(ist)\frac{1}{\omega}\exp(-t/\tau)\sin(\omega t)\theta(t) \\
&= -\frac{1}{\omega}\int_{0}^{\infty} dt \exp(ist-(t/\tau))\sin(\omega t) \\
&= \frac{-1}{2i\omega}[[(1/\tau)-is-i\omega]^{-1}+((1/\tau)-is+i\omega)^{-1}] \\
&= [((1/\tau)-is)^2-(i\omega)^2]^{-1} \\
&= [(s+(1/\tau))^2-\omega^2]^{-1}.
\end{aligned} \tag{1.36}$$

Considered as a function of the complex variable s, $G(s)$ is analytic in the upper half-plane and has poles at

$$s = \pm\,\omega - (i/\tau).$$

For τ large, these are close to the real axis at the natural frequencies of the system. The displacement from the real axis depends on τ the time for the exponential decay of the Green's function.

There are many important physical systems which behave as simple harmonic oscillators and for which the discussion of this section is directly relevant. In particular, the basic excitations of the lattice of a solid are phonons which behave as simple harmonic oscillators. We shall discuss these excitations in some detail in Chapter 7.

§1.4 Single-particle Green's functions

Whenever we come to evaluate explicitly the Green's functions obtained in §1.2 in a many-body problem, we usually find that we are quickly led to consider similar functions with a simpler structure, the single-particle Green's functions. To illustrate this, consider the following Green's function

$$G(\boldsymbol{r}, t; \boldsymbol{r}', t') = (i/\hbar)\langle 0|\,[\rho(\boldsymbol{r}, t), \rho(\boldsymbol{r}', t')]\,|0\rangle\,\theta(t-t') \quad (1.37)$$

where $\rho(\boldsymbol{r}, t)$ is the operator for the particle density at $\boldsymbol{r}$ at time t and $|0\rangle$ is the ground state. Let us evaluate G for a system of dynamically independent particles whose single-particle normalized eigenstates are $\phi_m(\boldsymbol{r})$ with energies ϵ_m. The Green's function can be evaluated using either Slater determinantal wave functions or second quantization. As we have to use the latter method in most of this book we shall use that method here. For convenience the main properties of the second quantized operators and states are listed in Appendix A. For a derivation of the properties readers are referred to an advanced text on quantum theory, for example Ziman (1969).

In these terms

$$\rho(\boldsymbol{r}) = \psi^+(\boldsymbol{r})\psi(\boldsymbol{r})$$

and

$$\psi(\boldsymbol{r}) = \sum c_n \phi_n(\boldsymbol{r})$$

where c_n destroys the state n and

$$c_n(t) = c_n \exp(-i\epsilon_n t/\hbar).$$

Hence, for $t > t'$,

$$G(\boldsymbol{r}, t, \boldsymbol{r}', t') = (i/\hbar)\langle 0|\,[\rho(\boldsymbol{r}, t), \rho(\boldsymbol{r}', t')]\,|0\rangle$$

$$= (i/\hbar) \sum_{k,l,mn} \phi_k^*(\boldsymbol{r})\phi_l(\boldsymbol{r})\phi_m^*(\boldsymbol{r}')\phi_n(\boldsymbol{r}')$$

$$\times \langle 0|\,[c_k^+(t)c_l(t), c_m^+(t')c_n(t')]\,|0\rangle.$$

Since the average requires only that the expectation value of the product $c_k^+(t)c_l(t)c_m^+(t')c_n(t')$ be taken in independent particle states, the average will be non-zero only when the product of operators has the net effect of not creating or destroying any particles. In other words, the product $c_l(t)c_n(t')$ must destroy the states which the product $c_k^+(t)c_m^+(t')$ creates. This requires that the pair of labels l, n is the same (apart from order) as the pair k, m. Hence, for a non-zero average

$$l = k, \qquad n = m \quad \text{or} \quad l = m, \qquad n = k.$$

Therefore

$$G(\boldsymbol{r}, t; \boldsymbol{r}', t') = (i/\hbar) \sum_{l,n} |\phi_l^*(\boldsymbol{r})|^2 \, |\phi_n(\boldsymbol{r}')|^2 \langle 0| \, [c_l^+(t)c_l(t), c_n^+(t')c_n(t')] \, |0\rangle$$

$$+ (i/\hbar) \sum_{l,n} \phi_n^*(\boldsymbol{r})\phi_l(\boldsymbol{r})\phi_l^*(\boldsymbol{r}')\phi_n(\boldsymbol{r}')$$

$$\times \langle 0| \, [c_n^+(t)c_l(t), c_l^+(t')c_n(t')] \, |0\rangle \, .$$

The first commutator is zero. The second term is zero if $n = l$ and otherwise depends upon $\langle 0|c_n^+(t)c_l(t)c_l^+(t')c_n(t')|0\rangle$.

If $n \neq l$, the degrees of freedom n and l are independent and one has

$$\langle 0|c_n^+(t)c_l(t)c_l^+(t')c_n(t')|0\rangle = \langle 0|c_n^+(t)c_n(t')|0\rangle \, \langle 0|c_l(t)c_l^+(t')|0\rangle,$$

$$G(\boldsymbol{r}, t, \boldsymbol{r}', t') = (i/\hbar) \, \{\langle 0|\psi^+(\boldsymbol{r}, t)\psi(\boldsymbol{r}', t')|0\rangle \, \langle 0|\psi(\boldsymbol{r}, t)\psi^+(\boldsymbol{r}', t')|0\rangle$$
$$- \langle 0|\psi^+(\boldsymbol{r}', t')\psi(\boldsymbol{r}, t)|0\rangle \, \langle 0|\psi(\boldsymbol{r}'t')\psi^+(\boldsymbol{r}, t)|0\rangle\}\theta(t - t') \tag{1.38}$$

and this leads us naturally to consider objects such as

$$F(\boldsymbol{r}, t; \boldsymbol{r}', t') = \langle 0|\psi(\boldsymbol{r}, t)\psi^+(\boldsymbol{r}', t')|0\rangle.$$

This new function has a simple interpretation because $\psi^+(\boldsymbol{r}, t)$ is the operator which creates a particle at the position $\boldsymbol{r}$ at time t (see Appendix A). If we denote

$$|\boldsymbol{r}, t\rangle = \psi^+(\boldsymbol{r}', t')|0\rangle$$

then $|\boldsymbol{r}, t\rangle$ describes a state which comprises the ground state with an extra particle at $\boldsymbol{r}$ at time t. Similarly, $|\boldsymbol{r}', t'\rangle$ has an extra particle at $\boldsymbol{r}'$ at time t. Therefore

$$\langle 0|\psi(\boldsymbol{r}, t)\psi^+(\boldsymbol{r}', t')|0\rangle = \langle \boldsymbol{r}t|\boldsymbol{r}'t'\rangle$$

is the probability amplitude for finding a particle at $\boldsymbol{r}$ at time t, given that a particle is placed at $\boldsymbol{r}'$ at the time t'. It therefore describes in quantum mechanical terms the way a particle travels from $\boldsymbol{r}'t'$ to $\boldsymbol{r}, t$. Similarly, $\langle 0|\psi^+(\boldsymbol{r}, t)\psi(\boldsymbol{r}', t')|0\rangle$ describes the travel of a "hole" from $\boldsymbol{r}', t'$ to $\boldsymbol{r}, t$.

The new functions are not themselves Green's functions because that term is reserved for functions which are singular or which have singular derivatives when the arguments become equal. It is easy, however, to construct Green's functions, known as single-particle

Green's functions from them. For example, we can write down the Green's functions

$$g_R(\boldsymbol{r}, t; \boldsymbol{r}', t') = -i\langle 0|[\psi(\boldsymbol{r}, t), \psi^+(\boldsymbol{r}', t')]|0\rangle\, \theta(t-t')$$

and

$$g_A(\boldsymbol{r}, t; \boldsymbol{r}', t') = i\langle 0|[\psi(\boldsymbol{r}, t), \psi^+(\boldsymbol{r}', t')]|0\rangle\, \theta(t'-t). \quad (1.39)$$

There are two things to be learnt from this example. The first is that the basic building blocks for Green's functions in many-body problems are likely to be the single-particle Green's functions. The second is that with the Green's functions we have so far constructed, there is not a simple relation between the Green's function $G(\boldsymbol{r}, t; \boldsymbol{r}'t')$ and g_A or g_R. Two important steps which have advanced the theory and made Green's functions a useful tool are:

(i) the discovery of Green's functions in which the relationship between those involving many operators can, at least for independent particles, be simply related to those involving fewer operators; and

(ii) the discovery of relations between these (formal) Green's functions and the observable ones.

The procedure in any calculation is to establish an approximation for calculating the former Green's functions and then to use (ii) to calculate the observable Green's functions from them. Section 2.2 is concerned with (ii). Most of the remainder of the book is concerned with the approximations.

Although we have come to single-particle Green's functions as basic building blocks, there are occasions when they are the observable functions. We give two examples.

(a) As we have seen, the single-particle Green's function describes the behaviour of an extra particle added to the system. An experiment where such behaviour is directly observed is one where particles can tunnel quantum mechanically through a barrier from one material to another. The results of such experiments can be given in terms of single-particle Green's functions (cf. §8.7).

(b) The thermodynamic properties including the density of states in energy of systems can be related to single-particle Green's functions (see §2.8).

§1.5 Correlation functions

The operator

$$\rho(\boldsymbol{r}) = \psi^+(\boldsymbol{r})\psi(\boldsymbol{r})$$

is the density operator whose expectation value $\langle|\psi^+(r)\psi(r)|\rangle$ is just the mean density at $\boldsymbol{r}$. In a macroscopic system it will be the actual density at $\boldsymbol{r}$. If the state, $|\ \rangle$, contains no particles at $\boldsymbol{r}$ the mean value is zero, otherwise it is finite. The integral $(1/N)\int \rho(\boldsymbol{r})dv$ integrated over a small volume Δv then has a mean value $P(\boldsymbol{r})\Delta v$ which is the probability of finding a particle in the small volume Δv. $P(\boldsymbol{r})$ is the probability density and

$$P(\boldsymbol{r}) = \frac{1}{N}\langle|\rho(\boldsymbol{r})|\rangle; \qquad \int P(\boldsymbol{r})\Delta v = 1$$

where the integral is over the whole volume. Now consider the operator

$$\rho(\boldsymbol{r})\rho(\boldsymbol{r}').$$

Since $\rho(\boldsymbol{r})$ and $\rho(\boldsymbol{r}')$ commute the product is Hermitian. Its expectation value $\langle|\rho(\boldsymbol{r})\rho(\boldsymbol{r}')|\rangle$ is zero in any state, $|\ \rangle$, for which there is no particle at either $\boldsymbol{r}$ and $\boldsymbol{r}'$ and is finite otherwise. The integral product

$$(1/N^2)\int \rho(\boldsymbol{r})dv \int \rho(\boldsymbol{r}')dv',$$

where the integrals are over small volume Δv, $\Delta v'$, then has a mean value $\int P(\boldsymbol{r}, \boldsymbol{r}')\Delta v\Delta v'$ and, for a macroscopic system,

$$P(\boldsymbol{r}, \boldsymbol{r}') = (1/N^2)\langle|\rho(\boldsymbol{r})\rho(\boldsymbol{r}')|\rangle$$

is the probability density that there is one particle at $\boldsymbol{r}$, and another at $\boldsymbol{r}'$.

If there were no correlation between the probability of finding a particle at $\boldsymbol{r}$, and the probability of finding a particle at $\boldsymbol{r}'$, we should have

$$P(\boldsymbol{r}, \boldsymbol{r}') = P(\boldsymbol{r})P(\boldsymbol{r}'),$$

that is,

$$\langle\rho(\boldsymbol{r})\rho(\boldsymbol{r}')\rangle = \langle\rho(\boldsymbol{r})\rangle\langle\rho(\boldsymbol{r}')\rangle. \tag{1.40}$$

This is true for a system of dynamically and statistically independent particles. For the usual systems of interest interactions between particles are important and (1.40) is not satisfied. The difference between the two sides

$$\langle\rho(\boldsymbol{r})\rho(\boldsymbol{r}')\rangle - \langle\rho(\boldsymbol{r})\rangle\langle\rho(\boldsymbol{r}')\rangle \tag{1.41}$$

therefore represents the correlation between the positions of the particles. For this reason

$$\langle\rho(\boldsymbol{r})\rho(\boldsymbol{r}')\rangle \tag{1.42}$$

is called the density–density correlation function. This is a slight misnomer because (1.41) represents the correlation, but it is very commonly used.

Particles in a quantum system always obey either Fermi or Bose statistics and are, therefore, never statistically independent. The density–density correlation function for such a system is therefore never zero.

The definition of a correlation function is very easily generalized to the case of other commuting observables. If A and B commute,

$$C_{AB} = \langle AB \rangle \tag{1.43}$$

is the correlation function between them. If A and B do not commute, it is not possible to measure them separately and so it is not strictly possible to talk about correlations between them. Nevertheless, for macroscopic systems the measurement of one may well not perturb the other by much and the correlation between them is practically measurable. Even when this is not the case it is possible formally to define correlation functions. Because of the non-commutability of the observables, the definitions are not unique and there is no universal agreement. Some authors call any expectation value of the form (1.43) a correlation function. Others (see Kubo 1957) generalize the function so that it is real and reduces to (1.43) for commuting observables. In this case,

$$C^s_{AB} = \tfrac{1}{2}\langle (AB + BA) \rangle = \tfrac{1}{2}\langle \{A, B\} \rangle . \tag{1.44}$$

When we allow observables which do not commute, there is no need for them to depend on the same time. We can therefore consider correlation functions

$$C_{AB}(t, t') = \langle A(t) B(t') \rangle \tag{1.45}$$

and

$$C^s_{AB}(t, t') = \tfrac{1}{2}\langle \{A(t), B(t')\} \rangle = \mathrm{Re}\, C_{AB}(t, t'). \tag{1.46}$$

The Green's function which appears in the linear response can be written in terms of correlation functions of the type in equation (1.45). This Green's function is

$$G(t, t') = (i/\hbar)\langle [B(t), M(t')] \rangle\, \theta(t - t') \tag{1.28}$$

$$= (2/\hbar)\, \mathrm{Im}\, C_{BM}(t, t')\theta(t - t'). \tag{1.47}$$

Notice a basic difference between Green's functions and correlation functions. The former (or their time derivatives) are discontinuous at $t = t'$ whereas the latter are not. The response of the system can be expressed equally well in terms of either but, in practice, it is usually easier to derive the Green's functions. The results of X-ray and neutron scattering experiments are frequently expressed in terms of the correlation functions (see Chapter 7).

§ 1.6 Semiquantitative considerations

Consider a one component system of particles of mass M interacting with each other through a two-body potential $V(r)$ which depends only on the separation r of the particles. If the density of particles is ρ, the order of magnitude of the mean separation of the particles r_s is given by

$$r_s \sim \rho^{-1/3}.$$

The mean potential energy is then of order $V(r_s)$. At high temperatures, the system will be classical and the mean kinetic energy of the particles will be $\sim kT$. At sufficiently high temperatures this will be greater than $V(r_s)$ and the system will be in the gaseous state. The behaviour is dominated by the kinetic energy and $V(r_s)$ can be treated by some kind of perturbation theory.

At very low temperatures, the classical kinetic theory is negligible. If the zero-point energy (the quantum contribution to the kinetic energy) is also negligible the behaviour is dominated by the potential energy and the system becomes a solid. In equilibrium, the total potential energy is a minimum and this determines the relative positions of the particles and therefore of r_s. Between these extreme temperatures, the system is in a state where the kinetic and potential energies are comparable.

In a quantum fluid such as liquid ^{4}He, the kinetic energy due to the zero-point motion is very important. Each atom is confined to a region of dimensions r_s, so its zero-point energy is $\sim \hbar^2/2mr_s^2$. We expect that this will be comparable with $V(r_s)$ in equilibrium.

Thus we see that for a one-component system we have ready-made starting points for problems at low and high temperatures when the system is in the solid or gaseous state but that at intermediate temperatures, especially when the system is in the liquid state, a good starting point is not obvious.

When the system comprises more than one component, equilibrium is determined by the overall energy or free energy and this need not be simply related to the potential between the same components or the potentials between the components. For example, in a metal the equilibrium positions of the ions are determined by the effective potential between the ions. There is then no reason to suppose that the potential energy of the electrons dominates their kinetic energy and, indeed, it usually does not. It is to deal with quantum systems for which neither the kinetic energy nor the potential energy dominates that many-body theory and the use of Green's functions have been built up. They are also of use when the forces between particles are singular, either by being long-range as are electrostatic and magnetic forces, or by having hard cores as do the usual forces between atoms. Finally, there are problems notably at phase transitions where small changes in the total energy make large qualitative changes to the system. The general theory is appropriate for these problems too.

§1.7 Summary and plan of the book

The discussion of this chapter has been designed to show that both classical and quantum mechanical Green's functions are useful for yielding the linear response of a system to a class of perturbation. The Green's functions ensure that the solution already satisfies the appropriate boundary conditions. There is reason to believe that the basic building blocks of the theory will be single-particle Green's functions, but that a judicious choice of Green's function will be necessary for the theory to appear in a simple form.

In the next chapter a number of different Green's functions are defined as well as appropriate Fourier transforms. The Fourier transforms are shown to be related to a function $G(\omega)$ which is analytic in the lower and upper halves of the complex plane of the variable ω. It is shown that through this function the different Green's functions can be related to each other. A number of important properties of the Green's functions are derived as well as the equations of motion of the single-particle Green's functions. It is also shown how the thermodynamic potential is related to single-particle Green's functions. This enables the thermodynamic properties to be determined from the single-particle Green's functions.

In Chapter 3, a perturbation expansion for a single-particle temperature Green's function in powers of a two-particle interaction is derived from the equations of motion and it is shown how the expansion can be represented by Feynman diagrams. The expansion is partially summed in terms of a self-energy and rules for calculating the self-energy are given. Particular approximations to the series such as the Hartree, Hartree–Fock and random phase approximations are introduced. A perturbation expansion for the response function is also derived and it is shown that in order to preserve certain conservation laws, certain identities, the Ward identities, must be satisfied. This lays restrictions on approximations for response functions.

It is very much easier to understand Feynman diagrams than the proof of perturbation theory. Therefore, although a proof is given we have tried to write the text in such a way that the reader can omit the proof if he so wishes. This means that §2.9 and the indicated parts of §3.2 and 3.10 can be omitted without rendering the remainder incomprehensible.

In later chapters, the general methods are applied to the particular problems of dilute impurities in a metal, the electron gas, a Fermi liquid, electrons and phonons, superconductivity, superfluidity, magnetism, disordered solids and critical phenomena.

References

Green, G. (1828). *An Essay on the Application of Mathematical Analysis to the Theories of Electricity and Magnetism*, Nottingham.
Kubo, R. (1957). *J. Phys. Soc. Japan* **12**, 570.
Kubo, R. (1971). "Statistical Mechanics", §2.4 and 2.7. North Holland, Amsterdam.
Ziman, J. M. (1969). "Elements of Advanced Quantum Theory", Chapters 1 and 3. Cambridge University Press, Cambridge.

Problems

1. Show that the classical Green's function which satisfies the equation

$$\nabla^2 G(\boldsymbol{r}, \boldsymbol{r}') - K^2 G(\boldsymbol{r}, \boldsymbol{r}') = -\delta(\boldsymbol{r} - \boldsymbol{r}')$$

and which tends to zero as $\boldsymbol{r}, \boldsymbol{r}'$ tend separately to infinity is

$$G(\boldsymbol{r}, \boldsymbol{r}') = \frac{\exp(-K|\boldsymbol{r} - \boldsymbol{r}'|)}{4\pi|\boldsymbol{r} - \boldsymbol{r}'|}.$$

2. A system of non-interacting fermions in free space is acted upon by a small time dependent perturbation.

$$H'(t) = \sum_{k,q} \{c_{k+q}^{+} c_k \phi_q \exp[i\omega_q t + (t/\tau)] + c_k^{+} c_{k+q} \phi_q^{*} \exp[-i\omega_q t + (t/\tau)]\}.$$

Write down the explicit form for the Heisenberg operator $\hat{H}'(t)$. Hence use Kubo's formula (1.24) to show that the mean spatial Fourier component of the number density,

$$n_q = \sum_k c_{k+q}^+ c_k ,$$

induced by the perturbation after it has been switched on for a time long compared with τ is $\langle n_q \rangle$ given by

$$\langle n_q \rangle = \sum_k (f_k - f_{k+q}) \times \left[\frac{\exp[i\omega_{-q}t + (t/\tau)]}{\omega_{-q} + \epsilon_k - \epsilon_{k+q} - i/\tau} + \frac{\exp[-i\omega_q t + (t/\tau)]}{\epsilon_k - \epsilon_{k+q} - \omega_q - i/\tau} \right] .$$

Chapter 2

Formal Matters

§2.1 Double-time Green's functions

The procedure we adopt to solve problems using Green's functions is to derive the equations satisfied by the appropriate Green's functions and then to find the solutions which satisfy the right boundary conditions. As we indicated in the last chapter the spatial boundary conditions cause no great difficulty but the temporal ones require a little care. In order to apply them correctly we need to know some of the formal properties of these Green's functions and this chapter is devoted to that purpose. For the reader not interested in proofs and derivations and also for easy reference, the main results are summarized at the end of this chapter in §2.10.

As was shown in the last chapter, an important Green's function is the retarded one which relates cause and effect. This does, as it must, satisfy the principle of causality, namely that cause always precedes effect. This principle is, of course, universal, as true in classical physics as in quantum physics, and it leads in all these cases to some useful and interesting properties of the temporal Fourier transforms of the causal functions in the complex plane of the "frequency" variable ω. These are the properties we study in succeeding sections and they include the Kramers–Krönig relations and the fluctuation–dissipation theorem, discussed in §§2.5 and 2.6.

In the final sections of this chapter we derive the equations of motion satisfied by the Green's functions and the relationship between the thermodynamic potential and Green's functions. We can use the latter to determine the thermodynamic properties once the Green's functions are known.

We begin by defining double-time Green's functions. By a double-time Green's function, we mean one which depends upon two operators, $A(t)$ and $B(t')$ (which may themselves be composites of simpler operators) which depend on the different times t and t'. Experience has shown that four Green's functions are particularly useful and important. These are defined as follows:

(1)
$$G_R(t, t') = -(i/\hbar)\langle[A(t), B(t')]_\epsilon\rangle\, \theta(t-t')$$
$$\equiv -(i/\hbar)\,\mathrm{Tr}\,\{\rho_0[A(t), B(t')]_\epsilon\}\, \theta(t-t'). \qquad (2.1)$$

This is the retarded Green's function which arises naturally in discussions of the linear response of a system and which has already been defined apart from the subscript ϵ. In fact, the notation means

$$[A(t), B(t')]_\epsilon = A(t)B(t') + \epsilon B(t')A(t). \qquad (2.2)$$

If A and B involve only Bose operators, ϵ is taken to be (-1) and the definition is the same as before, i.e.

$$[A(t), B(t')]_{-1} = [A(t), B(t')].$$

If A and B involve Fermi fields, two cases only are of importance. Either (a) A and B are even functions of the Fermi fields, that is, both are invariant when the signs of all the Fermi fields are changed simultaneously, or (b) both A and B are odd functions of the Fermi fields in which case they both change sign when all the Fermi fields change sign. In case (a) ϵ is again chosen to be (-1) and the former definition holds. In linear response theory, if A and B involve Fermi operators case (a) will hold. In case (b), ϵ is chosen to be $+1$ and

$$[A(t), B(t')]_{+1} = A(t)B(t') + B(t')A(t) = \{A(t), B(t')\}.$$

This case will not arise in linear response theory but it is important for discussions of single-particle properties (see §2.3).

(2)
$$G_A(t, t') = (i/\hbar)\langle[A(t), B(t')]_\epsilon\rangle\, \theta(t'-t) \qquad (2.3)$$

This is the advanced Green's function and agrees with the definition given in §1 when ϵ is -1.

(3)
$$G_c(t, t') = -(i/\hbar)\langle T\{A(t)B(t')\}\rangle \qquad (2.4)$$
$$\equiv -(i/\hbar)\langle A(t)B(t')\rangle\, \theta(t-t') + (i/\hbar)\epsilon\langle B(t')A(t)\rangle\, \theta(t'-t).$$

This is the causal Green's function. The parameter ϵ is again $+1$ or -1 according to the same rules as in (1). It has been found most useful for computation when the system is at absolute zero. In this book, we concentrate on properties at finite temperatures and do not make much use of this function.

(4)
$$\begin{aligned} G(\tau,\tau') &= -(1/\hbar)\langle T\{A(-i\tau)B(-i\tau')\}\rangle \\ &\equiv -(1/\hbar)\langle A(-i\tau)B(-i\tau')\rangle\,\theta(\tau-\tau') \\ &\quad + (\epsilon/\hbar)\langle B(-i\tau')A(-i\tau)\rangle\,\theta(\tau'-\tau). \end{aligned} \tag{2.5}$$

This is the temperature Green's function with ϵ defined as before. It has been found most useful for approximations and computations of the properties of condensed systems at finite temperatures. This Green's function plays a central role in this book.

In the following discussion, we treat temperature T, or more strictly $\beta = (k_B T)^{-1}$, as an imaginary time. It reduces the superficial complexity of formulae if energy and frequency are treated as having the same dimensions. It is normal practice to achieve this by using units in which $\hbar = 1$ and we shall follow this practice throughout the remainder of the book. This means that we do not distinguish between angular frequency and energy nor between wave vector and momentum. It is always simple to use a dimensional argument to insert the appropriate powers of $\hbar$ at the end of a calculation to obtain the result in conventional units (see §8.6).

§2.2 Formal properties

All the double-time Green's functions depend only on the difference of the time arguments when the Hamiltonian does not depend on the time explicitly. This is obvious physically because the Green's functions are sums of correlation functions of two operators, and for a system which has the same dynamic properties at all times the correlation between two observables at different times will depend only on the time elapsed between them. Formally this can be proved by considering one correlation function

$$C(t,t') = \langle A(t)B(t')\rangle. \tag{2.6}$$

Written explicitly

$$\begin{aligned} C(t,t') &= Z^{-1}\,\mathrm{Tr}\,\{\exp(-\beta H)\exp(iHt)A\exp(-iHt) \\ &\quad \times \exp(iHt')B\exp(-iHt')\}. \end{aligned} \tag{2.7}$$

Using the fact that $\exp(-\beta H)$ and $\exp(iHt)$ commute and using the cyclic property of the trace of a product of operators, we find

$$\begin{aligned} C(t, t') &= Z^{-1}\{\mathrm{Tr}\exp(-\beta H)A\exp[iH(t'-t)]B\exp[-iH(t'-t)]\} \\ &= \langle A(0)B(t'-t)\rangle \\ &\equiv C(t-t') \end{aligned} \tag{2.8}$$

and hence the result that the Green's functions are functions only of time-differences.

We can Fourier transform the functions with respect to their time-differences and examine the properties of the transforms. For example, we begin by constructing the function

$$\begin{aligned} G(\omega) &= \int_{-\infty}^{\infty} dt\exp[i\omega(t-t')]\, G_R(t-t') \\ &= -i\int_{-\infty}^{\infty} dt\exp[i\omega(t-t')]\,\langle[A(t), B(t')]_\epsilon\rangle\,\theta(t-t') \\ &= -i\int_0^{\infty} dt\exp(i\omega t)\langle[A(t), B(0)]_\epsilon\rangle. \end{aligned} \tag{2.9}$$

Now, provided that the average in the integrand does not grow with time as fast as an exponential and that ω is a complex number with a positive imaginery part, the integral will converge. For then we have

$$G(\omega_1 + i\omega_2) = -i\int_0^{\infty} dt\exp[i\omega_1 t - \omega_2 t]\,\langle[A(t), B(0)]_\epsilon\rangle,$$

which converges for $\omega_2 > 0$. Hence, considered as a function of the complex variable ω, $G(\omega)$ converges everywhere in the upper half-plane of ω. Since the average in the integrand is the sum of two correlation functions and, for a macroscopic system in thermodynamic equilibrium, we do not expect correlations to grow in time, the average is expected to satisfy the necessary conditions in practice for $G(\omega)$ to be analytic in the upper half-plane. It must be admitted that although this has been shown for simple systems and simple approximations, it has never been proved for real systems, unless we take it for granted that a system for which the average continues to grow with time cannot be in equilibrium. The Fourier transform of $G(t-t')$ is $G_R(\omega) = \mathrm{Lim}_{\delta\to 0}\, G(\omega + i\delta)$. $G_R(\omega)$ gives the forced response of the system to a sinusoidal force of angular frequency ω. It is often called the admittance of the system.

It is sometimes convenient to write $G(\omega)$ in a spectral form which demonstrates its analytic property. Let us denote by $|m\rangle$ and E_m the eigenstates and corresponding energies of the Hamiltonian, H. Then

$$H\,|m\rangle = E_m\,|m\rangle. \tag{2.10}$$

Then $G(\omega)$ can be written explicitly in terms of these states and energies as follows.

$$\begin{aligned}
G(\omega) &= -i \int_0^\infty dt \exp(i\omega t) Z^{-1} \sum_m \langle m| \exp(-\beta H) \\
&\quad \times [A(t), B(0)]_\epsilon\, |m\rangle \\
&= -i \int_0^\infty dt \exp(i\omega t) Z^{-1} \sum_m \exp(-\beta E_m) \langle m| \\
&\quad \times \{A(t)B(0) + \epsilon B(0)A(t)\}\, |m\rangle \\
&= -i \int_0^\infty dt \exp(i\omega t) Z^{-1} \sum_{m,n} \exp(-\beta E_m) \\
&\quad \times \{A_{mn}B_{nm} \exp[i(E_m - E_n)t] + \epsilon B_{mn}A_{nm} \exp[i(E_n - E_m)t]\} \\
&= -i \int_0^\infty dt \exp(i\omega t) Z^{-1} \sum_{m,n} (\exp(-\beta E_m) + \epsilon \exp(-\beta E_n)) \\
&\quad \times A_{mn}B_{nm} \exp[i(E_m - E_n)t],
\end{aligned}$$

where we have used the notation

$$A_{mn} = \langle m|A(0)|n\rangle = \langle m|A|n\rangle. \tag{2.11}$$

If it is assumed that the order of integration and summation can be changed the integration over time can be performed to yield

$$G(\omega) = -\frac{1}{Z} \sum_{m,n} \left(\frac{\exp(-\beta E_m) + \epsilon \exp(-\beta E_n)}{E_n - E_m - \omega} \right) A_{mn} B_{nm}\,.$$

Since the energies are real numbers this can be written

$$\begin{aligned}
G(\omega) &= \frac{1}{Z} \int_{-\infty}^{\infty} dx \sum_{m,n} \left(\frac{\exp(-\beta E_m) + \epsilon \exp(-\beta E_n)}{\omega - x} \right) A_{mn} B_{nm} \\
&\quad \times \delta(x - E_n + E_m) \\
&= \int_{-\infty}^{\infty} dx \frac{A(x)}{(\omega - x)}.
\end{aligned} \tag{2.12}$$

This is known as the spectral form of $G(\omega)$ and $A(x)$ is the spectral function with the explicit form

$$A(x) = Z^{-1} \sum (\exp(-\beta E_m) + \epsilon \exp(-\beta E_n)) A_{mn} B_{nm}$$
$$\times \delta(x - E_n + E_m)$$
$$= Z^{-1}(1 + \epsilon \exp(-\beta x)) \sum_{m,n} \exp(-\beta E_m) A_{mn} B_{nm}$$
$$\times \delta(x - E_n + E_m). \tag{2.13}$$

It follows that

$$\int_{-\infty}^{\infty} dx \frac{A(x)}{1 + \epsilon \exp(-\beta x)} = \frac{1}{Z} \sum_{m,n} \exp(-\beta E_m) A_{mn} B_{nm} = \langle AB \rangle. \tag{2.14}$$

This is a sum rule for $A(x)$. Provided that the sums over m and n converge so that $A(x)$ exists, the form (2.12) shows that $G_R(\omega)$ is analytic in the upper half-plane of ω. We can also obtain directly from equation (2.13), the sum rule

$$\int_{-\infty}^{\infty} dx\, A(x) = \langle [A, B]_\epsilon \rangle. \tag{2.15}$$

Since we expect the right-hand side of this equation to be finite (if A and B are not singular operators), it follows that as $|\omega|$ becomes large

$$G(\omega) \sim \omega^{-1} \int_{-\infty}^{\infty} dx\, A(x) = \langle [A, B]_\epsilon \rangle / \omega.$$

Hence, as $|\omega|$ tends to infinity, $G(\omega)$ approaches zero as fast as ω^{-1}, if $\langle [A, B]_\epsilon \rangle$ is non-zero, or faster if the average is zero.

The analyticity of $G(\omega)$ in the upper half-plane stems directly from the fact that $G_R(t)$ is a retarded function. Conversely if $G(\omega)$ is analytic in the upper half-plane and $G_R(t)$ is obtained from it by means of equation (2.9), then $G_R(t)$ will be a retarded function. Hence, when we come to solve equations of motion later on we can ensure that the boundary condition is satisfied by finding $G(\omega)$ and then obtaining $G_R(t)$ from it.

In exactly the same way, from the advanced Green's function we can define $G'(\omega)$

$$G'(\omega) = \int_{-\infty}^{\infty} dt \exp[i\omega(t - t')]\, G_A(t - t')$$

and is a function of the complex variable ω which is analytic in the lower half-plane ($\text{Im}\, \omega < 0$). Furthermore, $G'(\omega)$ is given by equation (2.12) with exactly the same form for $A(x)$. There is, therefore, a single function

$$G(\omega) = \int_{-\infty}^{\infty} dx \frac{A(x)}{(\omega - x)} \tag{2.16}$$

which is analytic in the upper half-plane and equal to $G_R(\omega)$ as ω approaches the real axis from above and is analytic in the lower half-plane and equal to $G_A(\omega)$ as ω approaches the real axis from below. In general, $G(\omega)$ is not analytic everywhere on the real axis, but has a cut given by

$$\begin{aligned} G_R(\omega) - G_A(\omega) &= \lim_{\delta \to 0} [G(\omega + i\delta) - G(\omega - i\delta)] \\ &= \lim_{\delta \to 0} \int_{-\infty}^{\infty} dx\, A(x) \left[\frac{1}{\omega + i\delta - x} - \frac{1}{\omega - i\delta - x} \right] \\ &= -2\pi i \int_{-\infty}^{\infty} dx\, A(x)\delta(x - \omega) \\ &= -2\pi i\, A(\omega). \end{aligned} \tag{2.17}$$

The spectral function is thus determined by the difference between the retarded and advanced Green's functions. Notice that $G_R(\omega)$ and $G_A(\omega)$ can also be written in the forms

$$\begin{aligned} G_R(\omega) &= \lim_{\delta \to 0} \int_{-\infty}^{\infty} dx \frac{A(x)}{(\omega + i\delta - x)} \\ &= \int_{-\infty}^{\infty} dx \left[P \frac{A(x)}{\omega - x} - i\pi A(x)\delta(x - \omega) \right], \end{aligned} \tag{2.18}$$

$$G_A(\omega) = \int_{-\infty}^{\infty} dx \left[P \frac{A(x)}{\omega - x} + i\pi A(x)\delta(x - \omega) \right], \tag{2.19}$$

where the symbol P denotes the principal part of the integral.

In a similar way, it can be shown that the causal Green's function can be written in the spectral form,

$$G_C(\omega) = \int_{-\infty}^{\infty} dx\, A(x) \left\{ \frac{P}{\omega - x} - i\pi \left[\frac{1 - \epsilon \exp(-\beta x)}{1 + \epsilon \exp(-\beta x)} \right] \delta(x - \omega) \right\} \tag{2.20}$$

and so depends on the same spectral function $A(x)$. Once $A(x)$ is known all the Green's functions can be obtained from it.

The function in square brackets in $G_C(\omega)$ can be related to the distribution function for the appropriate particles. For fermions, $\epsilon = 1$ and

$$\frac{1-\epsilon\exp(-\beta x)}{1+\epsilon\exp(-\beta x)} = \frac{\exp(\beta x)-1}{\exp(\beta x)+1} = 1-\frac{2}{\exp(\beta x)+1} = 1-2f(x)\,, \tag{2.21}$$

where $f(x)$ is the distribution function for independent fermions. (But (2.20) is an exact result, true even if the fermions interact!) For bosons ϵ is (-1) and

$$\frac{1-\epsilon\exp(-\beta x)}{1+\epsilon\exp(-\beta x)} = \frac{1+\exp(\beta x)}{1-\exp(\beta x)} = \frac{2}{1-\exp(\beta x)}-1 = 2n(x)-1, \tag{2.22}$$

where $n(x)$ is the distribution functions for bosons.

The temperature Green's function can be obtained from the causal one by varying t, t' continuously and appropriately in the complex plane. However, the most useful representation of the temperature Green's function can be obtained directly as follows. As with the other Green's functions $G(\tau_1, \tau_2)$ is a function only of the difference $\tau = \tau_1 - \tau_2$. Choose τ so that $0 < \tau \leqslant \beta$. Then $-\beta < (\tau - \beta) \leqslant 0$, and

$$\begin{aligned} G(\tau-\beta) &= \epsilon\,\langle B(0)A(-i\tau+i\beta)\rangle \\ &= \epsilon Z^{-1}\,\mathrm{Tr}\,\{\exp(-\beta H)B\exp[H(\tau-\beta)]\,A\exp[H(\beta-\tau)]\}\,. \end{aligned} \tag{2.23}$$

If the cyclic property of a trace of a product of operators is used, this can be rewritten

$$\begin{aligned} G(\tau-\beta) &= \epsilon Z^{-1}\,\langle\exp(-\beta H)A\exp(-H\tau)B\exp(H\tau)\rangle \\ &= \epsilon\,\langle AB(i\tau)\rangle \\ &= -\epsilon G(\tau). \end{aligned} \tag{2.24}$$

This important condition satisfied by $G(\tau)$ can be used to obtain a Fourier series for $G(\tau)$ in the region $-\beta < \tau \leqslant \beta$ (Matsubara, 1955). For clarity consider the cases $\epsilon = \pm 1$ separately.

(a) $\epsilon = 1$

This applies to Fermi-type operators where

$$G(\tau-\beta) = -G(\tau). \tag{2.25}$$

Therefore, in the interval $-\beta < \tau \leqslant \beta$, $G(\tau)$ can be expanded in terms of the exponentials $\exp(-i\zeta_l\tau)$ where $\zeta_l = (2l+1)\pi/\beta$ and

$$\exp[-i\zeta_l(\tau-\beta)] = \exp[+i(2l+1)\pi]\exp(-i\zeta_l\tau) = -\exp(i\zeta_l\tau).$$

Hence

$$G(\tau) = \beta^{-1}\sum_l \bar{G}(\zeta_l)\exp(-i\zeta_l\tau) \tag{2.26}$$

and

$$\bar{G}(\zeta_l) = \int_0^\beta d\tau \exp(i\zeta_l\tau)G(\tau). \tag{2.27}$$

The resultant $G(\tau)$ is a function periodic in τ with period 2β.

(b) $\epsilon = -1$

This applies to Bose-type operators where

$$G(\tau-\beta) = G(\tau). \tag{2.28}$$

In this case, in the interval $-\beta < \tau \leqslant \beta$, $G(\tau)$ can be expanded in a series of exponentials $\exp(-i\zeta_l\tau)$ where $\zeta_l = 2\pi l/\beta$ and

$$\exp[-i\zeta_l(\tau-\beta)] = \exp(i2\pi l)\exp(-i\zeta_l\tau) = \exp(-i\zeta_l\tau).$$

Hence

$$G(\tau) = \beta^{-1}\sum_l \bar{G}(\zeta_l)\exp(-i\zeta_l\tau) \tag{2.29}$$

and

$$\bar{G}(\zeta_l) = \int_0^\beta d\tau \exp(i\zeta_l\tau)G(\tau) \tag{2.30}$$

The resultant function $G(\tau)$ is periodic in τ with period β and is equal to the temperature Green's function defined in equation (2.5) in the interval $-\beta < \tau \leqslant \beta$. Outside this interval the two functions differ. We find in practice, however, that we need them both only for values of τ in the interval. Formally, the two cases $\epsilon = \pm 1$ look the same, the only difference being in the allowed values of ζ_l.

We can again use the exact eigenstates and eigenvalues of H to find a spectral form for $\bar{G}(\zeta_l)$. If we substitute from equation (2.5) into equation (2.27) or (2.30), we have

$$\begin{aligned}\bar{G}(\zeta_l) &= -\int_0^\beta d\tau \exp(i\zeta_l\tau)Z^{-1} \\ &\quad\times \mathrm{Tr}\,\{\exp(-\beta H)\exp(H\tau)A\exp(-H\tau)B\} \\ &= -\int_0^\beta d\tau \exp(i\zeta_l\tau)Z^{-1} \\ &\quad\times \sum_{m,n}\exp(-\beta E_m)A_{mn}B_{nm}\exp((E_m-E_n)\tau)\end{aligned}$$

$$= -Z^{-1} \sum_{mn} \exp(-\beta E_m) A_{mn} B_{nm} \frac{\exp[(E_m - E_n + i\zeta_l)\beta] - 1}{E_m - E_n + i\zeta_l}$$

$$= -Z^{-1} \sum_{m,n} A_{mn} B_{nm} \frac{[\exp(-\beta E_m) + \epsilon \exp(-\beta E_n)]}{E_n - E_m - i\zeta_l}$$

$$= \int_{-\infty}^{\infty} dx \frac{A(x)}{i\zeta_l - x}, \tag{2.31}$$

where the spectral function $A(x)$ is that already defined in equation (2.13).

The same spectral function $A(x)$, therefore, determines all the Green's functions and one should expect that once one Green's function is known, the others can be determined from it. Formally, it would seem that one can obtain $G(\omega)$ (cf. equation (2.12)) from $\bar{G}(\zeta_l)$ simply by replacing ζ_l by $(-i\omega)$. This would certainly be true if one obtained $\bar{G}(\zeta_l)$ in the spectral form (2.31). In practice, however, this is not the case and some caution is required to obtain $G(\omega)$ from $\bar{G}(\zeta_l)$. The statement of the problem we need to solve is quite simple. We have a function $\bar{G}(\zeta_l)$ with values at the discrete points ζ_l. We wish to determine from it the function $G(\omega)$ which is analytic separately in the upper and lower half-planes of the space of the complex variable ω and which has the values $\bar{G}(\zeta_l)$ at the points $\omega = i\zeta_l$. Can we determine $G(\omega)$ uniquely? Stated in this form we can see that the problem does not have a unique solution. For if $G(\omega)$ is one solution, that is $G(\omega)$ is analytic in the upper and lower half-planes and

$$G(i\zeta_l) = \bar{G}(\zeta_l) \tag{2.32}$$

and $h(\omega)$ is any function analytic in the upper and lower half-planes, then

$$G'(\omega) = G(\omega) + (1 + \epsilon \exp(\omega\beta))h(\omega) \tag{2.33}$$

is also a solution. For $(1 + \epsilon \exp(\omega\beta))$ is also analytic in the lower and upper half-planes and

$$1 + \epsilon \exp(i\zeta_l\beta) = 1 - \epsilon^2 = 0. \tag{2.34}$$

Hence, $G'(\omega)$ is analytic and

$$G'(i\zeta_l) = G(i\zeta_l) = \bar{G}(\zeta_l).$$

There is, however, one further property of $G(\omega)$ which we have established, namely that $G(\omega)$ goes to zero at least as fast as ω^{-1} as $|\omega| \to \infty$. Formally, this can be stated as

$$\lim_{|\omega|\to\infty} \omega G(\omega) = C, \tag{2.35}$$

where C is a finite number (actually $\int_{-\infty}^{\infty} dx\, A(x)$) which may be zero. In that case, it can be shown from the theory of functions of a complex variable (Baym and Mermin 1961) that there is a unique function $G(\omega)$ which is analytic in the complex variable ω in the upper half-plane, which satisfies

$$G(\omega) = G(i\zeta_l) \qquad \text{for all } \zeta_l > 0, \tag{2.36}$$

and which has the behaviour of equation (2.35) as $|\omega| \to \infty$. (Note that only the existence of C, not its value, is required for the validity of the theorem). This means that if we have determined $\bar{G}(\zeta_l)$ and if, from it, we can find an analytic function $G(\omega)$ such that (2.35) and (2.36) are satisfied then we know that this is the required function $G(\omega)$. Since, in practice, $\bar{G}(\zeta_l)$ is given by a formula, there is usually no difficulty in finding an analytic function $G(\omega)$ which satisfies (2.36). The main skill or trickery needed is to ensure that (2.35) is satisfied as well. Of course, if $\bar{G}(\zeta_l)$ is found in the spectral form (2.31), substitution of $(-i\omega)$ for ζ_l automatically ensures that both conditions are satisfied. There is an obvious and similar theorem that enables $G(\omega)$ to be determined in the lower half-plane.

It is worth stressing that the simple relationship between $G(\omega)$ and $\bar{G}(\zeta_l)$ results partly from the judicious choice of constant factors in the definition of $G_R(t)$, $G(\tau)$ and of their Fourier transforms. Other choices of these constants could lead to functions $G(-i\zeta_l)$ and $\bar{G}(\zeta_l)$ which differ by a constant factor. This is a matter for which one should always be on the watch.

Since the relationship between the various Green's functions in the complex ω-plane will be referred to repeatedly, we illustrate the geography of the complex plane in Fig. 2.1. The diagram shows the regions where $G(\omega)$, $G_R(\omega)$, $G_A(\omega)$ and $\bar{G}(\zeta_l)$ exist.

In the analysis of this section we have used the density matrix

$$\rho = \exp(-\beta H)/Z,$$

which is suitable for describing a canonical ensemble. For interacting systems it is usually more useful to use the grand canonical ensemble, for which (Reif 1965)

$$\rho = \exp[-\beta(H-\mu N)]/\tilde{Z} = \exp(-\beta\tilde{H})/\tilde{Z},$$
$$\tilde{Z} = \mathrm{Tr}\exp(-\beta\tilde{H}),$$

where μ is the chemical potential.

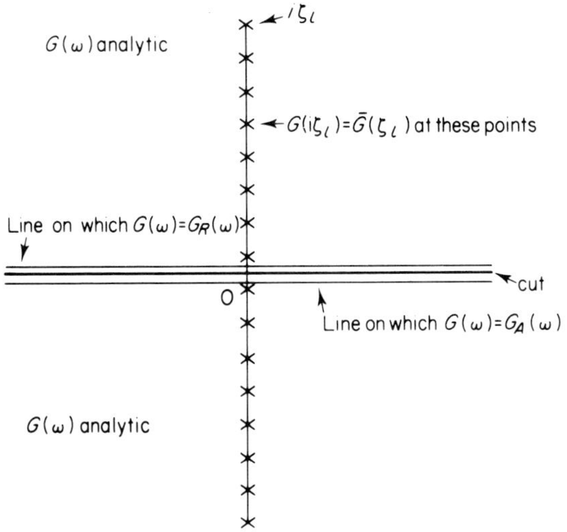

FIG. 2.1. The complex ω-plane, showing the region of analyticity of $G(\omega)$ and its relationship to $G_R(\omega)$, $G_A(\omega)$ and $\bar{G}(\zeta_l)$.

However, this requires no change in the formalism if all energies are measured from μ. For then the equations of motion are determined by $\tilde{H}$ not H. Thus all the preceding formulae are valid but with H replaced by $\tilde{H}$. Where confusion is unlikely we shall treat $\tilde{H}$ like H and not distinguish formally between the two.

§2.3 Single-particle Green's functions

(a) *General*

The relevance of single-particle Green's functions was discussed in §1.4. The single-particle temperature Green's function is defined by

$$G(\boldsymbol{r}\tau, \boldsymbol{r}'\tau') = \langle T\{\psi(\boldsymbol{r}, -i\tau)\,\psi^+(\boldsymbol{r}, -i\tau')\}\rangle \tag{2.37}$$

Note that the dependence of the operators on τ is given by,

$$\psi^+(\boldsymbol{r}, -i\tau) = \exp(H\tau)\,\psi^+(\boldsymbol{r})\exp(-H\tau),$$

$$\psi(\boldsymbol{r}, -i\tau) = \exp(H\tau)\,\psi(\boldsymbol{r})\exp(-H\tau).$$

Consequently, $[\psi(\boldsymbol{r}, -i\tau)]^+$ is equal to $\psi^+(\boldsymbol{r}, i\tau)$ and not to $\psi^+(\boldsymbol{r}, -i\tau)$ which the notation might suggest.

To avoid ambiguity, we introduce the notation,

$$\tilde{\psi}(\boldsymbol{r}, \tau) = \psi(\boldsymbol{r}, -i\tau), \qquad \tilde{\tilde{\psi}}(\boldsymbol{r}, \tau) = \psi^+(\boldsymbol{r}, -i\tau). \tag{2.38}$$

The spectral function is in this case

$$A(\boldsymbol{r},\boldsymbol{r}',x) = Z^{-1}(1+\epsilon\exp(-\beta x)) \tag{2.39}$$
$$\times \sum_{m,n} \exp(-\beta E_m)\psi(\boldsymbol{r})_{mn}\psi(\boldsymbol{r}')^+_{nm}\,\delta(x-E_n+E_m).$$

The diagonal part $A(\boldsymbol{r},\boldsymbol{r},\omega)$ is therefore real and we find

$$\begin{aligned}\operatorname{Im} G_R(\boldsymbol{r},\boldsymbol{r},\omega) &= \operatorname{Im}\int_{-\infty}^{\infty}\frac{A(\boldsymbol{r},\boldsymbol{r},x)}{\omega+i\delta-x}\,dx\\ &= -\pi\int_{-\infty}^{\infty} A(\boldsymbol{r},\boldsymbol{r},x)\,\delta(x-\omega)\,dx\\ &= -\pi A(\boldsymbol{r},\boldsymbol{r},\omega)\end{aligned} \tag{2.40}$$

and the diagonal part of the spectral function can be obtained immediately from the retarded Green's function.

The sum rule (2.15) now becomes

$$\int_{-\infty}^{\infty} dx\, A(\boldsymbol{r},\boldsymbol{r}',x) = \langle[\psi(\boldsymbol{r}),\psi^+(\boldsymbol{r}')]_\epsilon\rangle = \delta^3(\boldsymbol{r}-\boldsymbol{r}')$$

because of the commutation relations of the ψ's.

If the system is translationally invariant, one can go somewhat further than this. The correlation between two operators can depend only on the difference of positions. Thus $G(\boldsymbol{r},\boldsymbol{r}',\omega)$ is a function $G(\boldsymbol{r}-\boldsymbol{r}',\omega)$ of the difference $\boldsymbol{r}-\boldsymbol{r}'$. We introduce the space Fourier transform

$$\psi(\boldsymbol{k}) = \mathscr{V}^{-1/2}\sum \exp(i\boldsymbol{k}\,.\,\boldsymbol{r})c_{\boldsymbol{k}} \tag{2.41}$$

where V is the normalizing volume and $c_{\boldsymbol{k}}$ the destruction operator for a particle of momentum $\boldsymbol{k}$; since momentum is conserved in a translationally invariant system, averages of the form

$$\langle c_{\boldsymbol{k}}c^+_{\boldsymbol{k}'}\rangle$$

will be zero unless $\boldsymbol{k}'=\boldsymbol{k}$. Hence, if

$$G(\boldsymbol{r},\tau) = \frac{1}{\mathscr{V}}\sum_{\boldsymbol{k}}\exp(i\boldsymbol{k}\,.\,\boldsymbol{r})G(\boldsymbol{k},\tau), \tag{2.42}$$

we find

$$G(\boldsymbol{k},\tau) = -\langle T\{\tilde{c}_{\boldsymbol{k}}(\tau)\tilde{\tilde{c}}_{\boldsymbol{k}}(0)\}\rangle \tag{2.43}$$

and

$$A(\boldsymbol{k},x) = Z^{-1}(1+\epsilon \exp(-\beta x)) \times \sum_{mn} \exp(-\beta E_m)\, |c_{kmn}|^2\, \delta(x - E_n + E_m), \tag{2.44}$$

Again $A(\boldsymbol{k},x)$ is real and

$$\operatorname{Im} G_R(\boldsymbol{k},x) = -\pi A(\boldsymbol{k},x). \tag{2.45}$$

We shall see below that in a system of non-interacting particles $A(\boldsymbol{k}, \omega)$ is the density of states with momentum $\boldsymbol{k}$ and energy ω. Also in §8.7, it will be shown that the tunnelling current of particles, which we expect intuitively to be related to the density of states, is dependent on $A(\boldsymbol{k}, \omega)$. Thus in a translationally invariant system, the density of states with momentum $\boldsymbol{k}$ is *defined* to be

$$N_{\boldsymbol{k}}(\omega) = A(\boldsymbol{k}, \omega) = -(1/\pi) \operatorname{Im} G_R(\boldsymbol{k}, \omega).$$

Note that at the absolute zero of temperature

$$A(\boldsymbol{k}, \omega) = \sum_m |\langle m|c_{\boldsymbol{k}}^{+}|0\rangle|^2\, \delta(x - E_n + E_0).$$

Hence, $A(\boldsymbol{k}, \omega)$ simply counts the number of states with excitation energy ω and momentum $\boldsymbol{k}$ which are connected to the ground state through the addition of an extra particle. This is clearly what is meant by the density of states. The total density of states is thus

$$N(\omega) = (1/\mathscr{V}) \sum_{\boldsymbol{k}} A(\boldsymbol{k}, \omega) = -(1/\pi) \operatorname{Im} G_R(\boldsymbol{r}, \omega)|_{r=0}\,. \tag{2.46}$$

In a system that is not invariant under translations a local density of states is defined by

$$N(\boldsymbol{r}, \omega) = A(\boldsymbol{r}, \boldsymbol{r}, \omega) = -(1/\pi) \operatorname{Im} G(\boldsymbol{r}, \boldsymbol{r}, \omega) \tag{2.47}$$

with a total density of states

$$N(\omega) = \int N(\boldsymbol{r}, \omega)\, d^3r.$$

For the translationally invariant system the sum rule (2.15) becomes

$$\int_{-\infty}^{\infty} dx\, A(\boldsymbol{k},x) = \langle [c_{\boldsymbol{k}}, c_{\boldsymbol{k}}^{+}]_\epsilon \rangle = 1, \tag{2.48}$$

an exact result.

(b) *Independent particles*

To give some substance to the previous formal notions we consider a system of independent particles. In this case, the Hamiltonian is of the form

$$H = \sum_j \epsilon_i c_j^+ c_j \, , \tag{2.49}$$

where $c_j^\dagger$ creates a single particle in the state with normalized single-particle wave function $\phi_j(\boldsymbol{r})$ and single-particle energy ϵ_j. Hence

$$\psi(\boldsymbol{r}) = \sum_r c_r \phi(\boldsymbol{r}). \tag{2.50}$$

Because the different degrees of freedom i are independent,

$$\langle c_i c_j^+ \rangle = \langle c_i c_i^+ \rangle \, \delta_{ij} \, .$$

Hence

$$\begin{aligned} G_R(\boldsymbol{r}, \tau; \boldsymbol{r}', \tau') &= - \sum_j \phi_j(\boldsymbol{r}) \phi_j^*(\boldsymbol{r}') \, \langle [\tilde{c}_j(\tau), \tilde{\tilde{c}}_j(\tau')]_\epsilon \rangle \, \theta(t-t') \\ &= \sum_j \phi_j^*(\boldsymbol{r}) \phi_j^*(\boldsymbol{r}') \, G_R(j, t-t'). \end{aligned}$$

But, for independent particles

$$c_j(t) = c_j \exp(-i\epsilon_j t).$$

Therefore

$$\begin{aligned} G_R(j, t) &= -i \exp(-i\epsilon_j t) \, \langle [c_j, c_j^+]_\epsilon \rangle \, \theta(t) \\ &= -i \exp(-i\epsilon_j t) \theta(t). \end{aligned}$$

Hence

$$G(j, \omega) = 1/(\omega - \epsilon_j)$$

and

$$G(\boldsymbol{r}, \boldsymbol{r}', \omega) = \sum_j \frac{\phi_j(\boldsymbol{r}) \phi_j^*(\boldsymbol{r}')}{\omega - \epsilon_j} \, .$$

Therefore,

$$A(\boldsymbol{r}, \boldsymbol{r}', x) = \sum_j \phi_j(\boldsymbol{r}) \phi_j^* (\boldsymbol{r}') \delta(x - \epsilon_j). \tag{2.51}$$

In this case $A(\boldsymbol{r}, \boldsymbol{r}, x)$ clearly represents the local density of states at the point $\boldsymbol{r}$.

If the system is also translationally invariant,

$$\phi_j(\boldsymbol{r}) = \mathscr{V}^{-1/2} \exp(i\boldsymbol{k}\,.\,\boldsymbol{r})$$

and

$$A(\boldsymbol{k}, x) = \delta(x - \epsilon_k). \tag{2.52}$$

This is the density of states of a system of free particles and shows that a particle with momentum $\boldsymbol{k}$ definitely has the energy ϵ_k. The Green's function is

$$G(\boldsymbol{k}, \omega) = 1/(\omega - \epsilon_k) \tag{2.53}$$

and is an analytic function of ω except at the simple pole on the real axis $\omega = \epsilon_k$.

This case resembles that of the simple harmonic oscillator in §1.3.

(c) *Quasi-particles*

Many macroscopic systems, gases, liquids, and electrons in metals for example, behave as if they comprise nearly independent particles. How should we expect this behaviour to be reflected in the Green's functions? Let us, for simplicity, fix our attention on a translationally invariant system. Without interaction the Green's functions $G(\boldsymbol{k}, \omega)$ has a simple pole at $\omega = \epsilon_k$. However, the system will have other states with several particles excited which have the same momentum and energy as the single-particle state. If there is an interaction, we know from degenerate perturbation theory that these states will be coupled, the exact eigenstates being linear combinations of the original ones with energies spread out by the perturbation. Thus we expect $A(k, x)$ to change from a δ-function, to a function peaked near $x = \epsilon_k$ (which may be shifted from ϵ_k by the interaction) but with a finite spread. The stronger the interaction the greater the spread of energies and so the greater the width of $A(\boldsymbol{k}, x)$. However, the smaller the value of ϵ_k (energy measured from the chemical potential), the smaller the number of states that can be coupled to the original, that is the smaller the amount of phase space available for states to be coupled in, and so the smaller the width. We, therefore, expect that for ϵ_k small, $A(\boldsymbol{k}, x)$ is a sharply peaked function of x whose extent about the peak is increased with increasing interaction strength and is decreased as ϵ_k tends to zero.

We can be more specific about $A(\boldsymbol{k}, x)$ if we consider the time dependence of $G(\boldsymbol{k}, t)$. As we let $t \to \infty$, we expect correlations to decay, and ultimately to decay exponentially fast. Since for independent particles we have

$$G_R(\boldsymbol{k}, t) = -i \exp(-i\tilde{\epsilon}_k t)\theta(t),$$

we might expect that, for large times,

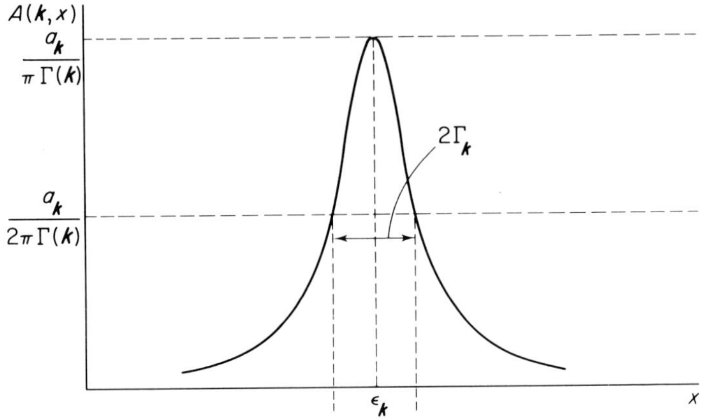

FIG. 2.2. The spectral function for a quasiparticle of energy $\epsilon_{\boldsymbol{k}}$, lifetime $\Gamma_{\boldsymbol{k}}^{-1}$.

$$G_R(\boldsymbol{k}, t) \sim -ia_k \exp(-i\tilde{\epsilon}_k t) \exp[-\Gamma(\boldsymbol{k})t]\,\theta(t). \qquad (2.54)$$

where $\Gamma(k)^{-1} > 0$ is the lifetime (cf. the damped harmonic oscillator, §1.3). Since the long time behaviour of $G_R(\boldsymbol{k}, t)$ determines the behaviour of $G_R(\boldsymbol{k}, \omega)$ for small, we expect that as $\omega \to 0$,

$$G_R(\boldsymbol{k}, \omega) \sim a_k[\omega + i\Gamma(\boldsymbol{k}) - \tilde{\epsilon}_k]^{-1}. \qquad (2.55)$$

Considered as a function of ω this has a pole at $\epsilon - i\Gamma(k)$. It is, therefore, analytic in the upper half-plane and this is as it should be.

For a translationally invariant system, equation (2.45) is generally valid. Hence

$$\begin{aligned} A(\boldsymbol{k}, x) &= \frac{1}{\pi} \operatorname{Im} \frac{a_k}{\tilde{\epsilon}_k - x - i\Gamma(\boldsymbol{k})} \\ &= \frac{1}{\pi} \frac{\Gamma(\boldsymbol{k}) \operatorname{Re} a_k + (\epsilon_k - x) \operatorname{Im} a_k}{(\tilde{\epsilon}_k - x)^2 + \Gamma(\boldsymbol{k})^2}. \end{aligned} \qquad (2.56)$$

From equation (2.44), we see that for fermions $A(\boldsymbol{k}, x)$ is never negative. Hence we must have a_k real and positive. Therefore

$$A(\boldsymbol{k}, x) = \frac{1}{\pi} \frac{a_k \Gamma(\boldsymbol{k})}{(\tilde{\epsilon}_k - x)^2 + \Gamma(\boldsymbol{k})^2}. \qquad (2.57)$$

This spectral function has a Lorentzian shape and is illustrated in Fig. 2.2.

From equation (2.57) we find

$$\int_{-\infty}^{\infty} A(\boldsymbol{k}, x)\, dx = a_k.$$

According to equation (2.48), this should imply $a_k = 1$. However, we have discussed only the long-time behaviour of $G(\boldsymbol{k}, t)$ and, therefore, the small x behaviour of $A(\boldsymbol{k}, x)$. There are expected to be other contributions to $A(k, x)$ for large values of x and these will also contribute to the integral. Hence a_k is often found to be different from unity.

The forms (2.56), (2.57) for $A(\boldsymbol{k}, x)$ show the spread of the spectral function about $x = \tilde{\epsilon}_k$ that one expects. A system for which (2.55) is a good approximation for $G(\boldsymbol{k}, \omega)$ is said to possess quasi-particles with energy ϵ_k and life-time $\Gamma(\boldsymbol{k})$. The approximation is best for $\tilde{\epsilon}_k$ small and, as $\tilde{\epsilon}_k \to 0$, we expect $\Gamma(\boldsymbol{k}) \to 0$. If one compares this form of $G(\boldsymbol{k}, \omega)$ with that before equation (2.51), one sees that a_k changes the weight of the wave function and it is therefore, often called the wave function renormalization. This quasi-particle picture is discussed in more detail in §§3.5 and 6.2

§2.4 Higher-order Green's functions

All the Green's functions defined in §2.1 can be generalized in many ways to cases where three or more operators are involved. The most useful generalization in the sense that they give the response of the system to some cause or that they arise naturally in computations are

$$G_R(t_1, t_2, t_3, \ldots, t_n) = (-i)^n \langle [\ldots[[A(t_1), B(t_2)], C(t_3)]\ldots] \rangle \times \theta(t_1 - t_2)\theta(t_2 - t_3)\ldots, \tag{2.58}$$

$$G_C(t_1, t_2, t_3, \ldots, t_n) = (-i)^n \langle T\{A(t_1)B(t_2)C(t_3)\ldots\}\rangle, \tag{2.59}$$

$$G(\tau_1, \tau_2, \tau_3, \ldots, \tau_n) = (-1)^n \langle T\{A(-i\tau_1)B(-i\tau_2)C(-i\tau_3)\ldots\}\rangle = (-1)^n \langle T\{\tilde{A}(\tau_1)\tilde{B}(\tau_2)\tilde{C}(\tau_3)\ldots\}\rangle, \tag{2.60}$$

where

$$\tilde{A}(\tau) = A(-i\tau) = \exp(H\tau)A\exp(-H\tau).$$

In equation (2.59), T is Wick's time-ordering operator which orders the operators according to the time label with the earliest time to the right. If the operators are Fermi-type operators (either Fermi operators or a sum of products of odd numbers of Fermi operators) the whole expression is multiplied by (-1) if the order of the operators is an odd permutation of that appearing in (2.59). In equation (2.60), T has a similar meaning except that it orders according to the "imaginary time" label τ. If $n/2$ of the operators are creation operators for particles and $n/2$ destruction operators, the function is called an $n/2$-particle Green's function.

The Green's functions (2.58) arise naturally when the non-linear response of the system is being studied. The Hall effect in a metal or semiconductor depends on a function of this kind. Green's functions of the type (2.59) are often used in the calculation of properties of systems in the ground state. Those of type (2.60) arise naturally in the computation of thermodynamic properties of macroscopic systems. We shall concentrate attention on the latter ones. The general theory of the functions (2.58) is quite complicated and not very revealing. Individual cases can be dealt with quite straightforwardly generalizing the methods of this book. (Applications of four-particle Green's functions are discussed in conference proceedings edited by Woods Halley (1978).)

Let us consider the functions (2.60). It is possible again to exploit the cyclic property of the trace to show that if $0 < \tau_i \leqslant \beta$

$$G(\tau_1, \tau_2, \ldots, \tau_i = 0, \tau_n) = -\epsilon\, G(\tau_1, \tau_2, \ldots, \tau_i = \beta, \ldots, \tau_n) \tag{2.61}$$

where all the τ's except the ith one are the same on the two sides of the equation. Take a special case to illustrate the proof of (2.61). Let A, B, C, D be Fermi-type operators and

$$G(\tau_1, \tau_2, \tau_3, \tau_4) = (-1)^4 \langle T\{A(\tau_1)B(\tau_2)C(\tau_3)D(\tau_4)\}\rangle.$$

Then for

$$0 < \tau_4 < \tau_2 < \tau_3 < \tau_1 = \beta,$$

$$G(\beta, \tau_2, \tau_3, \tau_4) = -(-1)^4 \langle A(\beta)C(\tau_3)B(\tau_2)D(\tau_4)\rangle.$$

The negative sign arises because the operators are Fermi operators and $ABCD$ is an odd permutation of $ABCD$. We also have that for

$$0 = \tau_1 < \tau_4 < \tau_2 < \tau_3 < \beta$$

$$G(0, \tau_2, \tau_3, \tau_4) = \langle C(\tau_3)B(\tau_2)D(\tau_4)A(0)\rangle.$$

But

$$\begin{aligned} G(\beta, \tau_2, \tau_3, \tau_4) &= -Z^{-1}\,\mathrm{Tr}\,[\exp(-\beta H)\exp(\beta H)A\exp(-\beta H) \\ &\quad \times C(\tau_3)B(\tau_2)D(\tau_4)] \\ &= -Z^{-1}\,\mathrm{Tr}\,[A\exp(-\beta H)C(\tau_3)B(\tau_2)D(\tau_4)] \\ &= -Z^{-1}\,\mathrm{Tr}\,[\exp(-\beta H)C(\tau_3)B(\tau_2)D(\tau_4)A(0)] \\ &= -G(0, \tau_2, \tau_3, \tau_4). \end{aligned} \tag{2.62}$$

This is equation (2.61) for the special case considered. In general, $G(\tau_1, \tau_2, \ldots)$ satisfies first-order differential equations in the variables $\tau_1, \tau_2, \ldots, \tau_n$. The conditions (2.61) are then boundary conditions sufficient to determine the Green's function throughout the domain

$$0 \leqslant \tau_i \leqslant \beta \qquad \text{for} \quad i = 1, \ldots, n.$$

The boundary condition can be satisfied automatically by expanding G in terms of the exponentials

$$\exp\left[-i \sum_{j=1}^{n} \zeta_{lj}\tau_j\right]$$

where the numbers ζ_{lj} are chosen such that

$$\exp(-i\zeta_{lj}\beta) = -\epsilon. \tag{2.63}$$

Hence, as in §2.2, $(\beta\zeta_{lj}/2\pi)$ are even integers for bosons and odd integers for fermions. The full expansion is

$$G(\tau_1, \tau_2, \ldots, \tau_n) = \beta^{-(n-1)} \sum_{\zeta_{lj}} \bar{G}(\zeta_{l1}, \zeta_{l2}, \ldots, \zeta_{ln}) \times \exp\left[-i \sum_{j=1}^{n} \zeta_{lj}\tau_j\right]. \tag{2.64}$$

When the Hamiltonian is independent of time G cannot depend on the origin of τ. Hence

$$G(\tau_1 + \alpha, \tau_2 + \alpha, \ldots, \tau_n + \alpha) = G(\tau_1, \tau_2, \ldots, \tau_n)$$

for all α. This will be true if, and only if, the allowed numbers ζ_{lj} are such that

$$\sum_{j=1}^{n} \zeta_{lj} = 0. \tag{2.65}$$

There are therefore only $(n-1)$ independent variables ζ_{lj}. In the case of the second-order Green's function we arrive back at the expansion given in §2.2.

§2.5 Kramers–Krönig relations

We have seen that the principle of causality leads to the conclusion that $G(\omega)$ is analytic in the upper half-plane of the complex variable ω. This in turn can be used to show that the real and imaginary parts of $G_R(\omega)$ are not independent but are related by (Kubo 1957),

$$\operatorname{Re} G_R(y) = \frac{1}{\pi} P \int_{-\infty}^{\infty} dx \, \frac{\operatorname{Im} G_R(x)}{x - y} \tag{2.66}$$

and

$$\operatorname{Im} G_R(y) = -\frac{1}{\pi} P \int_{-\infty}^{\infty} dx \, \frac{\operatorname{Re} G_R(x)}{x - y}. \tag{2.67}$$

These equations, which have a very wide validity, are known as Kramers–Krönig relations or dispersion relations. They enable the real part of $G_R(y)$ to be determined in terms of the imaginary part, if known, or vice versa. They are often used as a check on experimental results when both the real and imaginary parts of the admittance can be measured.

When the spectral function $A(x)$ is real, as for a translationally invariant system, equation (2.66) is consistent with the usual spectral form with

$$A(x) = -\pi^{-1} \operatorname{Im} G_R(x). \tag{2.68}$$

§2.6 Fluctuation-dissipation theorem

When the two operators A, B in the double-time Green's function (2.1) are identical and of the Bose type, the spectral function $A(x)$ is real and satisfies equation (2.68). Expansion of $A(x)$ in terms of the exact eigenfunctions as in equation (2.13) then leads to the relation (Kubo 1957).

$$-2 \operatorname{Im} G_R(\omega) = \pi(1 - \exp(-\beta\omega)) \langle B_\omega B_\omega^+ \rangle, \tag{2.69}$$

where

$$\langle B_\omega B_\omega^+ \rangle = \int dt \exp(i\omega t) \langle B(t) B^+(0) \rangle \tag{2.70}$$

is the mean-square fluctuation at angular frequency ω in the operator B. Since the dissipation of the energy of a system is often proportional to $\operatorname{Im} G_R(\omega)$, equation (2.69) relates dissipation to the mean-square fluctuation of an appropriate operator and is known as the fluctuation-dissipation theorem. As a special case the d.c. conductivity is related to the mean-square fluctuation in the current density.

§2.7 Equations of motion for Green's functions

How are Green's functions to be determined in a particular case? Save in very simple cases, such as that of independent particles, they are usually determined by some method of approximation and most of the remainder of this book is concerned with methods of approximation. Many approximations, usually the ones where the Green's function approach is of most importance, start from the equations of motion satisfied by Green's functions. We turn in this section to the derivation of these equations. Their form depends on the system, in particular on the Hamiltonian which describes the system. As an illustration, we consider here the case of a system of particles interacting through a two-body potential $V(\boldsymbol{r})$. The Hamiltonian is then (see Appendix A)

$$H = \int d^3\boldsymbol{r}\,\psi^+(\boldsymbol{r})h_0(\boldsymbol{r})\psi(\boldsymbol{r}) + \tfrac{1}{2}\int \psi^+(\boldsymbol{r})\psi^+(\boldsymbol{r}')V(\boldsymbol{r}-\boldsymbol{r}')\psi(\boldsymbol{r}')\psi(\boldsymbol{r})d^3\boldsymbol{r}d^3\boldsymbol{r}', \tag{2.71}$$

where $h_0(\boldsymbol{r})$ describes any one-body forces acting on the particles and includes the chemical potential, and $\psi(\boldsymbol{r})$ is the operator which destroys a particle at position $\boldsymbol{r}$. We have the commutation relations

$$\begin{aligned}[\psi(\boldsymbol{r}),\ \psi^+(\boldsymbol{r}')]_\epsilon &= \psi(\boldsymbol{r})\psi^+(\boldsymbol{r}') + \epsilon\psi^+(\boldsymbol{r}')\psi(\boldsymbol{r}) = \delta(\boldsymbol{r}-\boldsymbol{r}'),\\ [\psi(\boldsymbol{r}),\ \psi(\boldsymbol{r}')]_\epsilon &= 0.\end{aligned} \tag{2.72}$$

In the case of fermions $\psi(\boldsymbol{r})$ is a column vector with components $\psi_\sigma(\boldsymbol{r})$ for destroying particles with spin up and down. Equations (2.72) are then matrix equations, the first, written out in terms of its components, being

$$[\psi_\sigma(\boldsymbol{r}),\ \psi^+_{\sigma'}(\boldsymbol{r}')]_\epsilon = \delta_{\sigma\sigma'}\delta(\boldsymbol{r}-\boldsymbol{r}'). \tag{2.73}$$

Formally, we can deal with spin using equations (2.71) and (2.72) as well as the following conventions. The variable $\boldsymbol{r}$ is read as standing for both coordinate and spin label $(\boldsymbol{r}, \boldsymbol{\sigma})$ and an integral over $\boldsymbol{r}$ is interpreted as implying a sum over the spin label as well. Further, the δ-function $\delta(\boldsymbol{r}-\boldsymbol{r}')$ is interpreted as the product $\delta(\boldsymbol{r}-\boldsymbol{r}')\delta_{\sigma\sigma'}$. With these conventions the equations below also apply when the particles have spin.

For particles moving in an external scalar potential $U(\boldsymbol{r})$ and a vector potential $\boldsymbol{A}(\boldsymbol{r})$, $h_0(\boldsymbol{r})$ has the form

$$h_0(\boldsymbol{r}) = (1/2m)[-i\boldsymbol{\nabla} - (e/c)\boldsymbol{A}(\boldsymbol{r})]^2 - \mu + U(\boldsymbol{r}).$$

The operators A and B in the Green's function required are usually combinations of particle operators. Hence the Green's function is usually a combination of many-particle Green's functions and its τ-dependence will therefore depend on that of the field operators.

Let us start with their equations of motion. Heisenberg's equations and the commutation relations (2.72) yield

$$-\partial\tilde{\psi}/\partial\tau = [\tilde{\psi}, H]$$
$$= h_0(r)\psi(r,\tau) + \int d^3r'\,\tilde{\tilde{\psi}}(r',\tau)\,V(r-r')\psi(r',\tau)\psi(r,\tau), \tag{2.74}$$

$$\partial\tilde{\tilde{\psi}}/\partial\tau = h_0(r)\tilde{\tilde{\psi}}(r,\tau) + \int d^3r'\,\tilde{\tilde{\psi}}(r,\tau)\tilde{\tilde{\psi}}(r',\tau)V(r-r')\psi(r',\tau). \tag{2.75}$$

These equations can now be used to determine the derivatives with respect to τ of any Green's functions which depend on the particle operators.

Consider the important case of the single-particle temperature Green's function

$$G(r,\tau;r',\tau') = -\langle T\tilde{\psi}(r,\tau)\tilde{\tilde{\psi}}(r',\tau')$$
$$= -\langle\tilde{\psi}(r,\tau)\tilde{\tilde{\psi}}(r',\tau')\rangle\,\theta(\tau-\tau') + \epsilon\langle\tilde{\tilde{\psi}}(r',\tau')\tilde{\tilde{\psi}}(r,\tau)\rangle\,\theta(\tau'-\tau). \tag{2.76}$$

This depends on τ through the operator $\tilde{\psi}(\tau)$ as well as through the theta-function and both of these must be taken into account when differentiating. Hence

$$\frac{\partial}{\partial\tau}G(r,\tau;r',\tau') = -\langle[\tilde{\psi}(r,\tau), \tilde{\tilde{\psi}}(r',\tau')]_\epsilon\rangle\,\delta(\tau-\tau')$$
$$+ h_0(r)\langle T\tilde{\psi}(r,\tau)\tilde{\tilde{\psi}}(r',\tau')\rangle \tag{2.77}$$
$$+\int d^3r''\,V(r-r'')\langle T\tilde{\tilde{\psi}}(r'',\tau)\tilde{\psi}(r'',\tau)\tilde{\psi}(r,\tau)\tilde{\tilde{\psi}}(r',\tau')\rangle.$$

Because of the delta-function in the first term on the right-hand side, τ and τ' can be put equal in the remaining factor which, as a result of the commutation relation of $\tilde{\psi}$ and $\tilde{\tilde{\psi}}$, becomes $\delta(r-r')$. Hence

$$[(\partial/\partial\tau) + h_0(r)]\,G(r,\tau;r',\tau') = -\delta(r-r')\delta(\tau-\tau')$$
$$+\int d^3r''\,V(r-r'')G(r'',\tau,r,\tau;r',\tau',r'',\tau_2), \tag{2.78}$$

where

$$G(r_1,\tau_1,r_2,\tau_2 : r_3,\tau_3,r_4,\tau_4)$$
$$= \langle T\tilde{\psi}(r_1,\tau_1)\tilde{\psi}(r_2,\tau_2)\tilde{\tilde{\psi}}(r_4,\tau_4)\tilde{\tilde{\psi}}(r_3,\tau_3)\rangle$$

and the limit $\tau_2 \to 0$ is to be taken with $\tau_2 > \tau_1 > \tau$. In the same way we can find an equation for $\partial G(r,\tau;r',\tau')/\partial\tau'$. When the Hamiltonian does not depend on τ explicitly, G depends on the difference $(\tau-\tau')$ only and the second equation yields no new information.

Equation (2.78) by itself is not very helpful because although it gives the time derivative of $G(\boldsymbol{r}, \tau; \boldsymbol{r}', \tau')$ in terms of $G(\boldsymbol{r}'', \tau, \boldsymbol{r}'', \tau_1, \boldsymbol{r}, \tau_2; \boldsymbol{r}', \tau')$, the latter is unknown. One can, however, determine the equation satisfied by the derivative of the latter function. This equation introduces a new Green's function which depends on the product of six single-particle operators. The process can be continued to yield an infinite hierarchy of equations, successive equations introducing Green's functions containing two more single-particle operators in the product. The Green's functions have to be determined by the solution of the complete set of equations together with the boundary conditions. In practice this is done by introducing an approximation which reduces the infinite set of equations to a finite number, or, as in Chapter 3, by perturbation theory.

§2.8 The thermodynamic potential

We have shown that Green's functions arise naturally in the theory of the response of a many-body system to an external perturbation. More artificially, the equilibrium properties of the system can be related to appropriate Green's functions. This is not altogether surprising because the specific heat can be written in terms of the fluctuations in the energy of the system and fluctuations can, as we have seen, be written in terms of Green's functions.

All the thermodynamic functions can be derived from an appropriate thermodynamic potential. In many-body theory, it is convenient to work with varying numbers of particles and fixed chemical potential. We, therefore, choose a grand canonical ensemble and use the thermodynamic potential Ω defined by

$$\Omega = -kT \ln Z, \tag{2.79}$$

$$Z = \operatorname{Tr} \exp[-\beta H]. \tag{2.80}$$

Here H is the Hamiltonian with the term μN subtracted from it and the trace is taken over states with any numbers of particles. The entropy and specific heat are obtained from Ω through the thermodynamic relations.

$$S = -(\partial\Omega/\partial T)_{\mathscr{V},\mu}, \qquad C_V = T(\partial S/\partial T)_{\mathscr{V},\mu}. \tag{2.81}$$

One method to relate Ω to Green's functions is to vary the strength of the interaction. If the Hamiltonian is that defined in equation (2.71), we consider, first, the Hamiltonian

$$H(\lambda) = H_0 + \lambda V,$$

$$H_0 = \int d^3r\,\psi^+(r)h_0(r)\,\psi(r),$$

$$V = \frac{1}{2}\int d^3r\,d^3r'\,\psi^+(r)\psi^+(r')V(r-r')\psi(r')\psi(r). \qquad (2.82)$$

The Hamiltonian of the original problem is $H(1)$. The thermodynamic potential corresponding to $H(\lambda)$ is

$$\Omega(\lambda) = -kT\ln \mathrm{Tr}\exp[-\beta H(\lambda)].$$

Let us look first at the derivative of this potential,

$$\partial\Omega/\partial\lambda = Z^{-1}\,\mathrm{Tr}\exp[-\beta H(\lambda)]\,V = \langle V\rangle, \qquad (2.83)$$

the expectation value of the potential energy in the system $H_0 + \lambda V$. Since V involves the product of four field operators, it is possible to write the right-hand side as a two-particle Green's function. However, it is usually more convenient to exploit the equations of motion to write the right-hand side in terms of a single-particle Green's function. For the Hamiltonian $H(\lambda)$, the equation of motion (2.78) becomes

$$\begin{aligned}[(\partial/\partial\tau) + h_0(r)]\,G(r,\tau;r',\tau') &= -\delta(\tau-\tau')\delta(r-r')\\ &+\lambda\int d^3r''\,V(r-r'')\langle T[\tilde{\bar{\psi}}(r'',\tau)\tilde{\psi}(r'',\tau)\tilde{\psi}(r,\tau)\tilde{\bar{\psi}}(r',\tau')]\rangle.\end{aligned} \qquad (2.84)$$

We can obtain $\langle V\rangle$ from the last term by putting $r' = r$, $\tau' > \tau$, letting τ' tend to τ and integrating over r. In this way we obtain

$$\lambda\langle V\rangle = -\epsilon \lim_{\tau'\to\tau_+}\frac{1}{2}\int d^3r \lim_{r'\to r}\left\{\left[\frac{\partial}{\partial\tau} + h_0(r)\right]\times G(r,\tau;r',\tau') + \delta(\tau-\tau')(r-r')\right\}.$$

Using this result and integrating equation (2.83) over λ between 0 and 1, we obtain

$$\Omega(1) - \Omega(0) = -\frac{\epsilon}{2}\int_0^1\frac{d\lambda}{\lambda}\int d^3r\left\{\left[\frac{\partial}{\partial\tau} + h_0(r)\right]\times G(r,\tau;r',\tau') + \delta(\tau-\tau')\delta(r-r')\right\}\Bigg|_{r'=r,\tau'=\tau_+}. \qquad (2.85)$$

In this expression $\Omega(1)$ is the potential required and $\Omega(0)$ that for the non-interacting particles. Usually $\Omega(0)$ can be calculated quite easily and then equation (2.85) is an explicit formula for $\Omega(1)$. If the

Fourier transforms introduced in §2.2 are used, the equation can be rewritten

$$\Omega(1) - \Omega(0) = -\frac{\epsilon}{2\beta}\sum_l \int_0^1 \frac{d\lambda}{\lambda}\int d^3r\, \{[-i\zeta_l + h_0(\boldsymbol{r})] \times \bar{G}(\boldsymbol{r},\boldsymbol{r}',\zeta_l) + \delta(\boldsymbol{r}-\boldsymbol{r}')\exp(-i\zeta_l 0_-)\}|_{\boldsymbol{r}'=\boldsymbol{r}}, \tag{2.86}$$

where 0_+ is a small positive quantity $(\tau' - \tau)$ which tends to zero. If $h_0(\boldsymbol{r})$ is independent of position, being only a function of derivatives,

$$h_0(\boldsymbol{r}) = h(-i\boldsymbol{\nabla}),$$

the relation can also be written in terms of the spatial Fourier transforms of $\bar{G}$ (cf. equation (2.42)), as

$$\Omega(1) - \Omega(0) = -\frac{\epsilon}{2\beta}\sum_l \sum_{\boldsymbol{k}} \int \frac{d\lambda}{\lambda}\{[-i\zeta_l + h(\boldsymbol{k})]\, G(\boldsymbol{k},\zeta_l) + 1\} \times \exp(-i\zeta_l 0_-)|. \tag{2.87}$$

This form for the thermodynamic potential is often used in practice.

§2.9 An equation for a single-particle Green's function in an external field

It is sometimes convenient to define generalized Green's functions in the presence of external fields. This is of interest for itself as well as providing an alternative method for finding the effect of an external field. Since it is as easy to work with the generalized function as with the originals we shall base the general theory on these functions as far as possible. To be specific, consider the temperature Green's functions and choose the Hamiltonian to be given by

$$H(\tau) = H_0(\tau) + V,$$

$$H_0(\tau) = \int d^3r\, \psi^+(\boldsymbol{r}) h_0(\boldsymbol{r},\tau)\psi(\boldsymbol{r}),$$

$$h_0(\boldsymbol{r},\tau) = (1/2m)[-i\boldsymbol{\nabla} - (e/c)\boldsymbol{A}(\boldsymbol{r})]^2 - \mu + U(\boldsymbol{r},\tau), \tag{2.88}$$

where V is the two-body interaction given by equation (2.82). Further, we define (for $0 \leqslant \tau, \tau' \leqslant \beta$) a generalized Green's function by

$$G(\boldsymbol{r},\tau;\boldsymbol{r}',\tau';U) = -\operatorname{Tr}[TS(\beta)\psi_\tau(\boldsymbol{r})\psi^+_{\tau'}(\boldsymbol{r}')]/\operatorname{Tr} S(\beta), \tag{2.89}$$

where

$$S(\beta) = \exp\left\{-\int_0^\beta d\tau\, H(\tau)\right\}$$
$$\equiv \lim_{\Delta\tau\to 0_T} \prod_{\tau_r=0}^{\beta} [1-\Delta\tau\, H(\tau_r)]\,, \qquad \tau_r = r\Delta\tau,$$

and all the operators in the Green's function, including $H(\tau)$ and $S(\beta)$, are ordered according to time with the earliest on the right. The subscripts on the field operators indicate their position in the order. For example for $\tau > \tau'$, we have

$$G(\boldsymbol{r},\tau;\boldsymbol{r}'\tau';U) = -\mathrm{Tr}\left\{T\exp\left[-\int_\tau^\beta H(\tau'')d\tau''\right]\right.$$
$$\times\ \psi(\boldsymbol{r})T\exp\left[-\int_{\tau'}^{\tau} H(\tau'')d\tau''\right]$$
$$\left.\times\ \psi^+(\boldsymbol{r}')T\exp\left[-\int_0^{\tau'} H(\tau'')d\tau''\right]\right\}\Big/\mathrm{Tr}\,S(\beta).$$

When U is independent of τ, this definition of Green's function agrees with the previous one. Similar definitions can be given for the many-particle temperature Green's functions.

To obtain the equation for G rewrite it in the form

$$G(\boldsymbol{r},\tau;\boldsymbol{r}',\tau';U) = -Z^{-1}\,\mathrm{Tr}\,\{S(\beta)\tilde{\psi}(\boldsymbol{r},\tau)\tilde{\tilde{\psi}}(\boldsymbol{r}',\tau')\}\,\theta(\tau-\tau')$$
$$+\,\epsilon Z^{-1}\,\mathrm{Tr}\,\{S(\beta)\tilde{\tilde{\psi}}(\boldsymbol{r}',\tau')\tilde{\psi}(\boldsymbol{r},\tau)\}\,\theta(\tau'-\tau); \tag{2.90}$$

here

$$S(\beta) = T\exp\left[-\int_0^\beta H(\tau')d\tau'\right], \qquad Z = \mathrm{Tr}\,S(\beta)$$
$$\tilde{\psi}(\boldsymbol{r},\tau) = S(\tau)^{-1}\,\psi(\boldsymbol{r})S(\tau),$$
$$\tilde{\tilde{\psi}}(\boldsymbol{r}',\tau') = S(\tau)^{-1}\,\psi^+(\boldsymbol{r}')S(\tau).$$

It follows that $S(\tau)$, $S^{-1}(\tau)$ satisfy the equations

$$\partial S(\tau)/\partial\tau = -H(\tau)S(\tau), \qquad \partial S^{-1}/\partial\tau = S^{-1}H(\tau)$$

and $\tilde{\psi}(\boldsymbol{r},\tau)$ satisfies

$$\partial\tilde{\psi}(\boldsymbol{r},\tau)/\partial\tau = [H(\tau),\,\tilde{\psi}(\boldsymbol{r},\tau)]\,.$$

The last equation is the same as that in §2.7 except that H is replaced by $H(\tau)$, that is $U(\boldsymbol{r})$ is replaced by $U(\boldsymbol{r},\tau)$. The equation of motion for the new Green's function is, therefore, the same as that derived in §2.7 except that $U(\boldsymbol{r})$ is replaced by $U(\boldsymbol{r},\tau)$ or $h_0(\boldsymbol{r})$ is replaced by $h_0(\boldsymbol{r},\tau)$. Therefore,

$$[(\partial/\partial\tau) + h_0(\boldsymbol{r}, \tau)]\, G(\boldsymbol{r}, \tau; \boldsymbol{r}', \tau'; U) = -\delta(\tau - \tau')\delta(\boldsymbol{r} - \boldsymbol{r}')$$

$$+ \int d^3r''\, V(\boldsymbol{r} - \boldsymbol{r}'')\, \mathrm{Tr}\,\{T\, S(\beta)\tilde{\tilde{\psi}}(\boldsymbol{r}'', \tau)\tilde{\psi}(\boldsymbol{r}'', \tau)\tilde{\psi}(\boldsymbol{r}, \tau)$$

$$\times\, \tilde{\tilde{\psi}}(\boldsymbol{r}', \tau')\}/\mathrm{Tr}\, S(\beta) \qquad (2.91)$$

$$= -\delta(\tau - \tau')\delta(\boldsymbol{r} - \boldsymbol{r}') + \int d^3r''\, V(\boldsymbol{r} - \boldsymbol{r}'')G(\boldsymbol{r}'', \tau, \boldsymbol{r}, \tau_-; \boldsymbol{r}', \tau', \boldsymbol{r}'', \tau_+).$$

Similar equations can be derived for the higher-order Green's functions.

We can expand the one-particle Green's function to first order in the field U as follows. Since the T operator in equation (2.89) orders the other operators according to time, the operators can be treated as c-numbers as long as they keep the time label. Thus, if we put

$$H_{1\tau} = H(\tau) - \int d^3r\, \psi_\tau^+(r)U(r, \tau)\psi_\tau(r) \equiv H_1 - H'(\tau), \qquad (2.92)$$

then

$$S(\beta) = \exp\left[-\int_0^\beta d\tau\, H(\tau)\right]$$

$$= \exp\left\{-\int_0^\beta d\tau[H_{1\tau} + H'(\tau)]\right\}$$

$$\approx \exp\left\{-\int_0^\beta d\tau\, H_{1\tau}\right\}\left[1 - \int_0^\beta d\tau'\, H'(\tau')\right]. \qquad (2.93)$$

As with the field operators $H_{1\tau}$ does not depend explicitly on time but carries the time label to show its order. Then the first-order change in G, induced by the field U is

$$\delta G(r, \tau; r', \tau') = \mathrm{Tr}\left[T\, S_1(\beta) \int_0^\beta d\tau''\, H'(\tau'')\psi_\tau(r)\psi_\tau^+(r')\right] \Big/ \mathrm{Tr}\, T\, S_1(\beta)$$

$$- \frac{\mathrm{Tr}[T\, S_1(\beta)\psi_\tau(r)\psi_\tau^+(r')]\, \mathrm{Tr}[T\, S_1(\beta) \int_0^\beta d\tau\, H'(\tau)]}{[\mathrm{Tr}\, T\, S_1(\beta)]^2}$$

$$= \int_0^\beta d\tau''\, \langle T[\tilde{H}'(\tau'')\tilde{\psi}(r, \tau)\tilde{\tilde{\psi}}(r', \tau')\rangle$$

$$- \int_0^\beta d\tau''\, \langle T\tilde{H}'(\tau'')\rangle\, \langle T[\tilde{\psi}_\tau(r, \tau)\tilde{\tilde{\psi}}(r', \tau']\rangle, \qquad (2.94)$$

where now the averages are taken with respect to the system

described by the constant Hamiltonian H_1. When $\tau = \tau'$, the first term on the right-hand side of equation (2.94) is a double time Green's function from which the response of the system can be calculated (cf. §2.2 and the end of §3.2). The last term of (2.94) is often zero or easily calculated. Thus the response of the system can be calculated from δG, a method we use in §3.7, for example.

§2.10 Summary of results

The double-time temperature Green's function is defined by equation (2.5). The single-particle Green's function

$$G_1(\boldsymbol{r}, \tau; \boldsymbol{r}', \tau') = -\langle T\{\tilde{\psi}(\boldsymbol{r}, \tau)\tilde{\bar{\psi}}(\boldsymbol{r}', \tau')\}\rangle \tag{2.95}$$

satisfies the equation of motion (2.78) when there is a two-body interaction between the particles. This is only one of a hierarchy of equations. Thermodynamic properties can be obtained from the single-particle Green's function through equation (2.87).

In the absence of an external potential $G_1(\boldsymbol{r}, \tau; \boldsymbol{r}', \tau')$ is a function $G(\boldsymbol{r}, \boldsymbol{r}', \tau - \tau')$ of the "time" difference $(\tau - \tau')$ and not of τ and τ' separately. This Green's function satisfies the boundary conditions.

$$G_1(\boldsymbol{r}, \boldsymbol{r}', \beta) = -\epsilon G_1(\boldsymbol{r}, \boldsymbol{r}', 0). \tag{2.96}$$

The equations of motion together with the boundary condition (2.96) are sufficient for the unique determination of the single-particle Green's function.

The more general double-time Green's function, $G(\tau, \tau')$ defined by equation (2.5) is also (when the Hamiltonian and the operators do not depend explicitly on time) a function of the difference $(\tau - \tau')$ only. Further the analogue of equation (2.96) is satisfied.

$$G(\beta) = -\epsilon G(0). \tag{2.97}$$

This implies that $G(\tau)$ can be expanded in the form

$$G(\tau) = \beta^{-1} \sum_l \bar{G}(\zeta_l) \exp(-i\zeta_l\tau), \tag{2.98}$$

where

$$\zeta_l = (2l + 1)\pi/\beta \tag{2.99}$$

for Fermi-type operators, and

$$\zeta_l = 2l\pi/\beta \tag{2.100}$$

for Bose-type operators.

The retarded double-time Green's function $G_R(t, t')$ defined by equation (2.1) is the Green's function in terms of which the response of the system to perturbations is expressed. This function also depends only on the difference of its arguments and can be expressed as a Fourier integral,

$$G_R(t) = (2\pi)^{-1} \int_{-\infty}^{\infty} d\omega \exp(-i\omega t) G_R(\omega). \qquad (2.101)$$

Most of the practical methods for calculating Green's functions that are derived in this book, result in a calculation of $\bar{G}(\zeta_l)$ in the first instance. The determination of the required function $G_R(\omega)$ from this depends on the following important result: there exists a unique function $G(\omega)$ of the complex variable ω which is such that

(i) it is analytic in the upper half of the complex ω-plane

(ii) $$G(i\zeta_l) = \bar{G}(\zeta_l), \qquad (2.102)$$

(iii) $$\operatorname*{Lim}_{|\omega|\to\infty} \omega G(\omega) = \text{a finite constant}, \qquad (2.103)$$

and

(iv) $$\operatorname*{Lim}_{\delta\to 0} G(\omega + i\delta) = G_R(\omega), \qquad \text{for real } \omega. \qquad (2.104)$$

The properties (i), (ii), and (iii), determine $G(\omega)$ from $\bar{G}(\zeta_l)$ when the latter is known. Property (iv) determines $G_R(\omega)$ from $G(\omega)$. In this way $G(\omega)$ is determined once $\bar{G}(\zeta_l)$ is known. The applications discussed in this book illustrate how this determination can be made in practice.

References

General

Abrikosov, A. A., Gor'kov, L. P. and Dzyaloshinskii, I. Ye. (1965). "Quantum Field Theoretical Methods in Statistical Physics". Pergamon Press, Oxford.

Bonch-Bruevich, V. L. and Tyablikov, S. V. (1962). "The Green Function Method in Statistical Mechanics". North-Holland, Amsterdam.

Doniach, S. and Sondheimer, E. H. (1974). "Green's Functions for Solid State Physicists". Benjamin, New York.

Fetter, A. L. and Walecka, J. D. (1971). "Quantum Theory of Many Particle Systems". McGraw-Hill, New York.

Kadanoff, L. P. and Baym, G. (1962). "Quantum Statistical Mechanics". Benjamin, New York.

Parry, W. E. (1973) "The Many-Body Problem". Clarendon Press, Oxford.

Zubarev, D. N. (1960). *Usp. fiz. Nauk.* **71**, 71 (Translation: *Soviet Phys. Usp.* **3**, 320).

Special

Baym, G. and Mermin, D. N. 1961). *J. Maths. Phys.* **2**, 232.
Kubo, R. (1957). *J. Phys. Soc. Japan* **12**, 570.
Matsubara, T. (1955). *Prog. Theor. Phys.* **14**, 351.
Reif, F. (1965). "Fundamentals of Statistical and Thermal Physics," International Student Edition, Section 6.9. McGraw-Hill, New York.
Woods Halley, J. (1978). "Correlation Functions and Quasi-Particle Interactions in Condensed Matter". Plenum, New York.

Problems

1. Show that for non-interacting particles (bosons or fermions) the two-particle temperature Green's function is related to the one-particle Green's function through the equation

$$\begin{aligned} &G_2(r_1\tau_1, r_2, \tau_2; r_3, \tau_3, r_4, \tau_4) \\ &\equiv \langle T\,\tilde{\psi}(r_1, \tau_1)\tilde{\psi}(r_2, \tau_2)\tilde{\tilde{\psi}}(r_4, \tau_4)\tilde{\tilde{\psi}}(r_3, \tau_3)\rangle \\ &= G(r_1\tau_1; r_3\tau_3)G(r_2, \tau_2; r_4, \tau_4) - \epsilon G(r_1\tau_1; r_4, \tau_4)G(r_2\tau_2, r_3\tau_3). \end{aligned}$$

(Hint: Write $\tilde{\psi}$, $\tilde{\tilde{\psi}}$, respectively in terms of the destruction and creation operators for the independent particle states and then evaluate the averages.)

2. Prove the Kramers–Krönig relations as follows. For any contour C within and on which $G(\omega)$ is analytic, Cauchy's integral theorem states that if ω is a point within C,

$$G(\omega) = \frac{1}{2\pi i}\int_C \frac{du\, G(u)}{u - \omega}.$$

Choose for C the contour comprising the real axis from $-R$ to $+R$ and the semicircle Γ in the upper half-plane with this as diameter. Show that as $R \to \infty$, the integral on Γ tends to zero. Choose ω to be the point $y + i\delta\,(\delta \to 0)$. The real and imaginary parts of the resulting equation are the required relations.

3. Prove equation (2.69) by expanding the spectral function in terms of exact eigenfunctions.

4. A system comprises a harmonic oscillator of angular frequency Ω interacting with a bath of oscillators with angular frequencies ω_i through the Hamiltonian

$$H = \Omega a^+ a + \sum_i \omega_i b_i^+ b_i + \sum_i g_i(a^+ b_i + b_i^+ a).$$

Show that the Green's functions

$$G(\tau) = -\langle a(\tau)a^{+}(0)\rangle, \qquad F_i(\tau) = -\langle b_i(\tau)a^{+}(0)\rangle$$

satisfy the equations of motion

$$\left[\frac{\partial}{\partial\tau} + \Omega\right] G(\tau) = -\delta(\tau) - g_i F_i(\tau)$$

$$\left[\frac{\partial}{\partial\tau} + \omega_i\right] F_i(\tau) = -g_i G(\tau).$$

Using Fourier transforms solve these equations and show that

$$\bar{G}(\zeta) = \left\{ i\zeta_l - \Omega - \sum \left[g_i^2/(i\zeta_l - \omega_i)\right]\right\}^{-1}.$$

Show that the function $G(\omega)$ given by,

$$G(\omega) = \left\{\omega - \Omega - \sum \left[g_i^2/(\omega - \omega_i)\right]\right\}^{-1},$$

satisfies the condition (i), (ii), (iii) on page 51 and so is the correct continuation of $\bar{G}(\zeta)$. (Hint: To prove (i) show that the denominator does not vanish if the imaginary part of ω is not zero.) Hence, assuming a continuous spectrum of frequencies ω_i, find the spectral function for this problem. State conditions under which this describes a quasi-particle.

Chapter 3

General Approximations

§3.1 Perturbation theory with a single-particle potential

When the interaction between particles is weak an obvious first approximation to a Green's function to seek is an expansion in powers of the interaction. This approximation is known as perturbation theory. It is an approximation which can be carried out systematically and although one would expect it to be of practical use only for weak interactions, it illuminates a large number of other situations.

As an introduction to perturbation theory we consider first of all the case where there is no two-particle interaction but where there is a single-particle potential $U(r, \tau)$ (which may depend on "time", τ). This problem is of interest for its own sake (see Chapters 4 and 11) as well as for its illumination of the more general many-body problem. The equation for the Green's function $G(r_1\tau_1; r_2\tau_2; U)$ in the presence of the potential is

$$[(\partial/\partial\tau_1) + h_0(r_1) + U(r_1, \tau_1)]\, G(r_1, \tau_1; r_2, \tau_2; U) = -\delta(\tau_1 - \tau_2)\delta(r_1 - r_2), \tag{3.1}$$

where

$$h_0(r_1) = [-(1/2m)\nabla_1^2 - \mu + U_0(r)] \tag{3.2}$$

is an operator whose properties are assumed to be known. Thus we assume that we know the solution of the equation

$$[(\partial/\partial\tau_1) + h_0(\boldsymbol{r}_1)]\,G_0(\boldsymbol{r}_1,\tau_1;\boldsymbol{r}_2,\tau_2) = -\delta(\tau_1-\tau_2)\delta(\boldsymbol{r}_1-\boldsymbol{r}_2), \tag{3.3}$$

which is equation (3.1) when U is zero.

It is convenient to abbreviate a notation which is getting out of hand by showing the dependence of a function on $\boldsymbol{r}_1, \tau_1$, etc., by the numbers 1, etc. Thus we write

$$G(1, 2; U) \equiv G(\boldsymbol{r}_1, \tau_1; \boldsymbol{r}_2, \tau_2; U) \tag{3.4}$$

and also use the notation

$$\delta(1, 2) = \delta(\boldsymbol{r}_1-\boldsymbol{r}_2)\delta(\tau_1-\tau_2). \tag{3.5}$$

Then equations (3.1) and (3.3) can be rewritten

$$[(\partial/\partial\tau_1) + h_0(1) + U(1)]\,G(1, 2; U) = -\delta(1, 2), \qquad (3.1)'$$

$$[(\partial/\partial\tau_1) + h_0(1)]\,G_0(1, 2) = -\delta(1, 2). \qquad (3.3)'$$

Both equations have to be solved for $0 \leqslant \tau_1, \tau_2 \leqslant \beta$ subject to the standard boundary condition

$$G(0, 2; U) = -\epsilon G(\beta, 2; U), \tag{3.6}$$

where ϵ is 1 for fermions and -1 for bosons.

Now with $G_0(1, 2)$ a solution of $(3.3)'$ and (3.6), G also satisfies

$$G(1, 2, U) = G_0(1, 2) + \int_0^\beta d3\,G_0(1, 3)U(3)G(3, 2, U), \tag{3.7}$$

where we have used the notation

$$\int_0^\beta d3 = \int_0^\beta d\tau_3 \int_{\text{all space}} d^3r_3. \tag{3.8}$$

For, from (3.7) and (3.3)

$$[(\partial/\partial\tau_1)+h_0(1)]\,G(1,2;U) = -\delta(1,2)-\int_0^\beta d3\,\delta(1,3)U(3)G(3,2,U)$$

$$= -\delta(1,2)-U(1)G(1,2,U)$$

and this is equation (3.1). Further, since $G_0(1, 2)$ satisfies (3.6), any solution of (3.7) satisfies this condition. Thus equation (3.7) is an integral equation for G and the solution will automatically satisfy (3.6).

Now it is easy to find a formula for G in powers of U. If we write

$$G(1, 2; U) = \sum_{n=0} G_n(1, 2; U), \tag{3.9}$$

where G_n is of order n in U, $G_0(1, 2; U)$ is the solution with U zero and is, therefore, $G_0(1, 2)$. An iteration formula for $G_n (n > 0)$ can be found by substituting (3.9) in (3.7) and equating terms of the same power in U, thus

$$G_{n+1}(1, 2; U) = \int_0^\beta d3 G_0(1, 3) U(3) G_n(3, 2; U)$$
$$= \int_0^\beta d3 \int_0^\beta d4 \ldots \int_0^\beta d(2+n) G_0(1, 3) U(3) G_0(3, 4)$$
$$\times U(4) \times \ldots \times U(2+n) G_0(2+n, 2). \tag{3.10}$$

Hence

$$G(1,2;U) = G_0(1,2) + \int_0^\beta d1' G_0(1, 1') U(1') G_0(1', 2)$$
$$+ \int_0^\beta d1' d2' G_0(1, 1') U(1') G_0(1', 2') U(2') G_0(2', 2) + \ldots \tag{3.11}$$

It is convenient to express this result by means of diagrams and verbal pictures. Recall first that

$$G_0(1, 2) = -\langle\!\langle \tilde{\psi}(1) \tilde{\tilde{\psi}}(2) \rangle\!\rangle \theta(\tau_1 - \tau_2) + \epsilon \langle\!\langle \tilde{\tilde{\psi}}(2) \tilde{\psi}(1) \rangle\!\rangle \theta(\tau_2 - \tau_1)$$

and that $\tilde{\tilde{\psi}}(2)$ creates a particle at "2" and $\psi(1)$ destroys a particle at "1". Hence, the first term involves the creation of a particle at "2" and its later destruction at "1". We can therefore think of the first term as the propagator of a particle from "2" to "1" (cf. § 1.4). The second term involves the destruction of a particle at "1" and its later creation at "2". Hence the second term describes the propagation of a hole from "1" to "2". For many purposes, we find it convenient to ignore the difference between particles and holes and to say that $G_0(1, 2)$ describes the propagation of a particle from "2" to "1". This implies that for $\tau_2 > \tau_1$, $G_0(1, 2)$ describes the propagation of a hole from "1" to "2". In these terms, the propagation of a hole is equivalent to the propagation of a particle backwards in time. This idea is due to Feynman (1949) for real times and the same nomenclature has been carried over to Green's functions with imaginary times. Consistent with this idea of particle propagation, we represent $G_0(1, 2)$ by a continuous line directed from point 2 to a point "1" as shown in Fig. 3.1. Similarly, we regard $G(1, 2; U)$ as the propagator for a particle from "2" to "1" in the presence of the potential U. We illustrate this with a thick line directed from "2" to "1" as in Fig. 3.1(b).†

FIG. 3.1. The representation of a single-particle Green's function (a) with no external potential, (b) in the presence of a potential $U(r, \tau)$.

Now let us consider equation (3.11). The first term simply describes the propagation of the particle from 2 to 1 without U. The second term describes the propagation of the particle from 2 to $1'$ in the absence of U, a single scattering by the potential at $1'$ and the further propagation from $1'$ to 1 in the absence of U. To get the complete first-order contribution to G, we have to integrate over all the possible values of $1'$, i.e. over all the possible positions and times at which the single scattering can take place. If we denote a scattering by U by a cross, the first-order contribution to G is illustrated by Fig. 3.2.

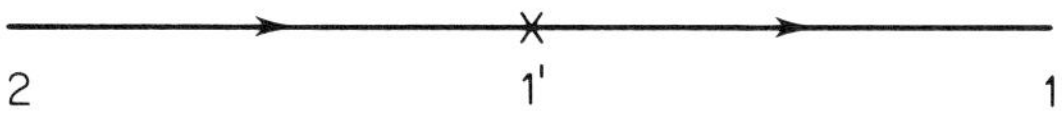

FIG. 3.2. The representation of the propagation of a particle from 2 to 1 with scattering by the potential at $1'$.

Conversely, given the diagram we can write down the corresponding contribution to the Green's function by means of the following prescription: (1) For each directed line from vertex b to vertex a, introduce a factor $G_0(a, b)$. (2) For the vertex $1'$ at which there is a cross, introduce a factor $U(1')$ and integrate over the variables $1'$ at the cross vertex.

These ideas are easily generalized to describe the nth order term in the expression (3.11). This term simply describes the propagation of the particle from 2 to 1 after n scatterings at $1', 2' \ldots, n'$ en route.

†There is some variety in the choice of the direction of the arrow on lines representing $G_0(1, 2)$ and $G(1, 2; U)$. Some authors direct it from "1" to "2" while others (as in this book) direct it from "2" to "1". This does not matter as long as one is completely consistent.

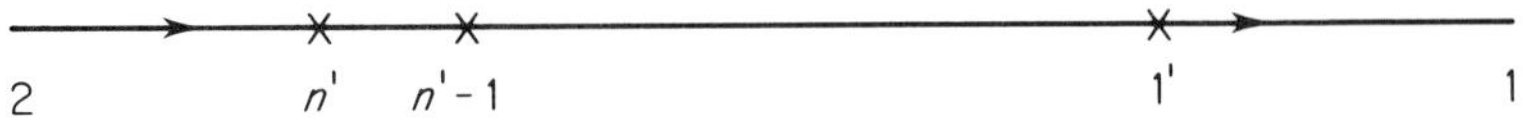

FIG. 3.3. Propagation from 2 to 1 with n' scatterings en route.

We have to integrate over all values of $1', 2' \ldots, n'$ to obtain the total nth order diagram. Again this is illustrated by the diagram of Fig. 3.3. To obtain the nth order contribution from Fig. 3.3 we generalize the previous rules to include a factor $U(i)$ for each cross vertex i and we integrate over all these variables i.

These rules allow one to represent equation (3.11) by the diagram of Fig. 3.4. From the figure we can immediately write down all the various contributions to G. The interpretation of the figure is also straightforward. In propagating from 2 to 1 the particle can suffer any number of scatterings by the potential. All possibilities contribute to the propagator.

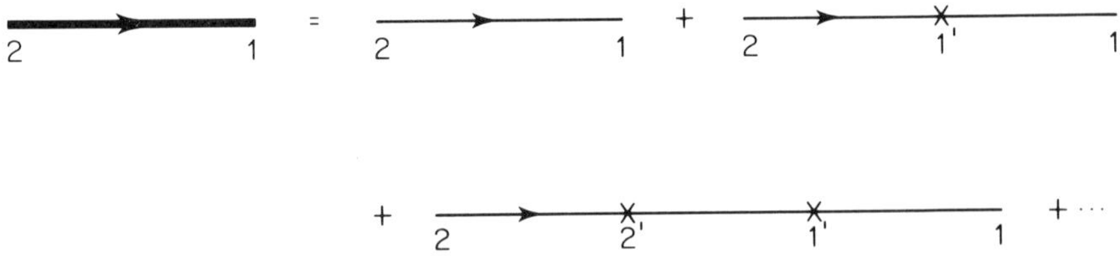

FIG. 3.4. Propagation of a particle from 2 to 1 in a potential U with any number of scatterings en route.

We also note that it is possible to represent the integral equation (3.7) by a figure, namely Fig. 3.5. The content of this diagram is that in propagating from 2 to 1 the particle can travel without scattering from 2 to 1 or it can travel first to any intermediate point 3, be scattered and then travel without scattering from 3 to 1.

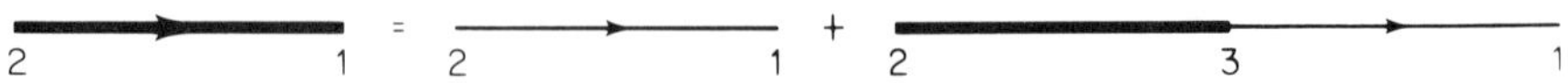

FIG. 3.5. The representation of equation (3.7).

These ideas of expanding a Green's function in powers of an interaction and representing the various terms by diagrams which can be pictured as scatterings of propagating particles are extended in the next section to the case when the particles interact with each other.

§3.2 Perturbation theory for interacting particles

As a model for this sort of problem, we consider the system for which the Hamiltonian is given by

$$\begin{aligned} H(\tau) &= H_0(\tau) + V, \\ H_0(\tau) &= \int d^3r\psi^+(r)h(r,\tau)\psi(r), \\ h(r,\tau) &= -(1/2m)\nabla^2 - \mu + U(r,\tau), \\ V &= \tfrac{1}{2}\int d^3r d^3r' \psi^+(r)\psi^+(r')V(r-r')\psi(r')\psi(r). \end{aligned} \quad (3.12)$$

This describes particles moving in an external potential $U(\boldsymbol{r},\tau)$ and interacting with each other through a two-body potential $V(\boldsymbol{r}-\boldsymbol{r}')$. It could, for example, describe electrons in a solid interacting through the Coulomb force.

Before trying to derive perturbation theory formally let us consider, in the light of the result of the last section, what we might expect the outcome of the two-particle interaction to be. Of course the integrand of V describes the destruction of particles at $\boldsymbol{r}$ and $\boldsymbol{r}'$ and the creation of particles at these points. This describes, therefore, the scattering of two particles, one at $\boldsymbol{r}$, one at $\boldsymbol{r}'$, the scattering amplitude being $V(\boldsymbol{r}-\boldsymbol{r}')$. We must add these two-particle scattering processes to the one-particle ones discussed in the last section.

Let us consider again the one-particle Green's function $G(1,2;U)$ and think of it as the propagator of a particle from 2 to 1. Without the two-particle interaction it is simply the propagator discussed in the last section and describes the propagation of the particle from 2 to 1 with any number of repeated scatterings by U in between. In this section we denote this propagator by $G_0(1,2)$.

What now is the effect of V? Presumably the particle can travel from 2 to 3 (the propagation described by $G_0(1,2)$) and there be scattered by the interaction V. However if this scattering is to take place it must scatter on another particle. Where is the other particle on which it might scatter? Now we must remember that we are dealing with a many-body problem and the Green's function involves an average over states containing many particles. The Green's function focuses attention on the propagation of one particle added to the medium of the other particles. It is therefore possible for the extra particle to be scattered by a particle in the medium at 4, say. We therefore show this particle in the medium as another line and the two-particle interaction by a dashed line. The propagation so far is illustrated in Fig. 3.6. However, the Green's function describes the overall propagation of the extra particle with the medium restored to its initial state. Hence the particle of the medium which has been excited must return to its initial state. This could happen at a second scattering so that the contribution to the propagator might appear as shown in Fig. 3.7.

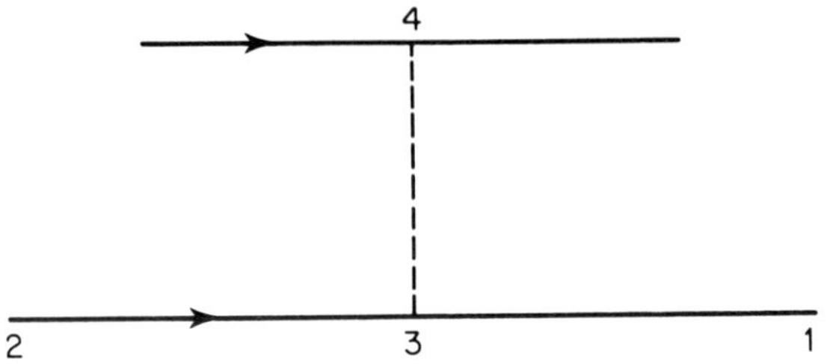

FIG. 3.6. The propagation of an extra particle from 2 to 1 with scattering at 3 on a particle of the medium at 4.

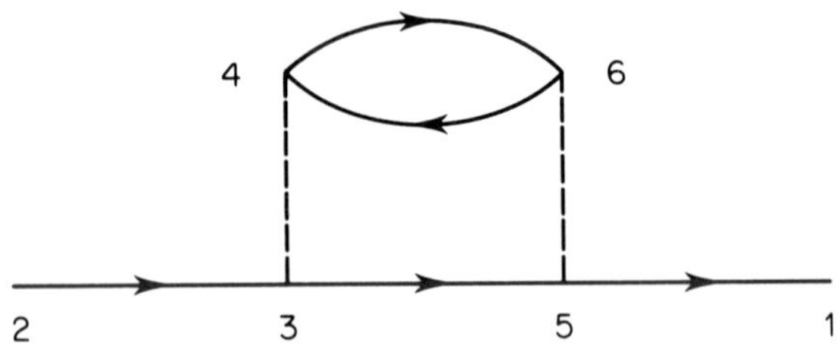

FIG. 3.7. A contribution to the propagator of a particle from 2 to 1 with scattering by the medium. A pair is created at 4 and annihilated at 6.

For fermions there is a natural language in which this can be described. The particle propagates from 2 to 3 where it is scattered by creating a particle–hole pair at 4. The extra particle then propagates from 3 to 5 where it is scattered again, simultaneously destroying the particle–hole pair at 6, the particle and hole having propagated from 4 to 6. We would expect, in analogy with the last section, that Fig. 3.7 would make a contribution to $G(1, 2)$ which is proportional to

$$\int d3d4d5d6G_0(1,5)G_0(5,3)G_0(3,2)G_0(4,6)G_0(6,4)V(3,4)V(5,6),$$

where

$$V(1,2) = V(r_1 - r_2)\delta(\tau_1 - \tau_2)$$

and the delta-function takes account of the instantaneous nature of the interaction. Since the scattering can take place at any points, the variables 3, 4, 5, 6 are integrated over appropriately.

Now the particle travelling through the medium can scatter continually off particles of the medium, and the particles of the medium, when excited, can scatter off each other as well as off the propagating particle. In order to calculate the complete propagator G we must take all of these possibilities into account. Each possibility makes its own contribution to G, and G is the sum of these contributions.

There are two simple contributions to G which are worthy of special mention. Consider again Fig. 3.6. The particle of the medium at 4 could be scattered into the same state so that at this stage the medium is unchanged. This is a limiting case where the particle at 4 propagates to 4. The contribution to G is then as illustrated in Fig. 3.8(a). This is the first term of the Hartree contribution to be discussed in more detail in § §3.6 and 3.7. Alternatively an exchange could take place. The extra particle could be scattered to the state originally occupied by the medium particle while the medium particle would propagate to 1. We should, therefore, have a particle propagating from 3 to 4. This is illustrated in Fig. 3.8(b). Figs. 3.8(a) and (b) constitute the first term of the Hartree–Fock contribution, discussed further in §3.6.

We turn now to a derivation of these results and of the exact rules for the contributions of the different diagrams. These are listed on p. 69 to which the reader not interested in the proof can easily jump. One of the advantages of Feynman diagrams is that it is very much easier to understand the perturbation expansion than its proof.

The usual proof is given in two stages. In the first the Green's function is expanded in powers of the two-particle interaction, each

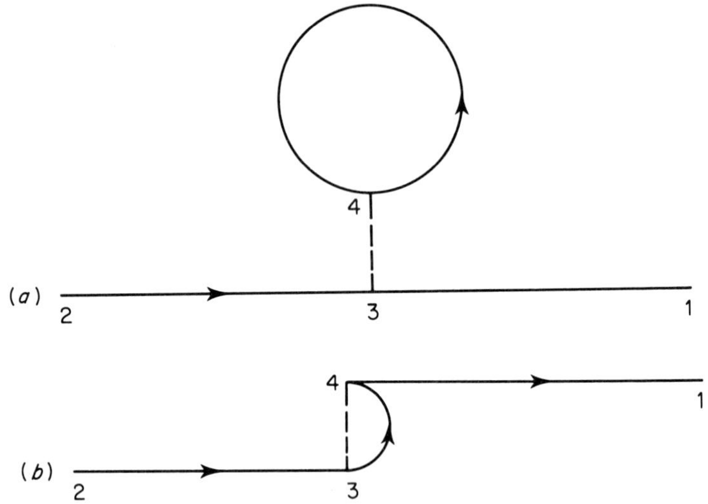

FIG. 3.8. Contributions to the one particle Green's function arising from scattering off one particle in the medium. In (a) the extra particle continues to propagate after the scattering. In (b) the two particles exchange.

term in the expansion involving a many-particle Green's function of the non-interacting system. In the second stage, Wick's theorem is used to express these many-particle Green's function in terms of single-particle Green's functions of the non-interacting system and so to relate the terms of the expansion to Feynman diagrams.

We follow this scheme along the lines of Ambegaoker (1969) and derive first the expansion for the one-particle Green's function (2.99) which can be rewritten

$$\begin{aligned} G(1,2) = & -Z^{-1}\,\mathrm{Tr}\,[S(\beta)S(\tau_1)^{-1}\psi(r_1)S(\tau_1) \\ & \times S(\tau_2)^{-1}\psi(r_2)^{+}S(\tau_2)]\,\theta(\tau_1-\tau_2) \\ & + \epsilon Z^{-1}\,\mathrm{Tr}\,[S(\beta)S(\tau_2)^{-1}\psi^{+}(r_2) \\ & \times S(\tau_2)S(\tau_1)^{-1}\psi(r_1)S(\tau_1)]\,\theta(\tau_2-\tau_1), \end{aligned} \tag{3.13}$$

where

$$S(\tau) = T\exp\left[-\int_0^{\tau} H(\tau')d\tau'\right] \equiv \lim_{N\to\infty} T\prod_{r=0}^{N}\exp\,[-Hr\tau/N)(\tau/N)]$$

$$= \lim_{N\to\infty}\exp[-H(\tau)\tau/N]\ldots\exp[-H(\tau/N)\tau/N]\exp[-H(0)\tau/N], \tag{3.14}$$

$$S(\tau)^{-1} = \bar{T}\exp\left[\int_0^{\tau} H(\tau')d\tau'\right], \tag{3.15}$$

and $\bar{T}$ orders operators with the earliest to the left. Thus for $\tau_1 > \tau_2$

$$S(\tau_1)S(\tau_2)^{-1} = T \exp\left[-\int_{\tau_2}^{\tau_1} H(\tau')d\tau'\right]. \tag{3.16}$$

Then $S(\tau)$ satisfies the equations

$$dS(\tau)/d\tau = -H(\tau)S(\tau), \qquad S(0) = 1. \tag{3.17}$$

If the operator $S_0(\tau)$ is defined by similar equations but with $V = 0$, then

$$dS_0(\tau)/d\tau = -H_0(\tau)S_0(\tau), \qquad S_0(0) = 1. \tag{3.18}$$

We now define an operator $U(\tau)$ by

$$S(\tau) = S_0(\tau)U(\tau). \tag{3.19}$$

Then from equations (3.19), (3.18) and (3.17), U satisfies the equations

$$dU(\tau)/d\tau = -\hat{V}(\tau)U(\tau), \qquad \hat{V}(\tau) = S_0(\tau)^{-1}VS_0(\tau), \qquad U(0) = 1 \tag{3.20}$$

and $\hat{V}(\tau)$ satisfies the Heisenberg equation of motion (in imaginary time),

$$d\hat{V}(\tau)/d\tau = [\hat{H}_0(\tau), \hat{V}(\tau)], \qquad \hat{H}_0(\tau) = S_0(\tau)^{-1}H_0(\tau)S_0(\tau). \tag{3.21}$$

Equation (3.20) describes the development of the system in what is usually called the interaction representation. The solution of equation (3.20) is, by iteration,

$$\begin{aligned} U(\tau) &= T \exp\left\{-\int_0^{\tau} \hat{V}(\tau')d\tau'\right\} \\ &= \sum_{n=0}^{\infty} ((-1)^n/n!) \int_0^{\tau} d\tau_n \int_0^{\tau} d\tau_{n-1} \ldots \\ &\quad \times \int_0^{\tau} d\tau_1\, T[\hat{V}(\tau_n) \ldots \hat{V}(\tau_2)\hat{V}(\tau_1)]. \end{aligned} \tag{3.22}$$

Moreover, for $\tau_1 > \tau_2$

$$U(\tau_1)U(\tau_2)^{-1} = T \exp\left[-\int_{\tau_2}^{\tau_1} \hat{V}(\tau')d\tau'\right]. \tag{3.23}$$

If one now substitutes for $S(\tau)$ into equation (3.13), one finds, for $\tau_1 > \tau_2$,

$$G(1, 2) = -Z^{-1}\,\mathrm{Tr}\left[S_0(\beta)T\left\{\exp\left[-\int_{\tau_1}^{\beta}\hat{V}(\tau)d\tau\right]\right\}\hat{\psi}(1)\right.$$
$$\left.\times T\left\{\exp\left[-\int_{\tau_2}^{\tau_1}\hat{V}(\tau)d\tau\right]\right\}\times\hat{\psi}(2)^+T\left\{\exp\left[-\int_0^{\tau_2}\hat{V}(\tau)d\tau\right]\right\}\right], \tag{3.24}$$

where

$$\hat{\psi}(1) = S_0(\tau_1)^{-1}\psi(r_1)S_0(\tau_1). \tag{3.25}$$

Hence, equation (3.24) can be rewritten, for $\tau_1 > \tau_2$ as

$$G(1, 2) = -\frac{\mathrm{Tr}\,[S_0(\beta)TU(\beta)\hat{\psi}(1)\hat{\psi}(2)]}{\mathrm{Tr}\,[S_0(\beta)TU(\beta)]}, \tag{3.26}$$

where the operators are in the Heisenberg representation for the non-interacting particles. Expression (3.26) can similarly be shown to be valid for $\tau_1 < \tau_2$. It can also be written as

$$G(1, 2) = -\frac{\langle TU(\beta)\hat{\psi}(1)\hat{\psi}(2)^+\rangle}{\langle TU(\beta)\rangle}, \tag{3.27}$$

where the averages are taken with respect to the non-interacting system.

If the expansion (3.22) for U is used, the numerator of equation (3.27) can be written as

$$-\sum_{n=0}((-1)^n/2^n n!)\int_0^{\beta} d\tau_n' \ldots d\tau_1' \langle T[\hat{V}(\tau_n')\ldots\hat{V}(\tau_1')\hat{\psi}(1)\hat{\psi}(2)^+]\rangle \tag{3.28}$$

$$= \sum_{n=0}((-1)^n/2^n n!)\int_0^{\beta} dn'' \ldots d1''\,dn' \ldots d1'$$
$$\times V(n'', n')\ldots V(2'', 2')V(1'', 1')$$
$$\times G_{0,2n+1}(1, 1', 1'', 2', 2'', \ldots, n', n'', 1'_+, 1''_+, \ldots n'_+, n''_+,), \tag{3.29}$$

where $G_{0,n}$ is the n-particle Green's function for the independent particle defined by

$$G_{0,n}(1, 2, \ldots, n; 1', 2', \ldots, n')$$
$$= (-1)^n\langle T[\hat{\psi}(1)\hat{\psi}(2)\ldots\hat{\psi}(n)\hat{\psi}(n')^+\ldots\hat{\psi}(2')^+\hat{\psi}(1')^+]\rangle$$

and

$$V(1, 2) = V(r_1 - r_2)\delta(\tau_1, \tau_2). \tag{3.31}$$

There exists a similar expansion of the denominator,

$$\langle TU(\beta)\rangle = \sum_{n=0} ((-1)^n/2^n n!) \int_0^\beta dn'' \ldots d1''\, dn' \ldots d1'$$
$$\times V(n'', n') \ldots V(1'', 1')$$
$$\times G_{0,2n}(1', 1'' \ldots n', n''; 1'_+, 1''_+ \ldots n'_+, n''_+). \quad (3.32)$$

In the many-particle Green's functions in equations (3.29) and (3.32), the creation operators are given a slight increase in time (denoted by i_+), so that equal time operators appear in the correct order. Equations (3.27), (3.31) and (3.32) provide the required expansion for $G(1, 2)$.

The next step is to show that for independent particles the n-particle Green's function can be written in terms of the 1-particle Green's functions. We note that $G_{0,}$ satisfies the boundary condition

$$G_{0,n}(1, 2, \ldots, n; 1', 2', \ldots, n')|_{\tau_1=0}$$
$$= -\epsilon G_{0n}(1, 2, \ldots, n; 1', 2', \ldots, n')|_{\tau_1=\beta}$$

and that for independent particles the equation of motion of the n-particle function is

$$[(d/d\tau_1) + h_0(1)]\, G_{0,n}(1, 2, \ldots, n; 1', 2', \ldots, n')$$
$$= -\sum_{i=1}^{n} (-\epsilon)^i \delta(1, i') G_{0,n-1}(2, 3, \ldots, n; 1', 2', \ldots, \not{i}', \ldots, n'); \quad (3.33)$$

here the right-hand side comes from differentiating the θ-functions implicit in the definition of the n-particle Green's function. This equation can now be integrated, if the one particle Green's function is used as in §3.1, to yield

$$G_{0n}(1, 2, \ldots, n; 1', 2', \ldots, n')$$
$$= \sum_{i=1}^{n} (-\epsilon)^i G_0(1, i') G_{0,n-1}(2, 3, \ldots, n; 1', 2', \ldots, \not{i}', \ldots, n'). \quad (3.34)$$

As a special case of this

$$G_{02}(1, 2; 1', 2') = G_0(1, 1')G_0(2, 2') - \epsilon G_0(1, 2')G_0(2, 1'). \quad (3.35)$$

By induction it follows that, for fermions,

$$G_{0,n}(1,2,\ldots,n;1',2',\ldots,n') = \begin{vmatrix} G_0(1,1') & G_0(1,2') & - & - & - & - \\ G_0(2,1') & G_0(2,2') & - & - & - & - \\ - & - & - & - & - & - \\ - & - & - & - & - & G_0(n,n') \end{vmatrix} \tag{3.36}$$

while for bosons,

$$G_{0,n}(1,2,\ldots,n;1',2',\ldots,n') = \sum G_0(1,\alpha_1)G_0(2,\alpha_2)\ldots G_0(n,\alpha_n), \tag{3.37}$$

where the sum is taken over all permutations $\alpha_1 \ldots \alpha_n$ of $1' \ldots n'$. Equations (3.36) and (3.37) constitute Wick's theorem for temperature Green's functions.

It follows from these results that there is a contribution to the Green's function of equation (3.29) from every pairing of the indices $1, 1', 1'', \ldots, n', n''$, with the indices $2, 1', 1'', \ldots, n', n''$. Each pairing can be represented by a Feynman diagram which contains two external vertices denoted by 1 and 2 and internal vertices labelled $1', 1'', 2', 2'', \ldots, n', n''$. The vertices i', i'' $(i = 1, \ldots, n)$ are joined by a dashed line to denote the interaction $V(i', i'')$. As in the previous section, the single-particle Green's function $G_0(a, b)$ is denoted by a continuous line directed from b to a. Since in equation (3.29), 1 labels a destruction operator and 2 a creation operator, a directed line will enter the vertex 1 and a directed line will leave the vertex 2. On the other hand each internal vertex labels both a creation and annihilation operator and so one directed line enters each internal vertex and one directed line leaves it. Each possible diagram drawn in this way contributes to the integrand of (3.29) a product of factors V for each dashed line and $G_0(a, b)$ for each continuous line. There is also an overall factor of ± 1 in the case of fermions which arises from the determinant (3.36) in Wick's theorem to which we shall return. In a similar way, one can draw diagrams to represent the various terms in the expansion of the denominator. These diagrams, however, have no external vertices. This means that all the vertices are connected in pairs by dashed lines and every vertex has one continuous line entering it and one leaving it. The diagrams in Figs. 3.7 and 3.8 all represent contributions to the numerator (3.29) as does that in Fig. 3.9(a). Fig. 3.9(b) represents a contribution to the denominator (3.32).

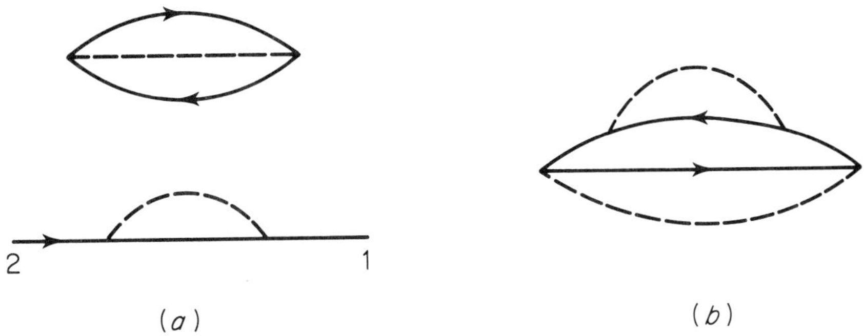

FIG. 3.9. Diagrams representing contributions to $G(1, 2)$; (a) is a disconnected contribution to $\langle TU(\beta)\hat{\psi}(1)\hat{\psi}(2)\rangle$ and (b) is a contribution to $\langle TU(\beta)\rangle$.

Now the overall sign of the diagrams in the case of fermions comes from the many-particle Green's function and is therefore associated with the continuous lines of the diagram. Because the continuous lines can end only at the points 1 and 2 in the diagrams that contribute to the numerator, there will always be one continuous line, the backbone, in the diagram proceeding from 2 to 1 (with possibly internal vertices lying on it), as well as closed continuous loops. The diagram of Fig. 3.9(a) contains one continuous loop as well as the backbone. Now it is not difficult to see that each continuous line makes its own contribution to the sign – the backbone contributes a factor $(-1)^r$ where r is the number of internal vertices on it and each closed loop contributes a factor $(-1)^{s+1}$ where s is the number of internal vertices on it. Hence, the overall factor is $(-1)^{l+2n} = (-1)^l$ where l is the number of closed loops. The overall factor for the diagrams contributing to the denominator is found in the same way.

The one-particle Green's function (3.27) is to be found from the ratio of expression (3.29) to (3.32). This can be obtained as follows. The diagrams that contribute to (3.29) fall into two classes, the connected ones which cannot be divided into two parts without cutting a line and the disconnected ones which are the remainder. Figures 3.7 and 3.8 illustrate connected diagrams while Fig. 3.9(b) is a disconnected one. We denote the sum of the connected diagrams by $-\langle TU(\beta)\hat{\psi}(1)\hat{\psi}(2)^{+}\rangle_C$.

Now consider the contribution to (3.29) of a disconnected graph. There is no overlap of variables of the connected part (containing the backbone) and the remainder. Thus the integral will split into two factors and can be written

$$(2^n n!)^{-1} CD$$

where $2n$ is the total number of internal vertices, C is the contribution of the connected part and D the contribution of the remainder. In fact if the connected part contains $2r$ internal vertices this is one contribution to

$$\langle T\hat{V}(\tau_n') \ldots \hat{V}(\tau_1')\hat{\psi}(1)\hat{\psi}(2)^+\rangle$$

of the form

$$\langle T\hat{V}(\tau_{\alpha_r}') \ldots \hat{V}(\tau_{\alpha_1}')\hat{\psi}(1)\hat{\psi}(2)^+\rangle_C \langle T\hat{V}(\tau_{\alpha_n}') \ldots \hat{V}(\tau_{\alpha_{r+1}}')\rangle.$$

Since all the contributions of this kind must be included

$$\begin{aligned}&\langle T\hat{V}(\tau_n') \ldots \hat{V}(\tau_1')\psi(1)\psi(2)^+\rangle \\ &= \sum \langle T\hat{V}(\tau_{\alpha_r}') \ldots \hat{V}(\tau_{\alpha_1}')\hat{\psi}(1)\hat{\psi}(2)^+\rangle_C \langle T\hat{V}(\tau_{\alpha_n}') \ldots \hat{V}(\tau_{\alpha_{r+1}}')\rangle\end{aligned} \tag{3.37}$$

where the sum is over all divisions of the variables $\tau_1, \ldots, \tau_n$ into two sets. Now when we integrate over the variables $\tau_1, \tau_2, \ldots, \tau_n$, each term which puts r operators V in the first bracket and $(n-r)$ in the second contributes the same amount to the integral. Since there are

$${}^nC_r = \frac{n!}{r!(n-r)!}$$

ways of choosing r operators V from the n of them, we find

$$\begin{aligned}&((-1)^n/2^n n!) \int_0^\beta d\tau_1' \ldots d\tau_n' \langle T\hat{V}(\tau_n') \ldots \hat{V}(\tau_1')\hat{\psi}(1)\hat{\psi}(2)^+\rangle \\ &= \sum_{r=0}^{n} ((-1)^r/2^r r!) \int d\tau_r' \ldots d\tau_1' \langle T\hat{V}(\tau_r') \ldots \hat{V}(\tau_1')\hat{\psi}(1)\hat{\psi}(2)^+\rangle_C \\ &\times ((-1)^{n-r}/2^{n-r}(n-r)!) \int d\tau_n' \ldots d\tau_{r+1}' \langle T\hat{V}(\tau_n') \ldots \hat{V}(\tau_{r+1}')\rangle.\end{aligned} \tag{3.38}$$

When we sum this expression over n we see that

$$\langle TU(\beta)\hat{\psi}(1)\hat{\psi}(2)^+\rangle = \langle TU(\beta)\hat{\psi}(1)\hat{\psi}(2)^+\rangle_C \langle TU(\beta)\rangle.$$

Hence

$$G(1, 2) = -\langle TU(\beta)\hat{\psi}(1)\hat{\psi}(2)^+\rangle_C. \tag{3.39}$$

This is the main result. In expanded form it is

$$\begin{aligned}G(1, 2) = -&\sum_{n=0}^{\infty} ((-1)^n/2^n n!) \int_0^\beta d\tau_n' \ldots d\tau_1' \\ &\times \langle T[\hat{V}(\tau_n') \ldots \hat{V}(\tau_1')\hat{\psi}(1)\hat{\psi}(2)^+]\rangle_C.\end{aligned} \tag{3.40}$$

There is one further simplification we can make. We note that permuting the labels $\tau'_1, \ldots, \tau'_n$ does not change the value of the integral in (3.40). Since there are $n!$ permutations, if for any diagram we keep only one set of labels for the imaginary times we should introduce a factor $n!$ which cancels the corresponding factor in the denominator. Similarly, if we look at the form (3.29) we see that interchanging any pair of labels i', i'' at the ends of a dashed line, does not change the value of the integral. Hence, if we keep only one set of labels for each dashed line we should multiply the result by the total number of interchanges of the pairs of labels; this is 2^n and cancels the corresponding factor in the denominator. From each class of diagrams which can be obtained one from the other by relabelling the vertices, we now include only one member. This means that we keep only topologically distinct diagrams.

We can now formulate a simple set of rules for calculating the nth order contribution to $G(1, 2)$. First draw all topologically distinct and connected diagrams which satisfy the following conditions:

(i) Each contains two external vertices 1 and 2.

(ii) Each contains n internal vertices.

(iii) The internal vertices are connected in pairs by dashed lines.

(iv) Continuous directed lines joining the vertices are drawn so that exactly one line is directed into and one away from each internal vertex, one line is directed away from vertex 2 and one directed towards vertex 1.

The contribution of each of these diagrams to $G(1, 2)$ is given according to the following rules:

(i) There is a factor $-V(i, j)$ for the dashed line connecting vertex i to vertex j.

(ii) There is a factor $G_0(a, b)$ for the continuous line directed from vertex b to vertex a.

(iii) There is a factor $(-\epsilon)$ for each closed continuous loop ($\epsilon = 1$ for fermions, -1 for bosons).

(iv) The product of factors is then integrated over the variables attached to the internal vertices.

Because of the labels i_+ in equations (3.29) and (3.32) any Green's function $G_0(a, b)$ with a and b at the same time is to be interpreted as $G_0(a, b_+)$. $G(1, 2)$ is the sum of the contributions from all these diagrams.

If the particles possess spin, the variable $\boldsymbol{r}$ includes the spin label (see §2.7) and integration over $\boldsymbol{r}$ includes a sum over spin. However, the interaction (3.12) conserves spin. Thus at each interaction vertex the spin label like the coordinate label is unchanged. With these conventions the rules we have given do not have to be changed.

Higher-order Green's functions can be expanded in powers of the interaction in the same kind of way, the proof closely following the one given above for the one-particle Green's function. For example, the two-particle Green's function defined by

$$G(1,2;3,4) = -\mathrm{Tr}\,[TS(\beta)\tilde{\psi}(1)\tilde{\psi}(2)\tilde{\tilde{\psi}}(4)\tilde{\tilde{\psi}}(3)]\,/\mathrm{Tr}\,[TS(\beta)], \tag{3.41}$$

can be expressed in the form

$$G(1,2;3,4) = \langle TU(\beta)\hat{\psi}(1)\hat{\psi}(2)\hat{\psi}(4)^{+}\,\hat{\psi}(3)^{+}\rangle_C, \tag{3.42}$$

where the right-hand side is the sum over the contributions from topologically distinct connected diagrams. In this case the diagrams all have the four external vertices 1, 2, 3, 4, and connected diagrams are those which cannot be separated into two parts without cutting a line, such that one part contains no external vertices. The vertices 3 and 4 have directed lines leaving them while the vertices 1 and 2 have directed lines entering them.

The contributions of individual diagrams are again calculated according to the rules (i) to (iv) except that in the case of fermions there is an additional factor. The diagrams fall into two classes. In the first class there is a continuous line from 3 to 1 (with perhaps internal vertices on it) and similarly a continuous line from 4 to 2. In the second class continuous lines run from 3 to 2 and from 4 to 1. In the case of fermions the latter contribution contains an extra factor (− 1) which arises from Wick's theorem.

Similar rules can be given for the n-particle Green's function, diagrams for which will contain $2n$ external vertices. The one- and two-particle Green's functions are the only ones needed in this book.

Response functions are usually related to two-particle Green's functions. We saw in §1.2 that the linear response of a quantum system is given by a retarded Green's function, equation (1.28), which depends on two operators $M(t)$ and $B(t)$. We further saw in §2.2 that the retarded Green's function can be obtained from the temperature one,

$$\mathrm{Tr}\,\rho_0\,[T\tilde{M}(\tau)\tilde{B}(\tau')]\,.$$

In many applications M and B are one-particle operators of the form

$$M(r_1) = \psi^+(r_1)m\psi(r_1), \qquad B(r_2) = \psi^+(r_2)m\psi(r_2),$$

where m and b may depend on $\boldsymbol{r}$ as well as on differential operators. In order to embrace interactions of external fields with currents the latter possibility must be included. The Green's function we are trying to calculate is linear in M and B. Hence our discussion can be generalized by adding together many terms of the kind we are considering in detail. Thus our choice for M and B is less restrictive than it might seem at first sight.

The quantity we have to calculate is then

$$\begin{aligned} \operatorname{Tr} \rho_0 [\tilde{\tilde{\psi}}(r_1, \tau_1)m(1)\tilde{\psi}(r_1,\tau_1)\tilde{\tilde{\psi}}(r_2, \tau_2)b(2)\psi(r_2, \tau_2)] \\ = m(1)b(2)G(1, 2; 3, 4)|_{3=1_+, 4=2_+}. \qquad (3.43) \end{aligned}$$

This can then be represented by all diagrams of the form shown in Fig. 3.10 where the square contains any number of internal vertices connected to each other and to the remainder of the diagram according to the usual rules. If m and b are differential operators m acts on the Green's function entering 1 and b on the Green's function entering 2. Examples of calculations based on this perturbation theory for the response function are given in §§4.3, 8.4 and 11.7. The random phase approximation discussed in §3.8 is a special case of this expansion.

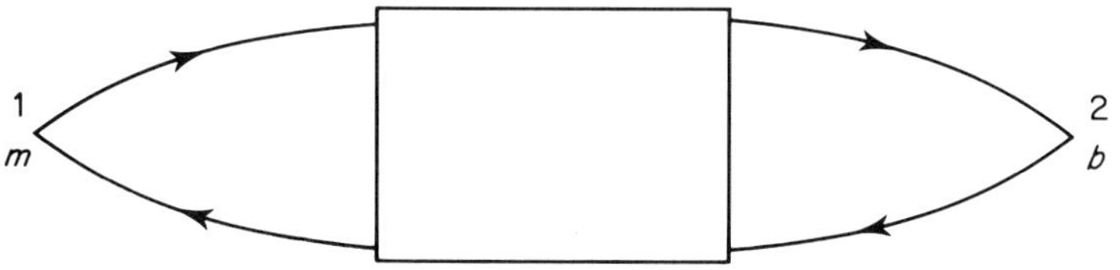

FIG. 3.10. A schematic representation of the diagrams which contribute to the response function $\langle T\hat{M}(1)\hat{B}(2)\rangle$. The box includes all diagrams which are connected to the remainder. The operators m and b act at the vertices 1 and 2 respectively.

§3.3 Time and translational invariance

Suppose that the Hamiltonian does not depend explicitly on time, $U(\boldsymbol{r}, \tau) = U(\boldsymbol{r})$. Then, as we have seen in §2.2, the Green's functions depend only on time-differences or τ-differences, not on the individual times themselves. It is then convenient to use appropriate Fourier transforms of the Green's functions. According to §2.2, for the temperature Green's functions we should use

$$G(1,2) = \beta^{-1} \sum_l \bar{G}(r_1, r_2, \zeta_l) \exp [i\zeta_l(\tau_2 - \tau_1)], \qquad (3.44)$$

where

$$\zeta_l = (2l+1)\pi/\beta, \qquad l = \text{integer, for fermions,}$$

and

$$\zeta_l = 2l\pi/\beta, \qquad l = \text{integer, for bosons.}$$

Also the contribution of a dashed line in a diagram is

$$V(1,2) = V(r_1 - r_2)\delta(\tau_1, \tau_2) = V(r_1 - r_2)\frac{\partial}{\partial \tau_1}\theta(\tau_1 - \tau_2),$$

where $\theta(\tau)$ is defined for $-\beta \leqslant \tau \leqslant \beta$ and is 1 for $0 \leqslant \tau \leqslant \beta$ and zero otherwise. This has the standard Fourier series

$$\theta(\tau) = \frac{i}{\beta} \sum_m \frac{\exp(-i\zeta_m \tau)}{\zeta_m}, \qquad \zeta_m = 2m\pi/\beta.$$

Hence

$$\delta(\tau) = \beta^{-1} \sum_m \exp(-i\zeta_m \tau), \qquad \zeta_m = 2m\pi/\beta, \qquad (3.45)$$

and

$$V(1,2) = \frac{V(r_1 - r_2)}{\beta} \sum_m \exp[-i\zeta_m(\tau_1 - \tau_2)]. \qquad (3.46)$$

Consider now the contribution to each diagram from a single Fourier component for each line. We can think of each line carrying an "energy", ζ_i. At an internal vertex three lines, one dashed and two continuous, meet. Suppose that the energies entering the vertex "i" are $\zeta_l, \zeta_m, \zeta_n$ (see Fig. 3.11). Then τ_i will appear in the contribution only in a factor.

$$\exp(i\zeta_l\tau_i)\exp(i\zeta_m\tau_i)\exp i(\zeta_n\tau_i) = \exp[i(\zeta_l + \zeta_m + \zeta_n)\tau_i].$$

The integral over τ_i can now be performed to yield

$$\int_0^\beta d\tau_i \exp[i(\zeta_l + \zeta_m + \zeta_n)\tau_i] = \beta\delta_{\zeta_l + \zeta_m + \zeta_n, 0}.$$

Hence, the total "energy" entering the vertex "i" is zero, that is, "energy" is conserved at each internal vertex. It follows that if an energy ζ_l leaves the external vertex 1, this energy ζ_l must enter the external vertex 2. The diagram, therefore, has an overall factor $\exp[i\zeta_l(\tau_2 - \tau_1)]$ and represents the Fourier transform, $\bar{G}_0(r_1 r_2, \zeta_l)$. If we fix the energy ζ_l leaving vertex "1" and arbitrarily fix the energies carried by the dashed lines, the energies of the remaining continuous lines are fixed by the energy conservation at each vertex.

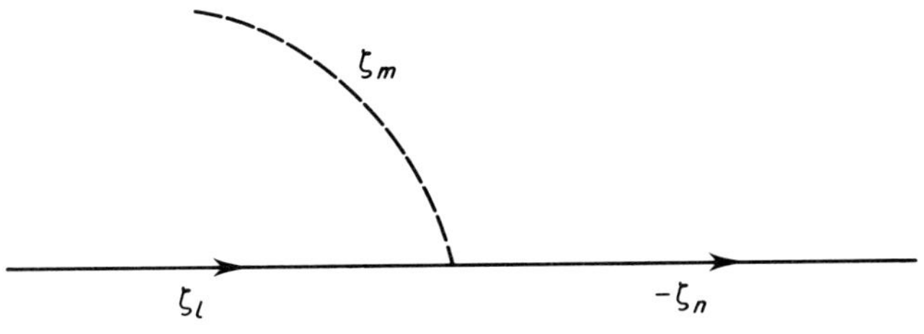

FIG. 3.11. A vertex showing the energies ζ_l, ζ_n, ζ_m entering.

But a diagram with $2n$ internal vertices has n dashed lines and, therefore, n arbitrary energies. To obtain the total contribution of such a diagram we must sum over the allowed values of these internal energies.

Our rules can now be expressed as a set of rules for calculating the contribution of an nth order diagram (which has $2n$ internal vertices) to $\bar{G}(r_1, r_2, \zeta_l)$. The diagrams are the same as before. The other rules are as follows:

(i) Ascribe an energy ζ_l to the lines leaving vertex 1 and entering vertex 2.

(ii) Ascribe an energy ζ_m to every other line in such a way that energy is conserved at each vertex.

(iii) Introduce a factor $\bar{G}_0(r_j, r_k, \zeta_m)$ from a directed line from vertex j to vertex k carrying energy ζ_m.

(iv) Introduce a factor $-V(r_j, r_k)$ for a dashed line joining vertices j and k.

(v) Introduce a factor β^{-n}.

(vi) Introduce a factor $(-\epsilon)$ for each closed loop.

(vii) Sum over all (n) remaining arbitrary energies ζ_m.

The factor β^{-n} arises as follows. There are $(2n + 1)$ continuous lines all together and each carries a factor β^{-1} arising from the definition of the "time" Fourier transform. Similarly, there are n dashed lines each carrying a factor β^{-1}. Each integration over the $2n$ internal times "τ" introduces a factor β. Finally, there is an overall factor β, coming from the transformation from $G(r_1, \tau_1; r_2, \tau_2)$ to $\bar{G}(r_1, r_2, \zeta_l)$. The product of these is

$$\beta^{-2n-1}\beta^{-n}\beta^{2n}\beta = \beta^{-n}.$$

If the system is translationally invariant $G(r_1, r_2, \zeta_l)$ cannot depend on the choice of origin and cannot depend on r_1 and r_2 separately, but only on their difference, $\boldsymbol{r}_1 - \boldsymbol{r}_2$. We can then introduce the spatial Fourier transforms

$$G(\boldsymbol{r}_1 - \boldsymbol{r}_2, \zeta_l) = \mathscr{V}^{-1} \sum_{\boldsymbol{k}} \exp(i\boldsymbol{k} \cdot \boldsymbol{r})\bar{G}(\boldsymbol{k}, \zeta_l), \qquad k_i = 2\pi n_i/\mathscr{V}^{1/3}$$

$$\to (2\pi)^{-3} \int d^3k \exp(i\boldsymbol{k} \cdot \boldsymbol{r})\bar{G}(\boldsymbol{k}, \zeta_l) \tag{3.47}$$

$$V(\boldsymbol{r}_1 - \boldsymbol{r}_2) = \mathscr{V}^{-1} \sum_{\boldsymbol{k}} \exp(i\boldsymbol{k} \cdot \boldsymbol{r})V(\boldsymbol{k}) \to (2\pi)^{-3} \int d^3k \exp(i\boldsymbol{k} \cdot \boldsymbol{r})V(\boldsymbol{k}), \tag{3.48}$$

and the treatment is completely analogous to that of the time transforms. Now each line carries a momentum $\boldsymbol{k}$ as well as energy ζ_l, and because of the integration over $\boldsymbol{r}$ at each vertex, momentum as well as energy is conserved at each vertex. The factor β is now replaced by $\beta\mathscr{V}$ or $\beta(2\pi)^3$. We still draw the same diagrams as before and have the following rules for the contribution of a diagram with $2n$ internal vertices to $G(\boldsymbol{k}, \zeta_l)$:

(i) Ascribe an energy ζ_l and momentum $\boldsymbol{k}$ (or wave vector $\boldsymbol{k}$) to the lines leaving vertex "1" and entering vertex "2" (these are now simply the lines entering and leaving the diagram – as the integrations over $\boldsymbol{r}_1, \tau_1, \boldsymbol{r}_2, \tau_2$ have now been performed there is no need to label the vertices "1" and "2").

(ii) Ascribe an energy ζ_m and momentum $\boldsymbol{k}'$ to every other line in such a way that energy and momentum are conserved at every vertex.

(iii) Introduce a factor $\bar{G}_0(\boldsymbol{k}, \zeta_m)$ for every continuous line carrying energy ζ_m, momentum $\boldsymbol{k}$.

(iv) Introduce a factor $V(\boldsymbol{k})$ for each dashed line carrying wave vector $\boldsymbol{k}$.

(v) Introduce a factor $[-\beta(2\pi)^3]^{-n}$ (or $(-\beta\mathscr{V})^{-n}$).

(vi) Introduce a factor $(-\epsilon)$ for each closed continuous loop.

(vii) Sum over all n remaining arbitrary energies ζ_m and integrate (or sum) over all remaining arbitrary momenta $\boldsymbol{k}'$.

For example the contributions of the diagrams of Fig. 3.12(a) and (b) to $\bar{G}_1(\boldsymbol{k}, \zeta)$ are, respectively,

$$\epsilon \frac{\bar{G}_0(\boldsymbol{k}, \zeta_l)}{(2\pi)^3\beta} \int d^3k' \sum_m \bar{G}_0(\boldsymbol{k}', \zeta_m)V(0)\bar{G}_0(\boldsymbol{k}, \zeta_l) \tag{3.49}$$

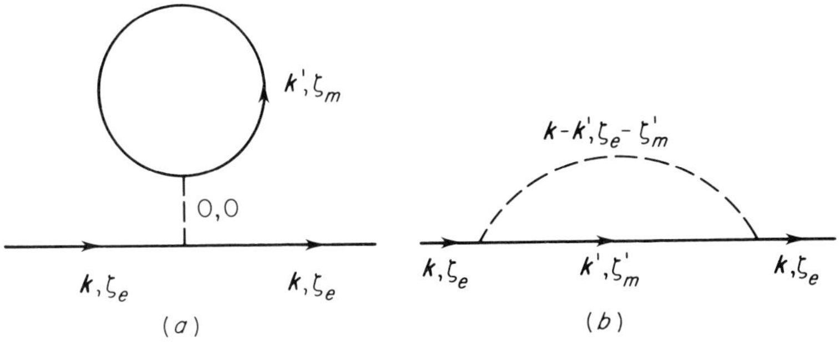

FIG. 3.12. The two first order contributions to the one-particle Green's function, showing the momentum and energy labels.

and

$$-\frac{\bar{G}_0(k, \zeta_l)}{(2\pi)^3\beta}\int d^3k' \sum_m V(k-k')\bar{G}_0(k', \zeta_m)\bar{G}_0(k, \zeta_l). \quad (3.50)$$

We will look at these expressions in the context of a particular example in the next section.

We have said in §§2.7 and 3.2, that we will deal with spin by treating the coordinate label as including spin. However, we cannot maintain this convention when we perform the spatial Fourier transform. As a consequence the spin labels should be shown explicitly. When the single-particle term of the Hamiltonian $h_0(r)$ is independent of spin, the previous results are easily generalized. Firstly, since the interaction conserves spin as well, the off-diagonal components $G(\uparrow, \downarrow)$ and $G(\downarrow, \uparrow)$ will be zero while the diagonal components $G(\uparrow, \uparrow)$ and $G(\downarrow, \downarrow)$ will be equal. Since spin is conserved at each interaction vertex, the only free spin labels over which we have to sum are attached to the closed loops, one for each loop. Further, the contribution from each loop is independent of spin. Hence, the contribution from each loop is simply doubled when spin is included. This means that the rules on p. 74 are unchanged except for (vi) which should now read:

(vi) Introduce a factor (-2ϵ) for each closed continuous loop.

As the particles with spin $\frac{1}{2}$ are fermions, this factor will be (-2) in practice. For spin $\frac{1}{2}$ particles then, the contribution Fig. 3.12(*a*) is doubled while that of Fig. 3.12(*b*) is unchanged.

Since the rules are referred to continually in this book they are summarized in Table 3.1.

TABLE 3.1. Rules for Perturbation Theory

Element of Graph	Contribution: Coordinate space	Contribution: Fourier space
k, ζ_l (2 → 1, solid line)	$G_0(1, 2)$	$G_0(\boldsymbol{k}, \zeta_l)$
q, ζ_m (2 - - - 1, dashed line)	$-V(1, 2) = -V(r_1 - r_2)\delta(\tau_1 - \tau_2)$	$-V(\boldsymbol{q})(\beta\mathscr{V})^{-1}$ or $-V(\boldsymbol{q})\beta^{-1}(2\pi)^{-3}$
Closed loop	$-\epsilon$†	$-\epsilon$†
(for fermions with spin $\frac{1}{2}$)	-2	-2

†Note that $\epsilon = 1$ for fermions, -1 for bosons.

§3.4 A system of weakly interacting fermions

To illustrate perturbation theory in practice, consider a system of free fermions interacting through a weak two-body force, described by the interaction Hamiltonian (2.92). The unperturbed part of the Hamiltonian is

$$H_0 = \int \psi^+(r)\left(\frac{-\nabla^2}{2m} - \mu\right)\psi(r)d^3r = \sum_k \left(\frac{k^2}{2m} - \mu\right) c_k^+ c_k. \quad (3.51)$$

As we have seen in §2.3, the spectral function in this case is

$$A(k, x) = \delta(x - \epsilon_k), \qquad \epsilon_k = (k^2/2m) - \mu$$

and, therefore

$$G_0(k, \zeta_l) = \int_{-\infty}^{\infty} \frac{dx\,\delta(x - \epsilon_k)}{i\zeta_l - x} = \frac{1}{i\zeta_l - \epsilon_k}. \quad (3.52)$$

(We are using the same symbol G for $G(r, \tau)$ and its Fourier transform $G(k, \zeta_l)$ as the independent variables make the meaning unambiguous.)

The first-order contributions G_1 to be added to G_0 to give G are those given in equations (3.49) and (3.50) with G_0 replaced by the expression in (3.52). We immediately come up against a problem. Since $G_0(k, \zeta_m)$ behaves like $1/(2m + 1)$ as m becomes large, the sums over m in both (3.49) and (3.50) diverge logarithmically. Does this mean that perturbation theory fails at the first test? Is the assumed expansion in powers of the interaction invalid? In fact, the answer to both these questions is no. We have simply been sloppy about the treatment of one point which probably looks pedantic. However, as we shall now show, a proper treatment yields finite results for (3.49) and (3.50).

In fact, in both terms the sums are

$$\beta^{-1} \sum_m G_0(k', \zeta_m) = G_0(k', \tau = 0).$$

However, one subsidiary rule tells us that, if the times of the two operators in the Green's function are equal, the destruction operator should be taken at a slightly earlier time to give the correct order. Hence, $G_0(k', \tau = 0)$ should be interpreted as $\lim_{\delta \to 0} G_0(k', -\delta)$. This is just $\epsilon\langle c_{k'}^+ c_{k'}\rangle$. Since $c_{k'}^+ c_{k'}$ is the number operator, we find

$$\epsilon\langle c_{k'}^+ c_{k'}\rangle = \epsilon f_{k'}$$

where $f_{k'}$ is the Fermi distribution function,

$$f_k = [\exp(\beta\epsilon_k) + 1]^{-1}.$$

The contributions of Figs. 3.8(*a*) and 3.8(*b*) can now be written as

$$\frac{G_0(k, \zeta_l)}{(2\pi)^3} \int d^3k' f_{k'} V(0) G_0(k, \zeta_l) \tag{3.53}$$

and

$$-\epsilon \frac{G_0(k, \zeta_l)}{(2\pi)^3} \int d^3k' V(k - k') f_{k'} G_0(k, \zeta_l). \tag{3.54}$$

Higher-order terms can be evaluated similarly. Only terms which involve the sum over a single Green's function, as in (3.49) and (3.50), require the introduction of δ and δ'. Other terms are convergent and lead to the correct result even if δ and δ' are taken to be zero at the outset. Further discussion of these results is delayed until after the next section.

§3.5 Self-energy

The perturbation series in the simple form in which we have obtained it is rarely of direct use. To understand this, consider again the problem of free particles interacting through a weak two-body potential. We discussed this problem in a qualitative way in §2.3(c). Without the interaction the retarded Green's function is

$$G_R(k, t) = -i \exp(-i\epsilon_k t)\theta(t).$$

With a weak interaction we expect that, for large times, this will be altered to

$$G_R(k, t) = -a_k \exp(-i\tilde{\epsilon}_k t) \exp[-\Gamma(k)t]\,\theta(t)$$

where $\tilde{\epsilon}_k$ differs from ϵ_k, as a result of the interaction, by a small amount and $\Gamma(k)$ is small and carries information about the way correlations decay in the interacting system. Now it is reasonable to expect that $(\tilde{\epsilon}_k - \epsilon_k)$ and $\Gamma(k)$ can be expressed as perturbation series in V which tend to zero as V tends to zero. Nevertheless, however small V is, there will be times sufficiently large (and we are interested in large times) for which $(\tilde{\epsilon}_k - \epsilon_k)t$ and $\Gamma(k)t$ are not small. This means that there are always times for which the perturbation series for $G_R(k, t)$ is not useful. What we really need to do is to extract from the series for $G(k, t)$ a perturbation series for $(\tilde{\epsilon}_k - \epsilon_k)$ and $\Gamma(k)$.

Similarly, we expect that the time Fourier transform should be of the form

$$G_R(k, \omega) \sim a_k/[\omega + i\Gamma(k) - \tilde{\epsilon}_k] \tag{3.55}$$

and we are interested in the poles of this function in the complex plane ω. This also requires an expansion of $\tilde{\epsilon}_k$ and $\Gamma(k)$ or of G_R^{-1} in powers of V rather than the expansion of G_R. The perturbation series we have obtained is useful because it is possible to sum the terms in such a way that we obtain a series for $G_R^{-1}(k, \omega)$.

Consider first the case with time and translation invariance so that one can work with the space and time transforms. The diagrams that contribute to $G(k, \zeta_l)$ can now be divided into several classes. In the first class take all diagrams (apart from G_0) which cannot be divided into two parts by cutting one continuous line (apart from the lines connected to the external vertices). Then we can represent the first class by the diagram of Fig. 3.13, where the black circle denotes the sum of diagrams which cannot be separated by cutting a continuous line. Some diagrams which contribute to this are illustrated in Fig. 3.13. The sum of the contributions to the black circle depends on k and ζ_l, is denoted by $\Sigma(k, \zeta_l)$ and, for reasons to become apparent shortly, is called the self-energy. The diagrams which contribute to $\Sigma(k, \zeta_l)$ are called proper self-energy parts. Into the second class, put all diagrams which can be separated into two parts by cutting one continuous internal line but which cannot be separated into more parts in this way. The sum of all these diagrams can be illustrated as in Fig. 3.14, where the black circles each stand for $\Sigma(k, \zeta_l)$.

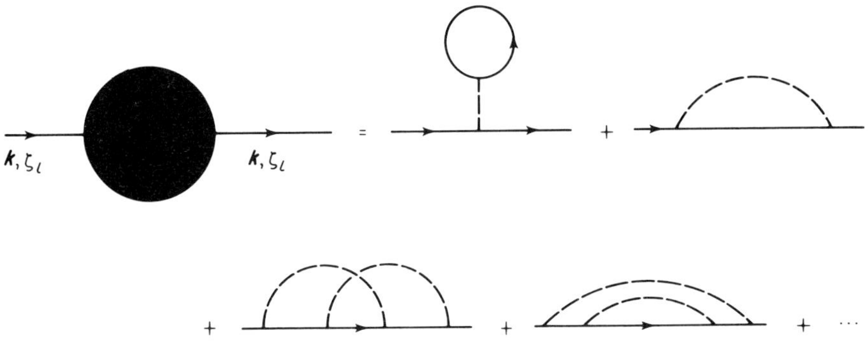

FIG. 3.13. Diagrams which contribute to the proper self-energy. The filled circle represents the total contribution to the self-energy.

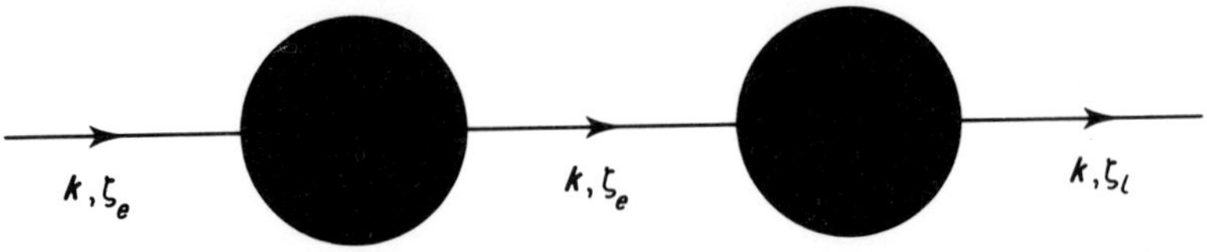

FIG. 3.14. Diagrams which can be separated into two parts by cutting one internal particle line only.

Each diagram contained in this representation belongs to this class and each part of such a diagram that is cut into two will belong to the first class. There is no double counting of diagrams.

This classification can be continued, the nth class being such that it can be divided into $(n + 1)$ parts by n cuts through n different continuous internal lines but it cannot be divided into $(n + 2)$ parts by $(n + 1)$ such cuts. The contribution of the nth class contains n black circles as illustrated in Fig. 3.15.

FIG. 3.15. Diagrams which can be separated into n parts by cutting exactly n different internal particle lines only.

Hence the total contribution to $G(k, \zeta_l)$ can be as illustrated in Fig. 3.16 and

$$\begin{aligned} G(k, \zeta_l) &= G_0(k, \zeta_l) + G_0(k, \zeta_l)\Sigma(k, \zeta_l)G_0(k, \zeta_l) \\ &\quad + G_0(k, \zeta_l)\Sigma(k, \zeta_l)G_0(k, \zeta_l)\Sigma(k, \zeta_l)G_0(k, \zeta_l) \\ &\quad + \dots \\ &= G_0(k, \zeta_l) + G_0(k, \zeta_l)\Sigma(k, \zeta_l)G(k, \zeta_l), \end{aligned} \tag{3.56}$$

and

$$G(k, \zeta_l) = [G_0(k, \zeta_l)^{-1} - \Sigma(k, \zeta_l)]^{-1}. \tag{3.57}$$

Equation (3.57) is known as Dyson's equation.

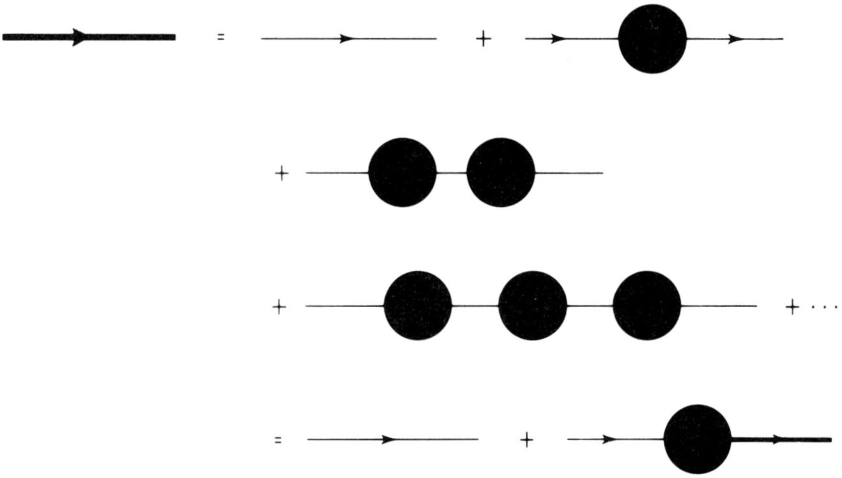

FIG. 3.16. The one-particle Green's function $G(1, 2)$ (the thick line here) represented in terms of proper self-energies.

This general result can be related to the Green's function given by equation (3.55) which was derived heuristically from the concept of a quasi-particle. From equation (3.57), one can obtain the Green's function analytic in the upper half-plane

$$G(k, \omega) = [\omega - \epsilon_k - \Sigma(k, -i\omega)]^{-1},$$

the poles, ω, of which satisfy,

$$\omega - \epsilon_k - \Sigma(k, -i\omega) = 0.$$

If this has a pole at $\tilde{\epsilon}_k - i\Gamma_k$ lying near the real axis ($\Gamma_k \ll |\tilde{\epsilon}_k|$), the real and imaginary parts can be determined in turn by

$$\tilde{\epsilon}_k - \epsilon_k - \text{Re}\,\Sigma(k, -i\tilde{\epsilon}_k + \delta) = 0$$

and

$$\Gamma_k = -\text{Im}\,\Sigma(k, -i\tilde{\epsilon}_k + \delta).$$

If this pole is isolated, the behaviour of the Green's function is given by

$$G(k, \omega) \approx a_k [\omega - \tilde{\epsilon}_k + i\Gamma(k)]^{-1},$$

where a_k is the residue at the pole and is given by

$$a_k^{-1} = 1 - \frac{\partial \Sigma}{\partial \omega}(\boldsymbol{k}, -i\tilde{\epsilon}_k + \delta).$$

This is the form of the Green's function for the existence of a good quasi-particle. The result is often written in terms of a lifetime, τ_k, defined by,

$$(2\tau_k)^{-1} = \Gamma(\boldsymbol{k}) = -\operatorname{Im} \Sigma(\boldsymbol{k}, -i\tilde{\epsilon}_k + \delta).$$

This lifetime plays an important role in transport phenomena (see Chapter 4).

From the expansion of $\Sigma(\boldsymbol{k}, \zeta_l)$ in powers of the interaction we can obtain expansions for $\tilde{\epsilon}_k$ and $\Gamma(\boldsymbol{k})$. Thus this expansion is a useful one. The reason for the name self-energy should now be apparent.

The first-order contributions to $\Sigma(\boldsymbol{k}, \zeta_l)$ are just those contained in Fig. 3.8 and already calculated in equation (3.49) and (3.50). We have

$$\Sigma(\boldsymbol{k}, \zeta_l) = (2\pi)^{-3} \int d^3 k' f_{k'} V(0) - \epsilon(2\pi)^{-3} \int d^3 k' V(\boldsymbol{k} - \boldsymbol{k}') f_{k'}. \tag{3.58}$$

The first term gives just the first-order Hartree correction to the energy of a free particle and the second the first-order Fock energy. To this order in V, Σ is real and we still find that Γ is zero. The first contribution to $\Gamma(k)$ comes in the next order where real scattering processes are introduced. We leave the discussion of these until later.

The classification of diagrams that we have just introduced will still be possible and useful even when the system is not time and space invariant. We will not be able to use Fourier transforms in the same way but, in the position and time representation, we shall be able to define a proper self-energy part as one which cannot be separated into two by cutting a single continuous line. The equation represented by Fig. 3.17 will then follow. Explicitly this equation is,

$$G(1, 2) = G_0(1, 2) + \int G_0(1, 1')\Sigma(1', 2')G(2', 2)\, d1'd2', \tag{3.59}$$

an integral equation for the Green's function. In the case of space and time invariance this equation, as we have seen, is easily solved when $\Sigma(1', 2')$ is known. In the general case, the solution of this equation can be difficult.

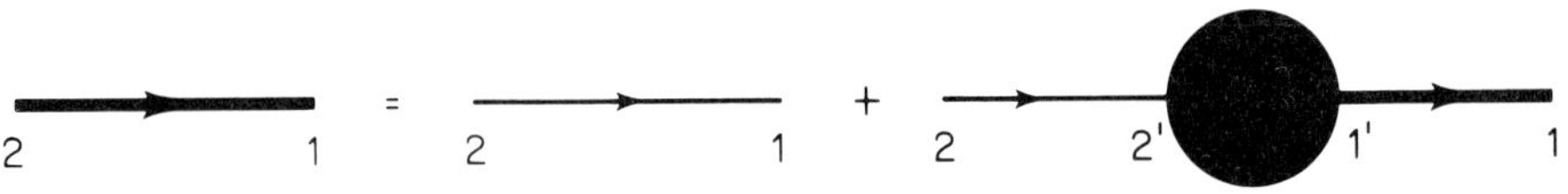

FIG. 3.17. Representation of the integral equation for $G(1, 2)$ (the thick line) with the proper self-energy as kernel.

§3.6 The Hartree–Fock approximation

We have already pointed out that, from its motivation and derivation, we expect perturbation theory to be useful when the interaction between particles is weak. In practice, this is rarely the case. The advantage of using Green's functions is that they suggest other possible starting points for approximations. Whatever the starting point, the Green's functions equations can then be used to generate higher-order approximations. For example, if the density of the system is low we might expect that correlations between large numbers of particles is negligible. This may be the case for other reasons as well, including the weakness of the interaction.

What approximations result if we neglect correlations between n and more particles? To answer this we have first to understand what we mean by correlations between particles. Let us start by considering two-particle correlations. We expect them to be contained in the two-particle Green's function

$$G_2(1, 2; 3, 4) = \langle T\{\tilde{\psi}(1)\tilde{\psi}(2)\tilde{\tilde{\psi}}(3)\tilde{\tilde{\psi}}(4)\}\rangle. \qquad (3.60)$$

Now, if there is no two-body interaction ($V = 0$), whatever the form of the one-particle potential, we have seen (§3.2) that

$$G_2(1, 2; 3, 4) = G(1, 4)G(2, 3) - \epsilon\, G(1, 3)G(2, 4). \qquad (3.61)$$

Thus even if there is no two-particle interaction G_2 is not zero. Hence not all of G_2 represents two-particle correlations. However, it is reasonable to take the difference between G_2 and (3.61) to be the two-particle correlation function $C(1, 2; 3, 4)$, that is,

$$C_2(1, 2; 3, 4) = G_2(1, 2; 3, 4) - G(1, 4)G(2, 3) + \epsilon\, G(1, 3)G(2, 4). \qquad (3.62)$$

Then C is certainly zero in any problem involving dynamically independent particles. Equation (3.62) can be represented by the diagram of Fig. 3.18. The diagram on the left contains all contributions to G_2 and the box which represents C_2 contains all diagrams which cannot be separated intwo two parts without cutting a line. This accords with a common sense view of the meaning of correlations between particles. This definition differs from that given in § 1.5 because we now subtract out the statistical correlation.

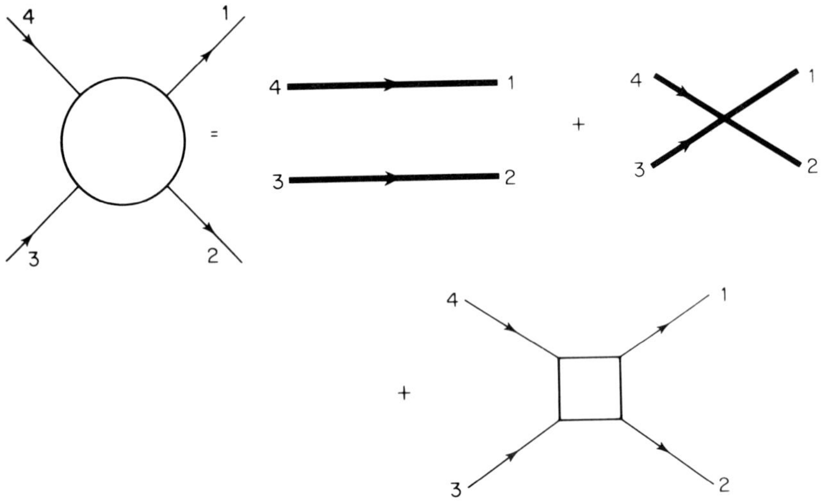

FIG. 3.18. The connection between the two-particle Green's function represented by the left of the equation and the two-particle correlation function, the last term. The thick lines represent $G(a, b)$.

In the same kind of way we can define a three-particle correlation function. This is probably most readily appreciated from the diagrams of Fig. 3.19. In that figure, the diagrams in (*a*) (of which there are 6) come from ignoring all correlations. The diagrams in (*b*) (of which there are 9) include only two-particle correlations. The diagram of (*c*) cannot be separated into two parts without cutting a line and, therefore, represents the three-particle correlation function. Hence

$$G_3(1, 2, 3; 4, 5, 6) = \sum \epsilon_P G(1, 6)G(2, 5)G(3, 4)$$

$$+ \sum \epsilon_P G(1, 6)C_2(2, 3; 4, 5) + C_3(1, 2, 3; 4, 5, 6), \qquad (3.63)$$

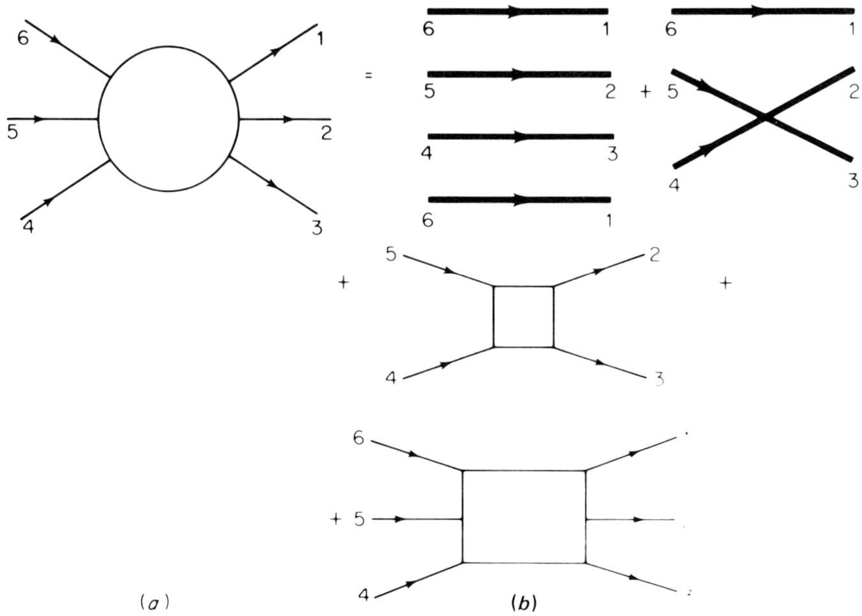

FIG. 3.19. The connection between the three-particle Green's function illustrated schematically on the left of the equation and the three-particle correlation illustrated by the last term. The latter cannot be separated into two parts without cutting a line. The thick lines represent $G(a, b)$.

where the sums are over distinct terms arising from permutations of the 1, 2, 3 vertices and of the 4, 5, 6 vertices. For fermions, there is also a factor of $\epsilon_P = -1$ resulting from odd permutations. Evidently one can define in a similar way a correlation function for n particles. It will again be represented by diagrams which cannot be separated into two parts without cutting a line.

As indicated in the introduction, for some systems we can expect to obtain a reasonable approximation if we ignore correlations of groups of more than n particles. Often, we can only tell what to choose for n and whether the approximation is satisfactory by putting it to the test.

The simplest approximation of this kind that we can make is to assume

$$C_n = 0, \qquad n \geqslant 2,$$

that is we ignore all correlations and assume that the system behaves like a set of independent particles, but not necessarily those of the bare Hamiltonian H_0. We seek a self-consistent solution for the independent particles which relies only on the vanishing of correlations. This approximation is the Hartree–Fock approximation.

Since the approximation neglects true two-particle correlations we expect it to be valid for systems of low number density ρ. To make this a quantitative statement we need to compare ρ with a parameter describing the potential. We would expect the correlations to be unimportant when the separation of the particles is much greater than the range of the potential a. Thus the Hartree–Fock approximation should be good when

$$\rho^{1/3} a \ll 1 .$$

When $\rho^{1/3} a$ is small the approximation can be made the start of an expansion in terms of this parameter. For long-range potentials such as the Coulomb interaction we should expect the Hartree–Fock approximation to fail. Note, however, that the validity of the approximation does not depend on the strength of the interaction.

Because C_2 is zero, G_2 is given by equation (3.61) and this can be substituted into equation (2.88) to yield the self-consistent equation for G,

$$[(\partial/\partial\tau) + h_0(1)]\, G(1,2) = -\delta(1,2) + \int d1' V(1,1')\{G(1,1')G(1',2) - \epsilon\, G(1,2)G(1',1'_+)\} \quad (3.64)$$

Equation (3.64) is the Hartree–Fock equation for the one-particle Green's function.

To understand what this equation means in terms of perturbation theory and corresponding diagrams, we integrate it using the function $G_0(1,2)$. Then, we find

$$G(1,2) = G_0(1,2) - \int_0^\beta d1' d2' G_0(1,1') V(1'_+, 2') \times \{G(1',2')G(2',2) - \epsilon\, G(1',2)G(2',2'_+)\} \quad (3.65)$$

This equation can be represented by the diagrams of Fig. 3.20 where now the self-consistent Green's function G replaces G_0 in the first-order approximation for Σ. For a translationally space and time invariant system,

$$\Sigma(k, \zeta_l) = \frac{\epsilon}{(2\pi)^3 \beta} \int d^3k' \sum_m V(0) \exp(i\zeta_m \delta')G(k', \zeta_m)$$

$$-\frac{1}{(2\pi)^3\beta}\int d^3k'\,V(k-k')\sum_m \exp(i\zeta_m\delta)G(k',\zeta_m). \qquad (3.66)$$

Σ is therefore independent of ζ_l and real. Hence, we obtain a correction to the energies of the particles but no contribution to $\Gamma(\boldsymbol{k})$. This is not surprising because we have neglected all scattering between the independent particles and so the particles cannot decay. Since G depends on $\Sigma(\boldsymbol{k})$, equation (3.66) is a self-consistent integral equation for $\Sigma(\boldsymbol{k})$. The first-order result (3.58) is obtained from this when G is replaced by G_0.

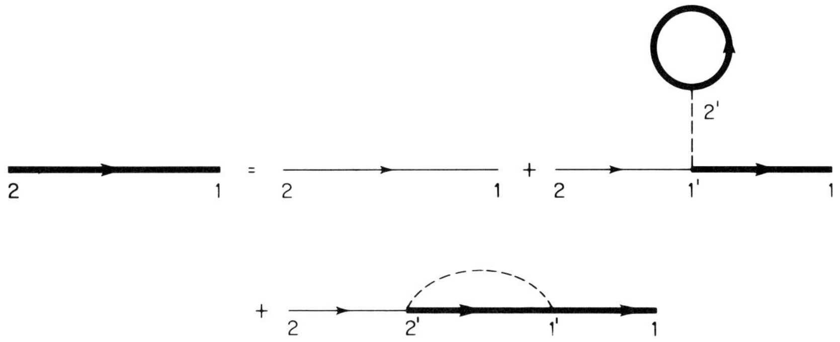

FIG. 3.20. Representation of the Hartree–Fock approximation to the one-particle Green's function (the thick lines).

Since the Hartree–Fock approximation does not assume that the interaction is weak, the result can differ substantially from perturbation theory. The self-energy $\Sigma(\boldsymbol{k})$ can be quite large and there is no unique way of finding corrections to $\Sigma(\boldsymbol{k})$. However, the lifetime of quasi-particles arises from the scattering of real quasi-particles, not the bare particles. We, therefore, expect that the leading corrections to Σ should come from the diagrams of Fig. 3.21, where the particle propagators are not the bare ones but either those calculated in the Hartree–Fock approximation or those calculated self-consistently to include the effect of the correction to Σ. These corrections are examined later in particular cases (§5.2).

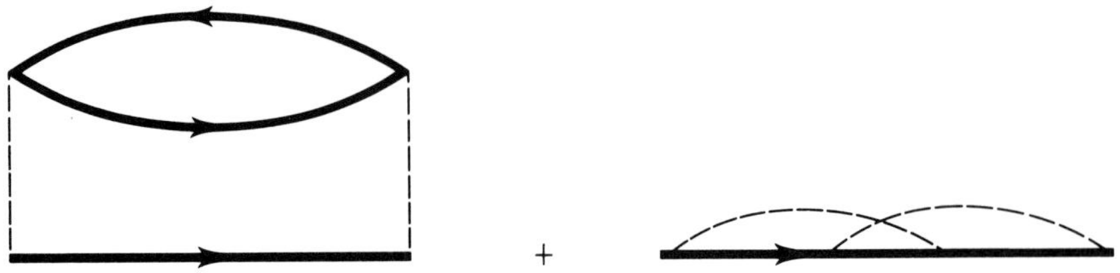

FIG. 3.21. Contributions to the self-energy which introduce a finite lifetime for quasi-particles. The thick lines represent $G(1, 2)$.

§3.7 The Hartree approximation

Historically the Hartree approximation preceded the Hartree–Fock approximation and was based on the very simple idea that, as a first approximation, each particle is acted upon by the mean potential due to the other particles. If the mean density of the other particles $\boldsymbol{r}$ at time t is $\rho(\boldsymbol{r}, t)$, this mean potential at $\boldsymbol{r}_1$ is

$$v(\boldsymbol{r}_1, t_1) = \int V(\boldsymbol{r}_1 - \boldsymbol{r}_1')\rho(\boldsymbol{r}_1', t_1)\, d^3 r_1' \equiv \int d1' V(1, 1')\rho(1'). \tag{3.67}$$

For a many-body system containing particles with wave functions spread out in space, it will make little difference if we include in ρ the contribution of the particle itself to the density and take ρ to be the total density. Then

$$\rho(1') = \langle \psi^+(1')\psi(1')\rangle = \epsilon G(1', 1'_+) \tag{3.68}$$

If the potential v is used in the equation for $G(1, 2)$ we obtain the equation

$$[(\partial/\partial\tau) + h_0(1) + v(1)]\, G(1, 2) = -\delta(1, 2) \tag{3.69}$$

with v to be determined self-consistently from

$$v(1) = \epsilon \int d1' V(1, 1')G(1', 1'_+). \tag{3.70}$$

Equations (3.69) and (3.70) together constitute the Hartree approximation. If equation (3.69) is compared with equation (3.64) we see that the former can be obtained from the latter by omitting the first integral from (3.64). This is the exchange term which also subtracts out the interaction of the particle with itself, included in the Hartree term. With reference to Fig. 3.20, the contribution of the Hartree term is given by the penultimate diagram of the figure. The exchange term comes from the last diagram.

Since the Hartree–Fock approximation sums a larger class of diagrams than the Hartree approximation, it might seem that it is a better approximation. However, this is not always the case, especially when long-range forces are involved (see §5.2). In such situations corrections to the exchange diagrams are very important and reduce their effect. In fact the Hartree approximation being based on a classical, self-consistent picture works well when classical considerations can give a good answer. Typically this occurs when macroscopic effects are paramount, for example in discussions of the long-wavelength collective modes such as plasma oscillations and sound waves, and of the effects of slowly varying external forces.

The Hartree approximation is also useful for starting an iterative solution of a problem. To initiate the Hartree approximation one needs to make a guess at the one-point function $v(1)$, the self-consistent potential. This is used to generate the Green's functions which can in turn be used to calculate an improved potential. The process can then be continued, at least in principle. Alternatively, the calculated Green's functions can be used to generate solutions of the Hartree–Fock equation or higher-order equations. To solve the Hartree–Fock equation itself by iteration requires, at the first stage, a guess for the two-point double time Green's function. It is usually less obvious what function to try for this.

§3.8 Random phase approximation(s) and screening

This method was introduced by Bohm and Pines (1953) in their studies of electrons interacting through Coulomb forces. The method was refined by Gell-Mann and Brueckner (1957) who showed how the perturbation diagrams could be summed systematically to yield improved approximations for the energy of an electron gas. Since then the method has been shown to be equivalent (in the first stage) to using the Hartree approximation for finding the first-order response of the electron gas to an external field $U(\boldsymbol{r}, t)$. This response may be needed for itself or it may be used to provide an approximation for a two-particle correlation function which, in its turn, can be used to provide an improved approximation for the free energy.

The effect of the external field is to change the local density of the particles and so to change the self-consistent field. Each particle is then acted upon by an effective field which is the sum of the external field and the self-consistent field due to all the particles. Through the self-consistent field the method takes some account of correlations between the particles and so approximates the response in a way which goes beyond an independent particle approximation.

If the response is required only to first order in the external field, it is possible to obtain the change in the Green's functions as solutions of linear inhomogeneous equations. As pointed out earlier (see §2.9) it is possible to use the method of temperature Green's functions if one uses an external field $U(\boldsymbol{r}, \tau)$ which depends on the imaginary time τ. The Hartree equation for the one-particle Green's functions is

$$[(\partial/\partial\tau) + h_0(1) + U(1) + v(1)]\,G(1,2) = -\,\delta(1,2), \quad (3.71)$$

where $v(1)$ is given by equation (3.70). To first order in U let us write

$$G(1,2) = G_0(1,2) + G_1(1,2), \qquad v(1) = v_0(1) + v_1(1) \quad (3.72)$$

where G_0, v_0 are independent of U. Then to first order in U, G_1 satisfies

$$[(\partial/\partial\tau) + h_0(1) + v_0(1)]\,G_1(1,2)$$
$$+\,\epsilon \int d1'V(1,1')G_1(1',1'_+)G_0(1,2) = -\,U(1)G_0(1,2). \quad (3.73)$$

The appropriate response function can then be obtained from $G_1(1,2)$. For example if $U(1)$ is an external potential, we may wish to know the change in number density, $\delta\rho(1)$, induced by it. This is simply

$$\delta\rho(1) = \epsilon\, G_1(1, 1_+). \quad (3.74)$$

According to §1.2, $\delta\rho(1)$ is related to $U(1)$ through a two-particle correlation function. Equation (3.74) therefore, provides an approximation for this correlation function. Because the correlation between the particles is allowed for in the change in the self-consistent field, the approximation goes beyond the independent particle approximation. Indeed, the two-particle function includes information concerning collective modes such as sound waves and plasmons. As we shall see in §5.2(c) this can lead to improved approximations for the free energy.

Equation (3.73) can also be written in the suggestive form

$$[(\partial/\partial\tau) + h_0(1) + v_0(1)]\,G_1(1,2) = -\,U_{\text{eff}}(1)G_0(1,2), \quad (3.75)$$

where

$$U_{\text{eff}}(1) = U(1) + \epsilon \int d1'V(1,1')G_1(1',1'_+) \quad (3.76)$$

and is an effective potential acting on the particles. The difference $U_{\text{eff}} - U$ is a shielding potential provided by the other particles and to be calculated self-consistently. In the case of long-range forces the shielding is substantial especially for slowly varying external fields. However, the interaction between any two particles in the system will be shielded in the same way by the other particles. Thus, the effective interaction between two particles is this shielded interaction and it is this shielded interaction which contributes to the exchange self-energy of a particle. Thus, when the shielding is substantial, the Hartree–Fock approximation for the exchange energy is poor and the Hartree approximation is a better one.

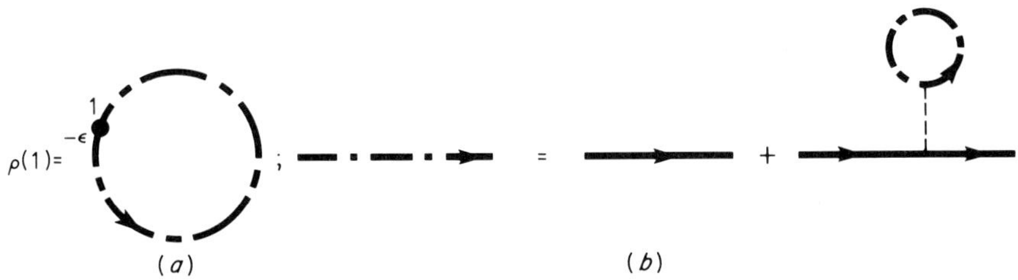

FIG. 3.22. Diagrams illustrating the calculation of the charge density in the Hartree approximation. The thick lines here represent $G_0(a, b; U)$, the broken thick line $G(a, b; U)$.

It is useful to look at the effect of shielding in terms of diagrams. Equation (3.68) is illustrated by Fig. 3.22(a) and the Hartree approximation for G by Fig. 3.22(b). To obtain $\delta\rho(1)$ to first order in U it is necessary to expand $G_0(1, 2, U)$ (the thick line) to first order in U as illustrated in Fig. 3.23. The expansion of the second term in 3.22(b) to first order in U is most easily accomplished by expanding in terms of $G_0(1, 2, U)$ (the thick lines) and then using Fig. 3.23. The result can be written as in Fig. 3.24 where the thick black lines now represent $G(1, 2; 0)$. Comparing the second term of Fig. 3.24(a)

FIG. 3.23. The expansion of the particle propagator $G_0(1, 2; U)$ (the thick line) to first order in the external field.

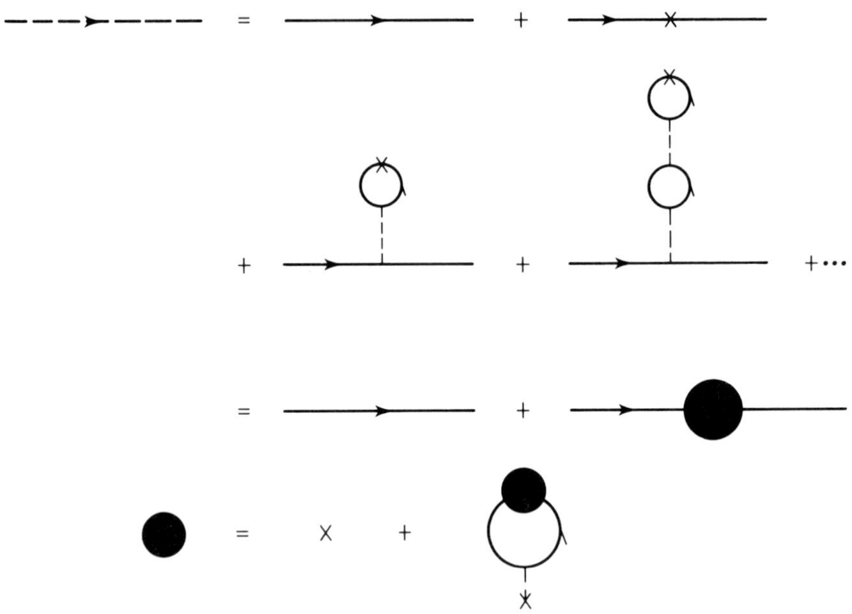

FIG. 3.24. The expansion of $G(1, 2; U)$ to first order in U and in terms of $G(1, 2; 0)$, here represented by thick black lines. The filled circle represents U_{eff} and the final equation represents the integral equation for this quantity.

with equation (3.75) we can see that the blob represents U_{eff}; Fig. 3.24(*b*) then represents equation (3.76). The second term of Fig. 3.24(*b*) thus represents the shielding of U by the particles of the system. The diagram for $\delta\rho(1)$ is now as illustrated in Fig. 3.25. The first diagram represents the direct effect of U, the others the shielding. These diagrams form a subset of those illustrated schematically in Fig. 3.10, which gives the total response to the field U.

The extension of the Hartree–Fock approximation can be dealt with in the same way. The diagrams for $\delta\rho(1)$ are illustrated in Fig. 3.26 where a new 3-point vertex function, $\Gamma(1, 2, 3)$, represented by the black triangle is introduced. The equations represented by Fig. 3.26 are

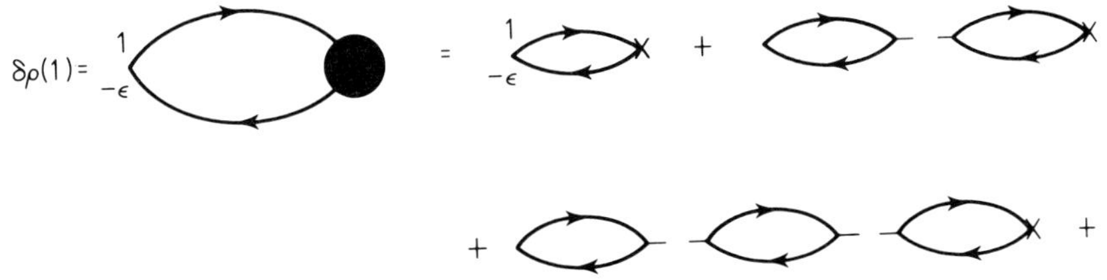

FIG. 3.25. The contribution to $\delta\rho$ linear in U in terms of the Green's function $G(1, 2; 0)$, represented by thick black lines.

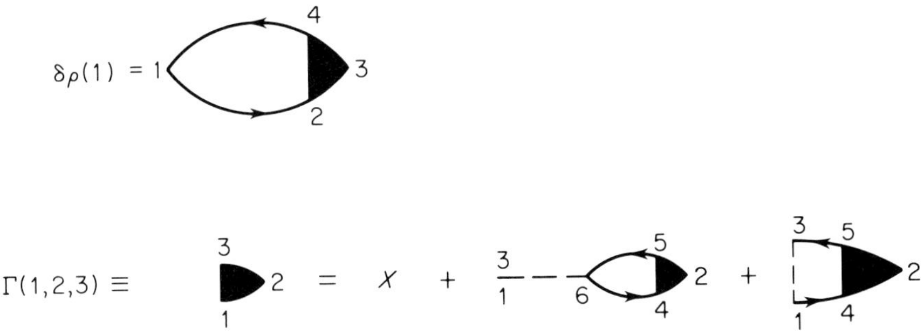

FIG. 3.26. The extension of the Hartree–Fock approximation to yield $\delta\rho(1)$ to first order in U. The thick black lines represent $G(1, 2; 0)$ calculated in the Hartree–Fock approximation.

$$\delta\rho(1) = \epsilon \int G(1, 3)\Gamma(2, 3, 4)G(4, 1)\, d2\, d3\, d4 \tag{3.77}$$

and

$$\begin{aligned}\Gamma(1, 2, 3) &= \delta(2, 3)\delta(1, 2)U(2) \\ &+ \delta(1, 3) \int V(6, 1)G(6, 4)G(5, 6)\Gamma(4, 2, 5)\, d4\, d5\, d6 \\ &+ \int V(1, 3)G(1, 4)G(5, 3)\Gamma(4, 2, 5)\, d4\, d5 \ldots\end{aligned} \tag{3.78}$$

§3.9 Conservation laws and Ward identities

The approximations described in the previous section have one further property of considerable importance; namely they preserve the general conservation laws which are satisfied by the original

Hamiltonian. For example, the Hamiltonian (3.12) on which we have concentrated attention conserves particle number and energy; in the case of particle number the conservation law is

$$\frac{\partial \rho(1)}{\partial t_1} + \operatorname{div}_1 \boldsymbol{j}(1) = 0, \qquad (3.79)$$

where $\rho(1)$ is the particle density and $\boldsymbol{j}$ the current density of particles. Both of these functions can be calculated from the one-particle Green's functions $G(1, 1'; U)$ or, to the first order in U, from two-particle Green's functions. The exact two-particle Green's functions ensure that (3.79) is satisfied to first order in U, but there is, in general, no guarantee that an approximate two-particle Green's function will satisfy (3.79). As many transport coefficients are related to conservation laws, one does not then obtain a consistent and reliable theory of these coefficients. This question has been considered in some depth by Baym and Kadanoff (1961) who lay down fairly general rules for obtaining approximations which preserve the conservation laws (sometimes called conserving approximations). Essentially they amount to using the same approximation for calculating the one-particle Green's functions in the presence of external fields as in their absence. The random phase approximations do this and so preserve the conservation laws of number, energy, and momentum. The reader is referred to the original paper by Baym and Kadanoff for further details.

In the limit of long wavelengths, the conservation laws imply relationships between an appropriate vertex function and the single-particle Green's functions. These identities are usually referred to as Ward identities after their discoverer (Ward, 1950) who first used them in elementary particle theory. As we shall not use these identities explicitly we refer the reader to Ward's paper and Schrieffer's (1964) book for further details.

References

General

Abrikosov, A. A., Gor'kov, L. P. and Dzyaloshinskii, I. Ye. (1965). "Quantum Field Theoretical Methods in Statistical Physics". Pergamon Press, Oxford.

Ambegaoker, V. (1969). "Superconductivity" (Ed. R. D. Parks), Vol. 1, Chapter 5. Dekker, New York.

Bonch-Bruevich, V. L. and Tyablikov, S. V. (1962). "The Green Function Method in Statistical Mechanics". North-Holland, Amsterdam.

Doniach, S. and Sondheimer, E. H. (1974). "Green's Functions for Solid State Physicists". Benjamin, New York.

Fetter, A. L. and Walecka, J. D. (1971). "Quantum Theory of Many Particle Systems". McGraw-Hill, New York.
Kadanoff, L. P. and Baym, G. (1962). "Quantum Statistical Mechanics". Benjamin, New York.
Parry, W. E. (1973). "The Many-Body Problem". Clarendon, Oxford.
Zubarev, D. N. (1960). *Usp. fiz. Nauk.* **71**, 71 (Translation: *Soviet Phys. Usp.* **3**, 320).

Special

Baym, G. and Kadanoff, L. P. (1961). *Phys. Rev.* **124**, 287.
Bohm, D. and Pines, D. (1953). *Phys. Rev.* **92**, 609, 626.
Feynman, R. P. (1949). *Phys. Rev.* **76**, 749, 769.
Gell-Mann, M. and Brueckner, K. (1957). *Phys. Rev.* **106**, 364.
Schrieffer, J. R. (1964). "Theory of Superconductivity". Benjamin, New York.
Ward, J. C. (1950). *Phys. Rev.* **78**, 182.

Problems

1. Write down the contributions to $\Sigma(\boldsymbol{k}, i\zeta)$ of the third and fourth diagrams shown explicitly in Fig. 3.13.
2. Draw all ten diagrams of the first order in the interaction which contribute to the two-particle Green's function when the interaction is given by equation (3.12). For the case of translational invariance and fermions show that the first order contribution to the two-particle correlation function is

$$C_2(k, \sigma_1, -k+p, \sigma_2; k', \sigma_3, p-k', \sigma_4) = G(k)G(p-k)G(k')G(p-k') \times [V(k-k')\delta_{\sigma_1\sigma_3}\delta_{\sigma_2\sigma_4} - V(p-k-k')\delta_{\sigma_1\sigma_4}\delta_{\sigma_2\sigma_3}].$$

3. Use equation (2.87) and the perturbation series for $G(\boldsymbol{k}, \zeta)$ to obtain the thermodynamic potential to first order in the interaction. Show that this agrees with the usual Hartree–Fock result.
4. By following the argument in §3.2, show that for Problem 4 of Chapter 2, it is possible to represent the terms in the perturbation expression of the one-particle Green's function in powers of g_i by diagrams and state their structure. (Note that the interaction either scatters a particle in the special oscillator state to one of the bath oscillator states or scatters in the opposite direction). Show that only one diagram contributes to the self-energy $\Sigma(\boldsymbol{k}, i\zeta)$ of the special oscillator and hence derive by perturbation theory the results already obtained by direct solution.

5. By following the argument in §3.2, write down the perturbation series for the one-particle Green's function of fermions interacting with a single classical magnetic impurity situated at the origin and described by the Hamiltonian

$$H = \int d^3r\,\psi^+(r)(-\tfrac{1}{2}\nabla^2 - \mu)\,\psi(r) + \tfrac{1}{2}J\,\psi^+(0)\mathbf{S}\cdot\boldsymbol{\sigma}\,\psi(0).$$

Here $\mathbf{S}$ is the impurity spin and the definition of the Green's function includes an average over the states of $\mathbf{S}$.

Chapter 4

Transport Coefficients of a Metal

§4.1 Introduction

We choose to begin the discussion of the application of Green's function techniques to physical problems by studying the transport coefficients of a metal. As we shall assume that the electrons in a metal are independent this is essentially a one-body problem. At the absolute zero of temperature a pure metallic crystal should have an infinite electrical conductivity because there is no mechanism for destroying the crystal momentum given to the electrons by an external electric field. In practice, however, the conductivity is finite in real crystals at finite temperatures because both the impurities that are present and the thermal lattice vibrations can and do destroy the crystal momentum and lead to a finite conductivity. This is a problem that was for a long time studied by means of the Boltzmann equation. However, the derived results seemed to be valid well beyond the apparent range of validity of the Boltzmann equation. Further there were and are many cases such as those of alloys and superconductors for which the Boltzmann equation was certainly not valid. Thus the Green's function approach was developed (Edwards 1958). This showed in the first place that, in the case of impurity scattering, the usual results were valid as long as $k_F l \gg 1$, where k_F is the Fermi wave vector and l the mean free path. This was outside the apparent range of the Boltzmann equation. (Actually it has been shown (Kadanoff and Baym 1962) that the Boltzmann equation in metals is valid for $k_F l \gg 1$ but their proof used the properties of Green's functions). Since the first results, Green's functions have been widely used to derive the transport properties of many different systems including many for which the Boltzmann equation is not valid. One advantage of Green's functions used in this way is that one can derive exact formal expressions for the transport coefficients in terms of Green's functions and these can then be approximated in controlled ways. Green's functions are also well suited to dealing with problems with impurities because they involve quantities which are close to the measurements. Although the Green's function approach is very different from that of the Boltzmann equation, and will seem unfamiliar, there are now well defined and successful procedures for using them which can quickly become familiar and can be used in a variety of problems.

In this book we shall deal only with the effect of impurities on transport phenomena. The same procedure can be used for discussing the effects of phonons but, because of the inelastic scattering, there are some technical complications. The interested reader may turn to Tewordt (1963) for further details. In this chapter we confine our attention to metals where the concentration of impurities is not too large. Indeed, we shall be assuming $k_F l \gg 1$. In this case, the main results can be derived analytically. In other cases such as alloys and semiconductors, the analysis cannot be carried so far analytically and the computer plays a more important role. We leave the discussion of such cases to Chapter 11.

As we have said, for the purposes of this chapter we assume that the electrons in a metal are non-interacting particles. Although, at first sight, this may seem far from the truth it is, for reasons given in Chapters 5 and 6, quite a good model. The electron energies will then have a band structure represented by the dependence of their energy $\epsilon(\boldsymbol{k})$ on crystal momentum $\boldsymbol{k}$. We measure energy from the chemical potential μ so that for a parabolic band

$$\epsilon(\boldsymbol{k}) = (k^2/2m) - \mu. \tag{4.1}$$

We shall concentrate on the conduction electrons which we shall assume form a degenerate gas of fermions. This means that the difference between the chemical potential and the band edge, the Fermi energy E_F is much greater than $k_B T$.

The Hamiltonian for the electrons is simply

$$H = \sum_{\boldsymbol{k},\sigma} \epsilon(\boldsymbol{k}) c^{+}_{\boldsymbol{k}\sigma} c_{\boldsymbol{k}\sigma}, \tag{4.2}$$

and the thermodynamic potential is

$$\Omega = \mathrm{Tr}\exp(-\beta H) = \prod_{k}\{1+\exp[-\beta\epsilon(\boldsymbol{k})]\}^2. \tag{4.3}$$

Other thermodynamic functions are the entropy per unit volume,

$$S = -2k_{\mathrm{B}}\sum_{\boldsymbol{k}}\{f[\epsilon(\boldsymbol{k})]\ln f[\epsilon(\boldsymbol{k})] + (1-f[\epsilon(\boldsymbol{k})])\ln(1-f[\epsilon(\boldsymbol{k})])\}, \tag{4.4}$$

and the specific heat (at constant volume and chemical potential) per unit volume,

$$C = 2k_{\mathrm{B}}\beta^2\sum_{\boldsymbol{k}}\epsilon(\boldsymbol{k})^2 f[\epsilon(\boldsymbol{k})]\{1-f[\epsilon(\boldsymbol{k})]\}. \tag{4.5}$$

In these expressions

$$f(\epsilon) = [\exp(\beta\epsilon)+1]^{-1} \tag{4.6}$$

and is the Fermi distribution function for the probability that a state of energy ϵ is occupied at temperature T. The total number of electrons present is

$$N = 2\sum_{\boldsymbol{k}} f[\epsilon(\boldsymbol{k})]. \tag{4.7}$$

For a given number of electrons equation (4.7) determines μ as a function of N and temperature. Our assumptions lead to

$$\mu(T)/\mu(0) = 1 + \mathrm{O}[(k_{\mathrm{B}}T/E_{\mathrm{F}})^2] \tag{4.8}$$

and changes of μ with temperature can be neglected.

The statistical factors in equation (4.5) ensure what is physically clear, namely that only electrons with an energy of the order of $k_{\mathrm{B}}T$ from the chemical potential contribute to the specific heat. Provided that band parameters do not vary rapidly near the chemical potential, for the specific heat per unit volume, C, we can write

$$C = 2k_{\mathrm{B}}\beta^2 N(0)\int_{-\infty}^{\infty} d\epsilon\,\epsilon^2 f(\epsilon)[1-f(\epsilon)], \tag{4.9}$$

where $N(0)$ is the density of states in energy per unit volume (for one spin) at the chemical potential $[\epsilon(\boldsymbol{k}) = 0]$ and is given by

$$N(0) = \mathscr{V}^{-1}\sum_{\boldsymbol{k}}\delta[\epsilon(\boldsymbol{k})]. \tag{4.10}$$

In $\boldsymbol{k}$-space, the ends of the wave vectors for which

$$\epsilon(\boldsymbol{k}) = 0 \tag{4.11}$$

lie on a surface, the Fermi surface.

As we have seen in §2.3(b), the single-particle Green's functions are given by the function, analytic in the upper and lower half-planes of ω,

$$G(\boldsymbol{k}, \omega) = [\omega - \epsilon(\boldsymbol{k})]^{-1}. \tag{4.12}$$

§4.2 The effect of scattering by impurities

We consider a model in which separate impurities are alike and where they act independently. We assume that they are distributed randomly in the metal but that the metal is homogeneous on a macroscopic scale. We also assume that they scatter the electrons elastically. Then they simply add to the Hamiltonian a potential term

$$\int d^3r\psi^+(\boldsymbol{r})\psi(\boldsymbol{r})U(\boldsymbol{r}) = \sum_\sigma \int d^3r\psi_\sigma^+(\boldsymbol{r})\psi_\sigma(\boldsymbol{r})U(\boldsymbol{r}), \tag{4.21}$$

where

$$U(\boldsymbol{r}) = \sum_{i=1}^{N_i} V(\boldsymbol{r} - \boldsymbol{R}_i) \tag{4.22}$$

and $V(\boldsymbol{r} - \boldsymbol{R})$ is the potential due to a single impurity situated at $\boldsymbol{R}$. The total number of impurities is N_i.

The problem is in the form discussed in §3.1. The equation satisfied by the temperature single-particle Green's function is

$$[(\partial/\partial\tau_1) + h_0(1) + U(\boldsymbol{r}_1)]\, G(1, 1') = -\delta(1, 1'), \tag{4.23}$$

where

$$\begin{aligned} h_0(1) &= \epsilon(-i\nabla_1), \\ G(1, 1') &= G_{\sigma\sigma}(1, 1') = \langle T\psi_\sigma(1)\psi_\sigma^+(1')\rangle, \qquad \sigma = \uparrow, \downarrow, \end{aligned} \tag{4.24}$$

and the solution of the unperturbed equation

$$[(\partial/\partial\tau_1) + h_0(1)]\, G_0(1, 1') = -\delta(1, 1') \tag{4.25}$$

has a Fourier transform given by equation (4.24). As discussed in §3.1, the solution of equation (4.23) is

$$\begin{aligned} G(1, 1') &= G_0(1, 1') + \int d2G_0(1, 2)U(2)G_0(2, 1') \\ &+ \int d2d3G_0(1, 2)U(2)G_0(2, 3)U(3)G_0(3, 1') + \ldots \end{aligned} \tag{4.26}$$

and is illustrated in Fig. 3.4.

Now we are not interested in the behaviour of the metal for one particular distribution of impurities. Indeed our measurements of such properties as the conductivity are made on a macroscopic scale on which the solid appears homogeneous in terms of the density of the impurities. Thus the measured quantities are usually spatial averages over regions containing large numbers of impurities. We can model this by taking an ensemble average over many metals containing impurities with the same macroscopic properties such as the same average density of impurities. Thus we should calculate the ensemble average

$$\langle G(\boldsymbol{r}, \boldsymbol{r}')\rangle$$

where the brackets indicate this average.

If this average is to be obtained from equation (4.26), it is necessary to find the averages

$$\langle U(2)\rangle, \qquad \langle U(2)U(3)\rangle, \qquad \langle U(2)U(3)U(4)\rangle$$

because $G_0(1, 2)$ is independent of the impurities. The result of the averaging will depend on what statistical properties we assume for the impurities. For concentrations that are not too large it is reasonable to assume that the positions of the impurities are independent and random. It is convenient to imagine that the metal is contained in a volume $\mathscr{V}$ with periodic boundary conditions. Then $U(1)$ can be expanded in the Fourier series

$$\begin{aligned} U(1) &= \sum_i V(\boldsymbol{r}_1 - \boldsymbol{R}_i) \\ &= \sum_{\boldsymbol{k}, i} v(\boldsymbol{k}) \exp\,[i\boldsymbol{k}\,.\,(\boldsymbol{r}_1 - \boldsymbol{R}_i)]. \end{aligned} \tag{4.27}$$

Hence

$$\begin{aligned} \langle U(1)\rangle &= \sum_{\boldsymbol{k}, i} v(\boldsymbol{k}) \exp\,(i\boldsymbol{k}\,.\boldsymbol{r}_1)\langle \exp\,(-\,i\boldsymbol{k}\,.\boldsymbol{R}_i)\rangle \\ &= \sum_{\boldsymbol{k}, i} v(\boldsymbol{k}) \exp\,(i\boldsymbol{k}\,.\boldsymbol{r}_1)\mathscr{V}^{-1} \int d^3R_i \exp\,(-\,i\boldsymbol{k}\,.\boldsymbol{R}_i) \\ &= \sum_{\boldsymbol{k}, i} v(\boldsymbol{k}) \exp\,(i\boldsymbol{k}\,.\boldsymbol{r}_1)\delta_{\boldsymbol{k},0} = N_i v(0). \end{aligned} \tag{4.28}$$

This is just the average of $U(\boldsymbol{r})$ over all space and represents a constant adjustment to all energies. Without loss of generality it can be put equal to zero.

The next average required is

$$\langle U(1)U(2)\rangle = \sum_{\boldsymbol{k}_1, \boldsymbol{k}_2} v(\boldsymbol{k}_1)v(\boldsymbol{k}_2) \exp\,(i\boldsymbol{k}_1\,.\boldsymbol{r}_1 + i\boldsymbol{k}_2\,.\boldsymbol{r}_2)$$

$$\times \sum_{i,j} \langle \exp(-i\mathbf{k}_1 . \mathbf{R}_i - i\mathbf{k}_2 . \mathbf{R}_j) \rangle. \tag{4.29}$$

If we use the fact that $v(0)$ has been taken to be zero, the average is non-zero only when $\mathbf{R}_j = \mathbf{R}_i$, $\mathbf{k}_2 = -\mathbf{k}_1$. Hence

$$\langle U(1)U(2) \rangle = N_i \sum_{\mathbf{k}} |v(\mathbf{k})|^2 \exp(i\mathbf{k} . \mathbf{r}_1 - \mathbf{r}_2) = N_i \langle v(1)v(2) \rangle, \tag{4.30}$$

where we have used the fact that for real potentials

$$v(-\mathbf{k}) = v^*(\mathbf{k}).$$

Also we write

$$v(1) = v(\mathbf{r}_1 - \mathbf{R}).$$

Similarly, we find that

$$\langle U(1)U(2)U(3) \rangle = N_i \langle v(1)v(2)v(3) \rangle. \tag{4.31}$$

When we have a product of four potentials the average is non-zero not only when all the potentials originate from the same impurity but also when two impurities each give rise to two potentials. Thus

$$\langle U(1)U(2)U(3)U(4) \rangle = N_i \sum_{\substack{\mathbf{k}_1, \mathbf{k}_2 \\ \mathbf{k}_3, \mathbf{k}_4}} v(\mathbf{k}_1)v(\mathbf{k}_2)v(\mathbf{k}_3)v(\mathbf{k}_4)$$

$$\times \exp(i\mathbf{k}_1 . \mathbf{r}_1 + i\mathbf{k}_2 . \mathbf{r}_2 + i\mathbf{k}_3 . \mathbf{r}_3 + i\mathbf{k}_4 . \mathbf{r}_4)\, \delta_{\mathbf{k}_1 + \mathbf{k}_2, -\mathbf{k}_3, -\mathbf{k}_4}$$

$$+ \sum_{\substack{\text{all} \\ \text{pairs}}} \langle U(1)U(2) \rangle \langle U(3)U(4) \rangle$$

$$= N_i \langle v(1)v(2)v(3)v(4) \rangle + N_i^2 \sum_{\substack{\text{all} \\ \text{pairs}}} \langle v(1)v(2) \rangle \langle v(3)v(4) \rangle. \tag{4.32}$$

The average of a general product of n U's can be written as a sum of products of averages of V's. For each way of grouping n v's there is a contribution to the sum. A term containing the product of r averages of v's will have a factor N_i^r. Thus the structure is

$$\langle U(1) \ldots U(2) \rangle = N_i \langle v(1) \ldots v(n) \rangle$$

$$+ N_i^2 \sum_r \langle v(\alpha_1) \ldots v(\alpha_r) \rangle \langle v(\alpha_{r+1}) \ldots v(\alpha_n) \rangle$$

$$+ N_i^3 \sum_{s \geqslant r} \langle v(\alpha_1) \ldots v(\alpha_r) \rangle \langle v(\alpha_{r+1}) \ldots v(\alpha_s) \rangle \langle v(\alpha_{s+1}) \ldots v(\alpha_n) \rangle$$

$$+ \ldots \tag{4.33}$$

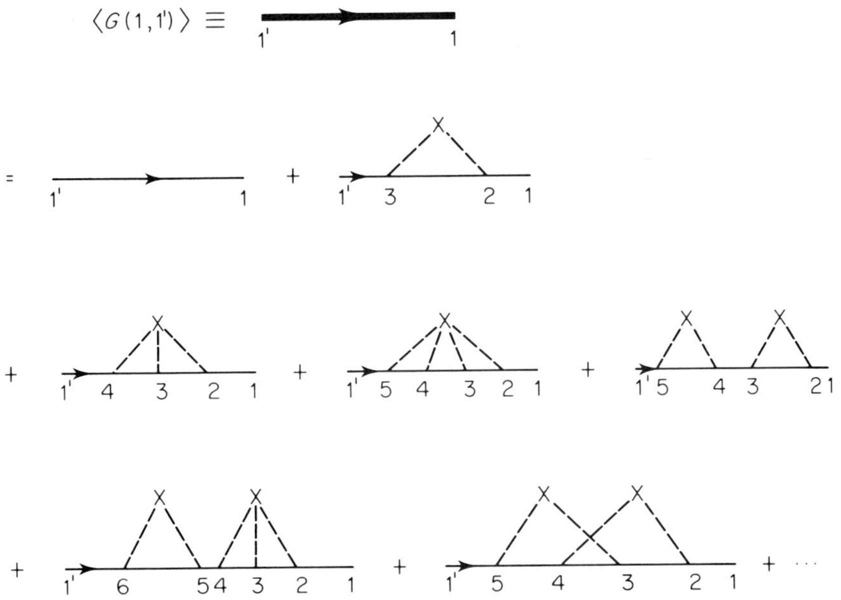

FIG. 4.1. The expansion of $G(1, 1')$ in terms of the impurity potential and $G_0(1, 1')$. A cross connected by dashed lines to points 1, 2, 3, . . . , represents a contribution $N_i\langle v(1)v(2)v(3)\ldots\rangle$.

The result for the average $\langle G(r, r')\rangle$ from equations (4.26) and (4.33) can now be easily illustrated in terms of diagrams as shown in Fig. 4.1. The contribution to $\langle G(r, r')\rangle$ from each part of Fig. 4.1 is given by the rules described in §3.2 with a straight line joining a, b representing $G_0(a, b)$, but with $N_i\langle v(\alpha_1)\ldots v(\alpha_r)\rangle$ represented by Fig. 4.2.

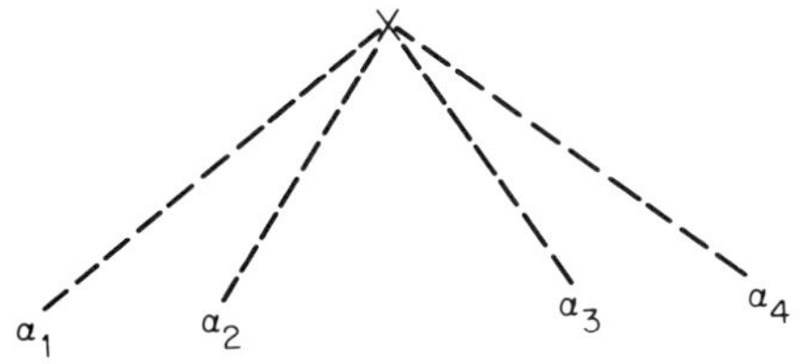

FIG. 4.2. The representation of the contribution $N_i\langle v(\alpha_1)v(\alpha_2)v(\alpha_3)v(\alpha_4)\rangle$.

Now one effect of the impurities should be to ensure that if the electrons suffer an impulse which gives them some momentum, this momentum will ultimately be destroyed. Thus, in real time, $\langle G(1, 1')\rangle$

should contain a sum of terms with decaying factors $\exp[-|t_1-t_1'|/\tau]$ where $1/\tau$ depends on the impurity potential and tends to zero as U tends to zero. Such a term is of infinite order in U and can only be obtained from Fig. 4.1 by summing an infinite number of terms. What we are saying in effect is that for sufficiently long times the impurity interaction is never weak. The situation is analogous to that discussed in §3.5 and can be resolved in a similar way by introducing the self-energy part of a diagram as one which cannot be cut into two parts by cutting only one electron line. Then one can write

$$\langle G(1,1')\rangle = G_0(1,1') + \int d2d3G_0(1,2)\Sigma(2,3)G(3,1'), \tag{4.34}$$

where the equation and Σ are illustrated in Fig. 4.3.

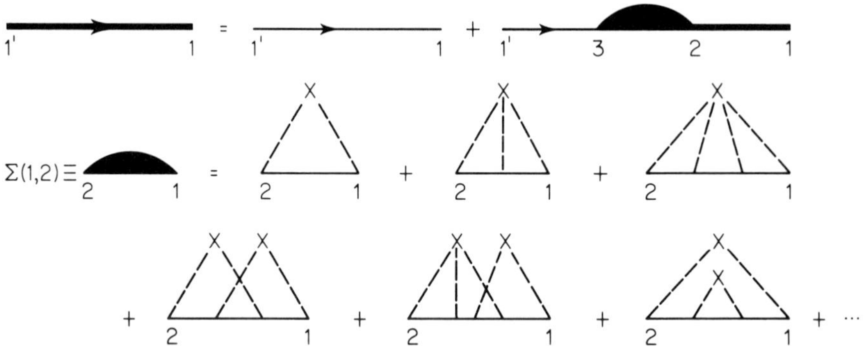

FIG. 4.3. The representation of the integral equation (4.34). The black area represents the self-energy Σ.

It is convenient to follow §3.3 and transform to momentum and energy variables. Then momentum $\boldsymbol{k}$ is carried by each dashed line to which a factor $v(\boldsymbol{k})$ corresponds and, because of relations like (4.30) and (4.31), the total momentum entering each impurity vertex is zero. Because the impurity potential is static, the scattering is elastic and each dashed line carries no energy while the electron always carries the same energy. Thus, using temperature Green's functions

$$\langle G(\boldsymbol{k},\omega_n)\rangle = G_0(\boldsymbol{k},\omega_n) + G_0(\boldsymbol{k},\omega_n)\Sigma(\boldsymbol{k},\omega_n)\langle G(\boldsymbol{k},\omega_n)\rangle \tag{4.35}$$

and Fig. 4.4 illustrates contributions to $\Sigma(\boldsymbol{k},\omega_n)$. Explicitly one obtains from equation (4.34) that

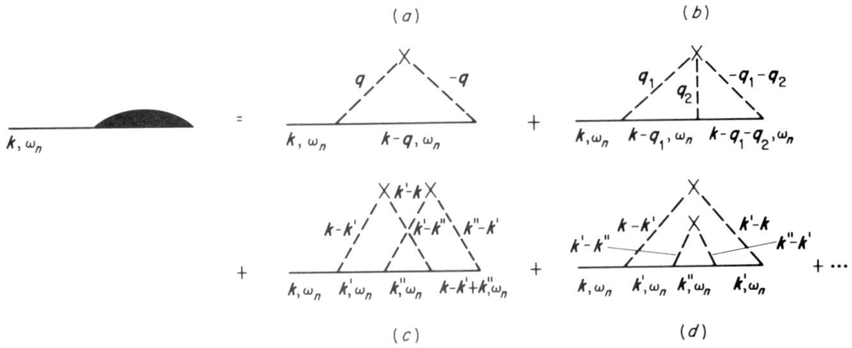

FIG. 4.4. Specific contributions to $\Sigma(k, \omega_n)$.

$$\langle G(k, \omega_n) \rangle = [G_0(k, \omega_n)^{-1} - \Sigma(k, \omega_n)]^{-1}. \tag{4.36}$$

The problem is reduced to obtaining $\Sigma(k, \omega_n)$.

In the first instance let us suppose that the scattering by the impurities is very weak. Then it is reasonable to approximate Σ by the first term illustrated in Fig. 4.36. Thus

$$\begin{aligned}
\Sigma(k, \omega_n) &= N_i \sum_q |v(q)|^2 G_0(k - q, \omega_n) \\
&= N_i \sum_{k'} \frac{|v(k - k')|^2}{i\omega_n - \epsilon(k')} \\
&= -N_i \sum_{k'} |v(k - k')|^2 \frac{[i\omega_n + \epsilon(k')]}{\omega_n^2 + \epsilon^2(k')} \\
&= \Sigma_0(k, \omega_n) + \Sigma_e(k, \omega_n),
\end{aligned} \tag{4.37}$$

where Σ_0 is odd in ω_n and Σ_e even in ω_n. Usually, as for example with transport phenomena, we are interested in electrons near the Fermi surface for which $\epsilon_k, \omega_n \ll E_F$. Let us evaluate Σ for such electrons.

Consider first Σ_e: If the integrand were confined to near the Fermi surface it would vanish because of the factor $\epsilon_{k'}$ which makes it antisymmetric about the Fermi surface. However, the integrand does not converge sufficiently rapidly for this argument to be satisfactory. Nevertheless, if we consider the difference in Σ_e between two values of k near the Fermi surface, the integrand will be sufficiently rapidly convergent for the argument to be valid.

Thus Σ_e is independent of $\boldsymbol{k}$ and ω_n for $\boldsymbol{k}$ near the Fermi surface and can be absorbed in a change in chemical potential. With this change, Σ_e can be formally neglected.

To evaluate Σ_0 we note that the integrand is sufficiently rapidly convergent for $\boldsymbol{k}'$ to lie always near the Fermi surface. In general $v(\boldsymbol{k}-\boldsymbol{k}')$ will not vary rapidly as $\boldsymbol{k}'$ varies near the Fermi surface. Thus we can write

$$\Sigma_0(\boldsymbol{k}, \omega_n) = -\frac{i\omega_n}{2\pi\tau_k}\int_{-\infty}^{\infty}\frac{d\epsilon'}{\omega_n^2+\epsilon'^2} = -\frac{i\omega_n}{|\omega_n|2\tau_k}, \tag{4.38}$$

where

$$\tau_k^{-1} = 2\pi N_i \sum_{\boldsymbol{k}'} |v(\boldsymbol{k}-\boldsymbol{k}')|^2\,\delta(\epsilon'). \tag{4.39}$$

As we shall see $2\tau_k$ is the lifetime for an electron of momentum $\boldsymbol{k}$ and its value as given by equation (4.39) is the same as is obtained by conventional perturbation theory in the Born approximation. For a spherically symmetric Fermi surface τ_k is independent of $\boldsymbol{k}$ for electrons near the Fermi surface.

The temperature Green's function is

$$\langle\!\langle G(\boldsymbol{k}, \omega_n)\rangle\!\rangle = [i\omega_n - \epsilon(\boldsymbol{k}) + i\omega_n/|\omega_n|2\tau_k]^{-1}. \tag{4.40}$$

In this case it is obvious that in the upper half-plane,

$$G(\boldsymbol{k}, \omega) = [\omega - \epsilon(\boldsymbol{k}) + i/2\tau_k]^{-1}, \tag{4.41}$$

while in the lower half-plane

$$G(\boldsymbol{k}, \omega) = [\omega - \epsilon(\boldsymbol{k}) - i/2\tau_k]^{-1}. \tag{4.42}$$

Thus $G_R(\boldsymbol{k}, \omega)$ is given by equation (4.41) with ω real. Furthermore

$$\begin{aligned} G_R(\boldsymbol{k}, t) &= \int_{-\infty}^{\infty}\frac{d\omega}{2\pi}\frac{\exp(-i\omega t)}{\omega-\epsilon(\boldsymbol{k})+i/2\tau_k} \\ &= -i\exp[-i\epsilon(\boldsymbol{k})t - t/2\tau_k]\,\theta(t). \end{aligned} \tag{4.43}$$

This confirms that impurity scattering does lead to a finite lifetime for electron states of definite momentum.

It is also revealing to look at the spatial Fourier transform of the Green's function,

$$G(\boldsymbol{r}, \omega) = \int d^3k\,\frac{\exp(i\boldsymbol{k}\,.\boldsymbol{r})}{\omega-\epsilon(\boldsymbol{k})+i/2\tau_k}.$$

For a spherically symmetric Fermi surface and $\omega, \tau^{-1} \ll E_F$, this can be written

$$G(r, \omega) = \frac{N(0)}{k_F r} \int_{-\infty}^{\infty} \frac{d\epsilon \sin kr}{\omega - \epsilon + i/2\tau},$$

with

$$k - k_F \approx m\epsilon/k_F.$$

Hence,

$$G(r, \omega) = \frac{N(0)}{2ik_F r} \int_{-\infty}^{\infty} d\epsilon \frac{\exp(ik_F r + im\epsilon r/k_F) - \exp(-ik_F r - im\epsilon r/k_F)}{\omega - \epsilon + i/2\tau},$$

$$= -[\pi N(0)/k_F r] \exp(ik_F r - r/2l),$$

$$l = v_F \tau = k_F \tau/m.$$

One can interpret this result as meaning that a propagating particle near the Fermi surface retains a memory of its initial state only over a distance of order l. This distance l is related to the lifetime τ in the same way as the transport mean free path is related to the transport lifetime to be defined later [equation (4.91)].

After this digression we return to the Fourier transform of the Green's function given by equation (4.42). A comparison of this equation with equation (2.17) shows that the spectral function is

$$A(k, \omega) = (i/2\pi)[(\omega - \epsilon(k) + i/2\tau_k)^{-1} - (\omega - \epsilon(k) - i/2\tau_k)^{-1}]$$

$$= (2\pi\tau_k)^{-1}\{[\omega - \epsilon(k)]^2 + i/4\tau_k^2\}^{-1}. \tag{4.44}$$

Thus the δ-function spectrum for free particles is spread out by the impurities into a Lorentzian. From equation (2.46) we see that the total density of states in energy per unit volume is

$$N(\omega) = (2\pi\mathscr{V})^{-1} \sum_k (\{[\omega - \epsilon(k)]^2 + i/4\tau^2\}\tau_k)^{-1}.$$

For a spherically symmetric Fermi surface this becomes

$$N(\omega) = [N(0)/2\pi\tau] \int d\epsilon\, [(\omega - \epsilon)^2 + i/4\tau^2]^{-1}, \tag{4.45}$$

where $N(0)$ is the density of states for the pure metal. Integration yields from equation (4.45)

$$N(\omega) = N(0),$$

and the density of states is unchanged. This remains true for a non-spherical Fermi surface provided that τ_k depends only on the direction of k for electrons near the Fermi surface. These results are to be expected when the scattering is elastic.

Before turning to the application of these results to the calculation of transport coefficients we consider corrections to the previous result from the terms so far neglected in $\Sigma(\mathbf{k}, \omega_n)$. First consider the contribution of all diagrams such as Figs. 4.4(*a*) and (*b*) containing one cross. This contribution can be written as

$$\Sigma(\mathbf{k}, \omega_n) = t(\mathbf{k}, \mathbf{k}, \omega_n) \tag{4.46}$$

where $t(\mathbf{k}', \mathbf{k}, \omega_n)$ is the t-matrix for scattering by an impurity at an imaginary energy ω_n defined by

$$t(\mathbf{k}' - \mathbf{k}, \omega_n) = v(\mathbf{k}' - \mathbf{k}) + \sum_{\mathbf{k}''} v(\mathbf{k}' - \mathbf{k}'')G(\mathbf{k}'', \omega_n)t(\mathbf{k}'' - \mathbf{k}, \omega_n). \tag{4.47}$$

Apart from its sign, t like Σ will not depend strongly on ω_n. The important part of Σ is

$$\Sigma_0(\mathbf{k}, \omega_n) = i \operatorname{Im} t(\mathbf{k}, \mathbf{k}, \omega_n) = -i\omega_n/|\omega_n|2\tau_{\mathbf{k}}, \tag{4.48}$$

where now

$$\tau_{\mathbf{k}}^{-1} = 2\pi N_i \sum_{\mathbf{k}'} |t(\mathbf{k}, \mathbf{k}')|^2 \delta(\epsilon') \tag{4.49}$$

and we have used the optical theorem for the t-matrix to derive (4.48) and (4.49). Thus the effect of these diagrams is to replace the Born approximation for scattering by the exact scattering cross section for a single impurity.

The remaining diagrams involve the interference of scattering by more than one impurity and should yield small contributions for low concentrations of impurities. If we compare the contributions of diagrams 4.4(*c*) and 4.4(*d*) we note that for their contributions to Σ_0 the momenta of the electron lines all lie near the Fermi surface. In the case of 4.4(*c*) this means that $(\mathbf{k}'' + \mathbf{k} - \mathbf{k}')$ lies near the Fermi surface. Since the spread of energies is $\sim 1/\tau$, this means that given $\mathbf{k}$ and $\mathbf{k}'$, $\mathbf{k}''$ is restricted to a solid angle $(E_F\tau)^{-1} \sim (k_F l)^{-1}$. Hence, 4.4(*c*) yields a contribution to Σ_0 which is smaller than that of 4.4(*d*) by a factor of this order. Thus for moderate mean free paths 4.4(*c*) and similar diagrams can be neglected. The effect of diagram 4.4(*c*) and its iterations is to change equation (4.38) to

$$\Sigma_0(\mathbf{k}, \omega_n) = -\frac{(i\omega_n + \Sigma_0)}{2\pi\tau_{\mathbf{k}}} \int \frac{d\epsilon'}{(\omega_n - i\Sigma_0)^2 + \epsilon'^2} \tag{4.50}$$

and this has the same solution, (4.38). Hence, neglecting only terms of relative order $(k_F l)^{-1}$, the one-particle Green's function is given by equation 4.40) with $\tau_{\mathbf{k}}$ defined by equation (4.49).

§4.3 The electrical conductivity

The conductivity tensor relates the current density J in the metal to the electric field E which induces it through Ohm's law,

$$J_\mu = \sigma_{\mu\nu} E_\nu . \tag{4.51}$$

As this is a linear relationship, the conductivity can be found by a proper application of Kubo's method (§1.2). The application of Kubo's formula requires a knowledge of the perturbing term in the Hamiltonian, H', and of the current operator, j_{op}. For the moment, we consider the general case of an electric field which may depend upon position and time and we represent it by the vector potential $A(r, t)$ where

$$E = -\partial A/\partial t. \tag{4.52}$$

For simplicity we specialize to the case of a parabolic band. Then, in the presence of the vector potential, the Hamiltonian for the electrons becomes (see Messiah 1961, and Appendix A)

$$H = \int \psi^+(r)\{(-i\nabla - eA)^2/2m + U(r)\}\psi(r)d^3r. \tag{4.53}$$

This expression ensures gauge invariance under the transformation

$$\psi \to \exp(i\phi)\psi, \qquad A \to A + e^{-1}\nabla\phi. \tag{4.54}$$

To first order in A (and therefore in E) the perturbation is

$$H' = -\int d^3rA \cdot j_1(r), \tag{4.55}$$

where

$$j_1(r) = (-ie/2m)[\psi^+\nabla\psi - (\nabla\psi^+)\psi]. \tag{4.56}$$

To obtain the operator for the total current density we note that, in the presence of a vector potential, the velocity of a classical particle v is related to its canonical momentum p by

$$v = p - eA.$$

This identification is consistent with the kinetic energy given in equation (4.53). It follows that the classical density of electric current carried by many particles is given by

$$j(r) = e\sum_i [p_i - eA(r_i)]\delta(r - r_i).$$

The Hermitian operator which corresponds to this in wave mechanics is

$$j(r) = \tfrac{1}{2}e \sum \{[p_i - eA(r_i)]\,\delta(r - r_i) + \delta(r - r_i)[p_i - eA(r_i)]\},$$

where care has been taken about the order of operators because p_i and r_i do not commute. In the occupation number representation, the corresponding operator is

$$\begin{aligned} j_{\rm op}(r) &= (e/2m)\{\psi^+(-i\nabla - eA)\psi + [(i\nabla - eA)\psi^+]\,\psi\} \\ &= j_1 - (e^2 A/m)\psi^+\psi. \end{aligned} \tag{4.57}$$

This expression is also Hermitian and invariant under the gauge transformation (4.54).

The application of Kubo's formula (§1.2) to the present problem shows that the average current induced by the external field is, to first order in the field,

$$\begin{aligned} J_\alpha(r, t) &= \langle j_{{\rm op}\,\alpha}(r)\rangle \\ &= (-e^2/m)A_\alpha\langle\psi^+\psi\rangle - \sum_\beta \int_0^\infty dt' \int d^3r' A_\beta(r', t')G^j_{R\,\alpha\beta}(1, 1'), \end{aligned} \tag{4.58}$$

where

$$G^j_{R\,\alpha\beta}(1, 1') = -i\langle[j_{1\alpha}(1), j_{1\beta}(1')]\rangle\theta(t_1 - t_1'). \tag{4.59}$$

The first term in equation (4.58) stems from the explicit dependence of $j_{\rm op}$ on A. This explicit dependence can be ignored in the second term as this is already first-order in A.

Now the expression (4.58) includes the transient response to switching on the field at $t = 0$ as well as the forced response. In general, it is the forced response which is measured and, provided that G^j includes a dissipative effect (as it does when impurities are present), it can be obtained by extending the lower limit of the integral in (4.58) to $-\infty$. Then, at any finite time the transients will have died away. In this case, one frequency component of A will give rise to the corresponding frequency component of J. Thus we obtain

$$J_\alpha(r, \omega) = (-ne^2/m)A_\alpha(r, \omega) - \sum_\beta \int d^3r' A_\beta(r', \omega)G^j_{R\,\alpha\beta}(1, 1'; \omega), \tag{4.60}$$

where we have introduced the number density of conduction electrons n. As was explained in §2.1, G^j_R can be obtained by analytic continuation from the temperature Green's function which we now proceed to calculate.

The temperature Green's function related to G^j_R is

$$G^j_{\alpha\beta}(1, 1') = -\langle T\tilde{j}_{1\alpha}(1)j_{1\beta}(1')\rangle. \tag{4.61}$$

If the expression (4.56) is used for $\boldsymbol{j}_1$, this can be rewritten as

$$G^j_{\alpha\beta}(1, 1') = (e^2/4m^2)(\nabla_{2'\beta} - \nabla_{1'\beta}) \times (\nabla_{2\alpha} - \nabla_{1\alpha})G_{2\sigma\sigma'}(2, 2'; 1, 1')\big|_{\substack{\tau_2'=\tau_{1-}',\, r_2'=r_1' \\ \tau_2=\tau_{1-},\, r_2=r_1}} \quad (4.62)$$

where

$$G_{2\sigma\sigma'}(2, 2'; 1, 1') = \langle T[\tilde{\psi}_\sigma(2)\tilde{\psi}_{\sigma'}(2')\tilde{\bar{\psi}}_{\sigma'}(1')\tilde{\bar{\psi}}_\sigma(1)]\rangle. \quad (4.63)$$

Now the particles are independent. Hence G_α can be written [cf. equation (3.38)]

$$G_{2\sigma\sigma'}(2, 2'; 1, 1') = G_{\sigma'\sigma'}(2', 1')G_{\sigma\sigma}(2, 1) - G_{\sigma\sigma'}(2, 1')G_{\sigma'\sigma}(2', 1). \quad (4.64)$$

If this expression is substituted into equation (4.62) the result can be represented by Fig. 4.5 where the thick black lines represent the exact one-particle Green's functions and the vertices represent the operators

$$(-ie/2m)(\nabla_2 - \nabla_1) \quad \text{and} \quad -(ie/2m)(\nabla_{2'} - \nabla_{1'})$$

which act before the limits,

$$1' \to 1, \qquad 2' \to 2,$$

are taken (cf. §3.10). In fact when substituted into equation (4.62) the first term gives a contribution to G^j of

$$\langle j_{1\beta}(1')\rangle\, \langle j_{1\alpha}(1)\rangle.$$

But $\langle \boldsymbol{j}(1)\rangle$ is simply the current density in the absence of an electric field and this is zero. Hence, only the second term contributes to (4.62).

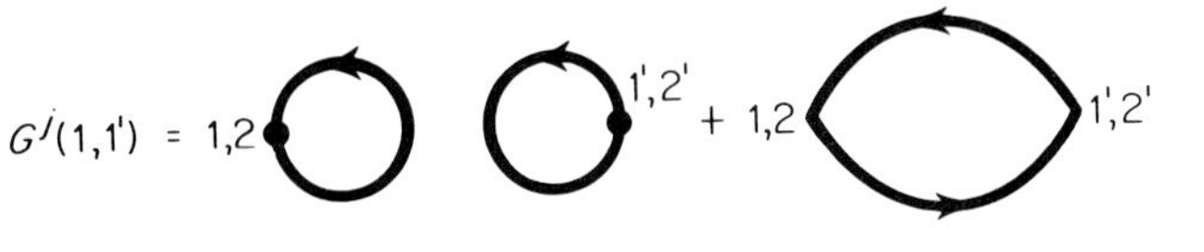

FIG. 4.5. The representation of the contributions to the current–current correlation function. The thick black lines represent the particle propagators in the presence of impurities.

As explained in the last section we are interested only in the ensemble average of $J(\boldsymbol{r}, \omega)$ over the positions of the impurities. Thus we require

$$\langle G_{\sigma\sigma'}(2, 1')G_{\sigma'\sigma}(2', 1)\rangle.$$

Now the same impurities are involved in the two Green's functions so when we perform the ensemble average there will be interference terms resulting. For the moment let us ignore these and suppose that the Green's functions are independent. Then

$$\langle G_{\sigma\sigma'}(2, 1')G_{\sigma'\sigma}(2', 1)\rangle = \langle G_{\sigma\sigma'}(2, 1')\rangle\langle G_{\sigma'\sigma}(2', 1)\rangle. \quad (4.65)$$

This means that the thick black lines in Fig. 4.5 now represent the average Green's functions displayed in Fig. 4.3.

After the spatial averaging, $G^j(1', 1)$ depends only on the coordinate difference $(\boldsymbol{r}_1' - \boldsymbol{r}_1)$ and it is convenient to use spatial Fourier transforms. One Fourier component of $\boldsymbol{J}(\boldsymbol{r}, \omega)$ depends on the corresponding Fourier component of $\boldsymbol{A}$. Hence

$$J_\alpha(\boldsymbol{q}, \omega) = -\sum_\beta K_{\alpha\beta}(\boldsymbol{q}, \omega)A_\beta(\boldsymbol{q}, \omega), \quad (4.66)$$

where

$$K_{\alpha\beta}(\boldsymbol{q}, \omega) = (ne^2/m)\delta_{\alpha\beta} + G^j_{R\,\alpha\beta}(\boldsymbol{q}, \omega) \quad (4.67)$$

and

$$G^j_{\alpha\beta}(\boldsymbol{q}, i\omega_\nu) = (2e^2/4m^2\beta\mathscr{V})$$

$$\times \sum_{\boldsymbol{k},\lambda} (2k_\alpha + q_\alpha)(2k_\beta + q_\beta)G(\boldsymbol{k}, \omega_\lambda')G(\boldsymbol{k} + \boldsymbol{q}, \omega_\lambda' + \omega_\nu), \quad (4.68)$$

$$\omega_\lambda' = (2\lambda + 1)\pi/\beta, \qquad \omega_\nu = 2\pi\nu/\beta,$$

and we have used the "time" Fourier transform of the temperature Green's functions. The factor 2 takes account of the sum over spins.

For many cases of interest, the wavelength of the field is very much less than the Fermi wavelength and ω is very much less than E_F. Then, the fields cause transitions only of electrons with wave vectors near the Fermi surface. Thus we have

$$q \ll k \quad \text{and} \quad \epsilon(\boldsymbol{k}) \equiv \epsilon_k \ll E_F. \quad (4.69)$$

We exploit these inequalities in a number of ways to evaluate (4.68).

We can first neglect q_α, q_β in the first two brackets of the left-hand side of equation (4.69). Also the vectors k_α, k_β can be replaced by $k_{F\alpha}, k_{F\beta}$, that is, vectors in the same direction as k_α and k_β but with fixed magnitudes k_F. If we introduce the spectral form of $G(\boldsymbol{k}, \omega_\lambda)$, we obtain

$$G^j_{\alpha\beta}(\boldsymbol{q}, i\omega_\nu) = \frac{2e^2N(0)}{m^2\beta}\int\frac{d\Omega}{4\pi}k_{F\alpha}k_{F\beta}\int d\epsilon_k$$

$$\times \sum_\lambda \int_{-\infty}^{\infty} dx dy \frac{A(\boldsymbol{k}, x)A(\boldsymbol{k} + \boldsymbol{q}, y)}{(i\omega_\lambda' - x)[i(\omega_\lambda' + \omega_\nu) - y]}, \quad (4.70)$$

where the spectral function $A(\boldsymbol{k}, x)$ is given by equation (4.44). The summation over λ can be performed by the method of Appendix B and leads to

$$G^j_{\alpha\beta}(\boldsymbol{q}, i\omega_\nu) = \frac{2e^2 N(0)}{m^2}\int\frac{d\Omega}{4\pi}k_{F\alpha}k_{F\beta}\int d\epsilon_k \times\int dxdy\,\frac{A(\boldsymbol{k},x)A(\boldsymbol{k}+\boldsymbol{q},y)[f(x)-f(y)]}{x-y+i\omega_\nu}. \tag{4.71}$$

The function obtained formally from this by replacing $i\omega_\nu$ by ω is analytic in the upper half-plane and satisfies condition (iii), §2.10, as $\omega\to\infty$. Hence it is $G^i_{\beta\alpha}(\boldsymbol{q},\omega)$ in the upper half-plane of ω. We can straightforwardly perform the integral over y in the term involving $f(x)$ and over x in the term involving $f(y)$ to obtain

$$G^j_{\alpha\beta}(\boldsymbol{q}, \omega) = \frac{2e^2 N(0)}{m^2}\int\frac{d\Omega}{4\pi}k_{F\alpha}k_{F\beta}\int d\epsilon_k \times\int dxf(x)\left[\frac{A(\boldsymbol{k},x)}{x-\epsilon_{k+q}+\omega+i/2\tau}-\frac{A(\boldsymbol{k}+\boldsymbol{q},x)}{\epsilon_k-x+\omega+i/2\tau}\right]. \tag{4.72}$$

The next simple integral to perform is that over ϵ_k. Unfortunately, however, the integrals do not converge for the orders of integration over ϵ_k and x to be interchanged. Instead we proceed as follows. Consistent with our inequalities (4.69), we note that

$$\epsilon_{k+q} = \epsilon_k + \boldsymbol{k}.\boldsymbol{q}/m + q^2/2m \approx \epsilon_k + \boldsymbol{v}_F.\boldsymbol{q}. \tag{4.73}$$

where $\boldsymbol{v}_F$ is a vector in the direction of $\boldsymbol{k}$ but with magnitude equal to the Fermi velocity k_F/m. Then ϵ_k appears in the integrand only in the combination $x-\epsilon_k$. Hence, the integrals over x and ϵ_k have the form

$$I = \int_{-\infty}^{\infty} d\epsilon_k \int_{-\infty}^{\infty} dxf(x)F(x-\epsilon_k), \tag{4.74}$$

where again in accord with (4.69) we extend the integration over ϵ_k to $\pm\infty$. If we now integrate by parts, first over ϵ_k, then over x, we find

$$\begin{aligned} I &= -\int_{-\infty}^{\infty} d\epsilon\,\epsilon\,\frac{d}{d\epsilon}\int_{-\infty}^{\infty} dxf(x)F(x-\epsilon) \\ &= \int_{-\infty}^{\infty} d\epsilon\,\epsilon\int_{-\infty}^{\infty} dxf(x)\,\frac{dF(x-\epsilon)}{dx} \\ &= -\int_{-\infty}^{\infty} d\epsilon\,\epsilon\int_{-\infty}^{\infty} dx\,\frac{df}{dx}F(x-\epsilon). \end{aligned} \tag{4.75}$$

Because of the factor df/dx, the convergence is now sufficient for the orders of integration to be interchanged and that over ϵ taken first. The integrals over ϵ and x are now elementary and lead to

$$G^{j}_{\alpha\beta}(\boldsymbol{q}, \omega) = \frac{2e^2 N(0)}{m^2}\int \frac{d\Omega}{4\pi} k_{F\alpha} k_{F\beta} \frac{v_F q\mu - i/\tau}{\omega - v_F q\mu + i/\tau}, \tag{4.76}$$

where

$$\mu = \boldsymbol{k}_F . \boldsymbol{q}/k_F q = \boldsymbol{v}_F . \boldsymbol{q}/v_F q. \tag{4.77}$$

Hence

$$G^{j}_{\alpha\beta}(\boldsymbol{q}, \omega) = \frac{2e^2 N(0)}{m^2}\int \frac{d\Omega}{4\pi} k_{F\alpha} k_{F\beta}\left[-1 + \frac{\omega}{\omega - v_F q\mu + i/\tau}\right]$$

and

$$\frac{2e^2 N(0)}{m^2}\int \frac{d\Omega}{4\pi} k_{F\alpha} k_{F\beta} = \frac{2e^2 N(0) k_F^2}{3m^2}\,\delta_{\alpha\beta}\int\frac{d\Omega}{4\pi}$$

$$= \frac{4e^2 N(0) E_F}{3m}\,\delta_{\alpha\beta}. \tag{4.78}$$

But, for the free-electron model we are using,

$$n = \tfrac{4}{3}N(0)E_F. \tag{4.79}$$

It follows from equations (4.66), (4.67), (4.76), (4.78) and (4.79) that

$$K_{\alpha\beta}(\boldsymbol{q}, \omega) = \frac{3ne^2}{mk_F^2}\int \frac{d\Omega}{4\pi} k_{F\alpha} k_{F\beta} \frac{\omega}{\omega - v_F q\mu + i/\tau} \tag{4.80}$$

and

$$\boldsymbol{J}(\boldsymbol{q}, \omega) = -\frac{3ne^2}{mk_F^2}\int \frac{d\Omega}{4\pi} \frac{\boldsymbol{k}[\boldsymbol{k}.\boldsymbol{A}(\boldsymbol{q}, \omega)]\,\omega}{\omega - v_F q\mu + i/\tau}. \tag{4.81}$$

But

$$\boldsymbol{E}(\boldsymbol{q}, \omega) = i\omega \boldsymbol{A}(\boldsymbol{q}, \omega).$$

Hence

$$\boldsymbol{J}(\boldsymbol{q}, \omega) = \frac{3ne^2 i}{mk_F^2}\int \frac{d\Omega}{4\pi} \frac{\boldsymbol{k}[\boldsymbol{k}.\boldsymbol{E}(\boldsymbol{q}, \omega)]}{\omega - v_F q\mu + i/\tau}. \tag{4.82}$$

Apart from the definition of τ to which we shall return in a moment this expression is the same as that derived from the Boltzmann equation in the relaxation time approximation (Ziman 1960) and used successfully in the analysis of experiments on the anomalous skin effect in metals.

The d.c. conductivity is obtained from equation (4.82) by considering the effect of a static uniform electric field. We therefore require only the Fourier components $\omega = 0$, $\boldsymbol{q} = 0$. It then follows that

$$\boldsymbol{J} = (ne^2\tau/m)\boldsymbol{E}, \tag{4.83}$$

with the usual formula for the conductivity

$$\sigma = ne^2\tau/m. \tag{4.84}$$

The calculation so far has ignored the interference between the two Green's functions occuring in the response when the ensemble average over the impurities has been taken. However, it is clear that there will be contributions from averages of the form

$$\langle v(\alpha_1)v(\alpha_2)\ldots v(\alpha_r)\rangle,$$

where some of the impurity potentials arise from the interation of one of the Green's functions and some from the other. If we use the representation of Fig. 4.2 for the averages, we can represent the effect of averaging on $G^j(1', 1)$ as illustrated in Fig. 4.6. In the complete sum the impurity potential can act any number of times on each line, and all possible averages must be included.

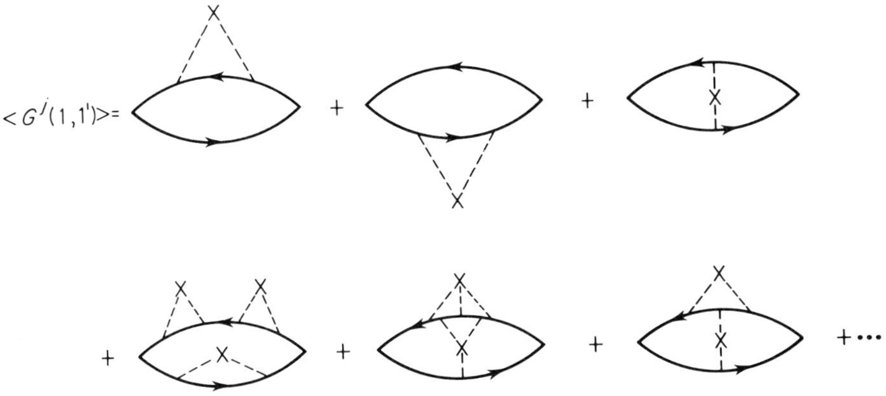

FIG. 4.6. Diagrams representing the averaged correlation function $G^j(1, 1')$.

The diagrams strongly resemble those that arise in perturbation theory when there is a two-body interaction and these are discussed in Chapter 3. The approximation we have used to obtain the average Green's function are analogous to the Hartree–Fock approximation discussed in §3.6. (There are no Hartree diagrams in the present problem.) As we have said in §3.8, we expect that, for consistency, the diagrams we should include in $\langle G^j\rangle$ are those given by the random phase approximation and illustrated in Fig. 4.7. In the diagram the heavy lines represent the average single-particle Green's

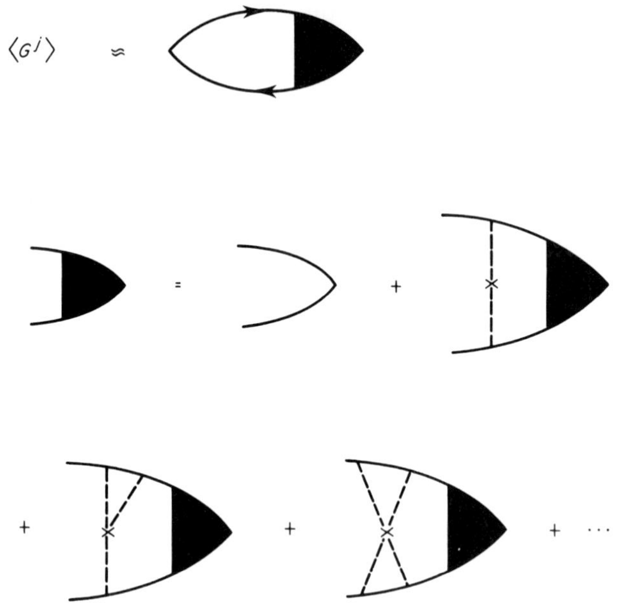

FIG. 4.7. The representation of the random phase approximation for $\langle G^j(1, 1')\rangle$. The thick black lines represent the average single-particle Green's function $\langle G\rangle$.

functions already calculated. The contribution of the first diagram (with no impurity shown explicitly) is the one already calculated. The remaining diagrams to be included all have one impurity site shown explicitly. All diagrams which can be displayed in this way are to be included. The diagrams being ignored all involve the interference of scattering from one impurity site with scattering from another. These will therefore all be a factor of at most $(k_F l)^{-1}$ smaller. It is therefore consistent to exclude them.

Let us consider first the contribution of the first two diagrams of Fig. 4.7 to $\langle G^j\rangle$. These correspond with the first diagrams used in evaluating $\langle G\rangle$ and are the most important if the scattering by a single impurity is weak. For these diagrams the contribution to G^j is given by

$$G^j_{\beta\alpha}(\boldsymbol{q}, i\omega_\nu) = (2e^2/4m^2\beta) \sum_{\boldsymbol{k},\lambda} (2k_\alpha + q_\alpha)\Gamma_\beta(\boldsymbol{k}, \boldsymbol{k}+\boldsymbol{q}) \times G(\boldsymbol{k}, \omega'_\lambda)G(\boldsymbol{k}+\boldsymbol{q}, \omega'_\lambda + \omega_\nu) \tag{4.85}$$

where the vertex part $\mathbf{\Gamma}$ satisfies the integral equation

$$\mathbf{\Gamma}(k, k+q) = (2k+q) + N_i \sum_{k'} |v(k-k')|^2 G(k', \omega'_\lambda) \times \mathbf{\Gamma}(k', k'+q) G(k'+q, \omega'_\lambda + \omega_\nu) \qquad (4.86)$$

This integral equation is the analogue of the Boltzmann equation and, indeed, the Boltzmann equation can be obtained from it. However, it is just as easy to work directly with equation (4.86). Note that $\mathbf{\Gamma}$ is also a function of ω'_λ and ω_ν but that these are parameters in equation (4.86).

Consider first the case where $ql \ll 1$. Then q is much smaller than all other wave vectors in the problem and can be neglected. Since $\mathbf{\Gamma}$ now depends only on the vector $\boldsymbol{k}$ it must be proportional to $\boldsymbol{k}$. Hence

$$\mathbf{\Gamma} = \gamma \boldsymbol{k}. \qquad (4.87)$$

If we now use the results that the vectors $\boldsymbol{k}$ and $\boldsymbol{k}'$ in equations (4.85) and (4.86) will lie near the Fermi surface and that only the component of $\mathbf{\Gamma}(\boldsymbol{k}')$ parallel to $\boldsymbol{k}$ will contribute to the sum in (4.86) we obtain as the equation for γ,

$$\gamma = 2 + (\gamma/2\pi\tau') \int d\epsilon' G(k', \omega'_\lambda) G(k', \omega'_\lambda + \omega_\nu) \qquad (4.88)$$

where

$$(\tau')^{-1} = \tfrac{1}{2} N_i N(0) \int d\Omega \, |v(k-k')|^2 \boldsymbol{k} \, . \, \boldsymbol{k}/kk'. \qquad (4.89)$$

Compared with τ^{-1} defined by equation (4.39), $(\tau')^{-1}$ has an extra angular factor in the integrand. Note that γ is independent of $\boldsymbol{k}$, but, like Γ, depends on ω'_λ and ω_ν. The integral in equation (4.88) is straightforward and yields (for $\omega_\nu > 0$)

$$\gamma = 2 + \frac{i\gamma}{\tau'} \frac{1}{i\omega_\nu + i/\tau'} \theta(-\omega'_\lambda)\theta(\omega'_\lambda + \omega_\nu). \qquad (4.90)$$

Whence

$$\gamma - 2 = \frac{(2i/\tau')\theta(-\omega'_\lambda)\theta(\omega'_\lambda + \omega_\lambda)}{i\omega_\nu + i/\tau_{\text{tr}}}, \qquad (4.91)$$

where

$$\tau_{\text{tr}}^{-1} = \tfrac{1}{2} N_i N(0) \int d\Omega \, |v(k-k')|^2 [1 - \boldsymbol{k} \, . \, \boldsymbol{k}'/kk'] \qquad (4.92)$$

and τ_{tr} is the usual relaxation time for transport processes.

The difference between $K^0_{\beta\alpha}$ determined previously and given by equation (4.80) and the corrected one, $K_{\beta\alpha}$, is (for $ql \ll 1$)

$$K_{\beta\alpha}(i\omega_\nu) - K^0_{\beta\alpha}(i\omega_\nu) = \frac{3ne^2}{2mk_F^2\beta}\int\frac{d\Omega}{4\pi}k_{F\alpha}k_{F\beta}\int d\epsilon\sum_\lambda G(\boldsymbol{k}, \omega'_\lambda)$$
$$\times G(\boldsymbol{k}, \omega'_\lambda + \omega_\nu)[\gamma(\omega'_\lambda) - 2].$$

The convergence here is sufficiently rapid for the order of integration over ϵ and summation over λ to be reversed. The integral over ϵ has been performed to obtain equation (4.90) from (4.88). Here it yields

$$K_{\beta\alpha}(i\omega_\nu) - K^0_{\beta\alpha}(i\omega_\nu) = \frac{2\pi ne^2}{m\beta}\sum_\lambda \frac{\theta(-\omega'_\lambda)\theta(\omega'_\lambda + \omega_\nu)i^2/\tau'}{(i\omega_\nu + i/\tau_{tr})(i\omega_\nu + i/\tau)}$$
$$= -\frac{ne^2}{m}\frac{\omega_\nu/\tau'}{(i\omega_\nu + i/\tau_{tr})(i\omega_\nu + i/\tau)}. \tag{4.93}$$

Again we can analytically continue by replacing ω_ν by $-i\omega$, and use equation (4.80) to obtain

$$K_{\beta\alpha}(\omega) = \frac{ne^2}{m}\left[\frac{i/\tau'}{(\omega + i/\tau_{tr})(\omega + i/\tau)} + \frac{\omega}{\omega + i/\tau}\right]$$
$$= \frac{ne^2}{m}\frac{\omega}{(\omega + i/\tau_{tr})}. \tag{4.94}$$

Thus, in the long-wavelength limit, the effect of the interference terms is to replace the lifetime τ by the transport relaxation time τ_{tr} and one obtains results identical to those obtained from the Boltzmann equation. For example it follows immediately from equation (4.93) that the d.c. conductivity is

$$\sigma = ne^2\tau_{tr}/m.$$

When the wavelength is not much greater than the mean free path, neither the Boltzmann equation nor the integral equation for the vertex part $\boldsymbol{\Gamma}$, equation (4.86), leads to the relaxation time approximation. Nevertheless, it seems (Pippard 1968) that as far as the conductivity is concerned, equation (4.82) with τ replaced by τ_{tr} provides a satisfactory basis for the analysis of the anomalous skin effect of metals. As it is written, equation (4.82) is valid only for metals with a spherical Fermi surface. It can be generalized, however, to take account of an arbitrary shape of Fermi surface. This is essential for a proper analysis of the anomalous skin effect in metals but would take us beyond the scope of this book.

The same methods can be applied to obtain other transport

coefficients of metals such as the thermal conductivity (Tewordt 1963), and the coefficient for sound absorption (Tsuneto 1961).

References

Edwards, S. F. (1958). *Phil. Mag.* **3**, 1020.
Kadanoff, L. P. and Baym, G. (1962). "Quantum Statistical Mechanics", Chapter 9. Benjamin, New York.
Messiah, A. (1961). "Quantum Mechanics", Vol. 1, p. 67. North-Holland, Amsterdam.
Pippard, A. B. (1978). *Proc. Roy. Soc.* **A305**, 291.
Tewordt, L. (1963). *Phys. Rev.* **129**, 657.
Tsuneto, T. (1961). *Phys. Rev.* **121**, 402.
Ziman, J. M. (1960). "Electrons and Phonons". Oxford University Press, Oxford.

Problems

1. Consider the scattering of electrons in metals as given by both non-magnetic and magnetic impurities, through an interaction Hamiltonian

$$H' = \sum_i \int d^3r\psi^+(\boldsymbol{r})V_1(\boldsymbol{r}-\boldsymbol{R}_i)\psi(\boldsymbol{r})$$
$$+ \sum_\beta \int d^3r\psi^+(\boldsymbol{r})\boldsymbol{S}_\beta\cdot\boldsymbol{\sigma}\,\psi(\boldsymbol{r})\,V_2(\boldsymbol{r}-\boldsymbol{R}_\beta)$$

here the locations of the non-magnetic and the magnetic impurities are denoted by $\boldsymbol{R}_i$ and $\boldsymbol{R}_\beta$, respectively, and $\boldsymbol{S}_\beta$ is the spin of the impurity at $\boldsymbol{R}_\beta$ (cf. Problem 5, Chapter 3). Presuming that the positions and spins of the impurities are randomly distributed (and ignoring the dynamics of impurity spins) show that to second order in the interaction, the odd part of the electron self-energy is still given by equation (4.38) but the lifetime is now defined by

$$\tau_k^{-1} = 2\pi N_i \sum_{k'} |V_1(\boldsymbol{k}-\boldsymbol{k}')|^2\,\delta(\epsilon')$$
$$+ 2\pi N_\beta S(S+1) \sum_{k'} |V_2(\boldsymbol{k}-\boldsymbol{k}')|^2\,\delta(\epsilon').$$

(Higher-order contributions to the self-energy from magnetic impurities lead to the Kondo effect and the resistance minimum in magnetic alloys (see Problem 3, Chapter 10). The effect of magnetic impurities on superconducting properties is discussed in some detail in §8.5).

2. Derive equation (4.90) from equation (4.85) for the case $\omega_\gamma > 0$.
3. Show that if only *s*-wave scattering by the impurities is significant (implying that $v(\boldsymbol{k})$ is approximately constant), there are no vertex corrections and the relation between $\boldsymbol{J}$ and $\boldsymbol{A}$ (even with $\boldsymbol{q} \neq 0$) is given by equation (4.81). (This equation is often used to discuss the anomalous skin effect in metals.)

Chapter 5

The Coloumb Gas

§5.1 Macroscopic considerations

In solid state physics the strongest interaction with which we have to deal is the Coulomb interaction between the charged particles, the electrons and the ions. In a metal for example, two electrons separated by the average atomic distance have a Coulomb energy of the order of 1 Rydberg, comparable with the kinetic energy or band energy of an electron at the Fermi surface. Furthermore, since the Coulomb potential falls off with distance only as r^{-1}, it is a long-range potential. Despite this, the electrons in a metal behave as if they are nearly independent particles, or at least nearly independent quasi-particles. A major problem for solid state physics is, therefore, to understand how this comes about. This chapter and the next are devoted to discussing the answer to this problem.

In this chapter we concentrate attention on the Coulomb gas of electrons, assuming that the ions of the lattice provide a static uniform background of positive charge sufficient to preserve overall charge neutrality. This is the jellium model of a metal. Despite its crudeness this is a useful model both for pedagogical purposes and for understanding the properties of metals; indeed, with only slight modifications it can provide valid results for real metals.

As a first step towards the understanding of the behaviour of this Coulomb gas, we consider a macroscopic treatment which should indicate how the gas behaves when subject to slow perturbations of long wavelength. This treatment is analogous to the usual electromagnetic treatment of dielectrics and magnets which does not depend on a knowledge of their microscopic properties. Because the Coulomb interaction is long-range, the macroscopic treatment is particularly important. The macroscopic results will subsequently be used as a guide to the microscopic theory.

The results of the macroscopic theory are not difficult to understand. If there is a small fluctuation in the charge density at any point, this will produce an electric field which will tend to restore charge neutrality. The electrons however possess inertia and so will possess kinetic energy when neutrality is restored. They therefore overshoot and a charge density oscillation is set up, the charge everywhere oscillating about the equilibrium value. This motion is called a plasma oscillation or, in its quantized form, a plasmon. The excitation energy of a plasmon in a metal is of the order of a few eV.

If an external charge which is either static or which oscillates slowly compared with the plasma frequency is introduced into the metal, the electrons can relax to neutralize the charge. However, the electrons possess kinetic energy and so a pressure. This means that they cannot be confined to very small regions and cannot neutralize a point or well localized charge completely. There is a screening distance depending on the kinetic energy of the electrons, over which the neutralization takes place. A second charge in the metal will be affected by the first charge only within this screening distance. Since any two electrons in the gas are two localized charges whose interaction is influenced by the remainder of the Coulomb gas, their interaction is also effectively screened. Thus there are two main effects of the long-range character of the Coulomb interaction in the gas, the existence of the plasma oscillations and the screening of the interaction between any two individual electrons. In the remainder of this section we shall show how these results may be derived from the macroscopic theory. In subsequent sections of the chapter we show how they result from the microscopic theory. In the next chapter we show how the residual part of the interaction between the electrons can be treated.

On the macroscopic scale we can use electrodynamics and hydrodynamics to describe the electrons. We shall only be interested in small departures from equilibrium and so shall use only linearized forms of these theories. Further, since we shall concentrate on charge oscillations and the velocities of the electrons are much less than the velocity of light, we can ignore magnetic fields. If then ρ is the number of electrons per unit volume of the electron fluid and $\mathbf{v}$ the local drift velocity, the linear hydrodynamic equation of motion is

$$m\frac{\partial(\rho\boldsymbol{v})}{\partial t} = -\boldsymbol{\nabla}p + e\rho\boldsymbol{E}; \tag{5.1}$$

here e and m are respectively the charge and mass of the electron, p the local pressure and $\boldsymbol{E}$ the local electric field.

We must also have conservation of charge so

$$\partial\rho/\partial t + \operatorname{div}(\rho\boldsymbol{v}) = 0. \tag{5.2}$$

Finally, the electric field is determined by the deviation of ρ from its equilibrium value ρ_0. Thus

$$\operatorname{div}\boldsymbol{E} = e(\rho-\rho_0)/\epsilon_0. \tag{5.3}$$

The pressure term takes account of the kinetic energy of the electrons as well as the short-range forces.

If we introduce

$$\delta\rho = \rho-\rho_0 \tag{5.4}$$

and treat $\delta\rho$ and $\boldsymbol{v}$ as first-order terms, the equations reduce to

$$m\frac{\partial\boldsymbol{v}}{\partial t} = -\frac{1}{\rho_0}\frac{\partial p}{\partial\rho}\boldsymbol{\nabla}\rho + e\boldsymbol{E},$$

$$\frac{\partial(\delta\rho)}{\partial t} + \rho_0\operatorname{div}\boldsymbol{v} = 0,$$

and

$$\operatorname{div}\boldsymbol{E} = e\delta\rho/\epsilon_0.$$

We have also assumed that the motion is sufficiently slow for local equilibrium to obtain so that p is a local function of ρ. For long-wavelength oscillations the derivative of p with respect to ρ should be taken at constant entropy. It is possible to eliminate div $\boldsymbol{v}$ from the three equations to obtain

$$m\frac{\partial^2\delta\rho}{\partial t^2} - \frac{\partial p}{\partial\rho}\nabla^2\delta\rho + \frac{e\rho_0}{\epsilon_0}\delta\rho = 0. \tag{5.5}$$

This is an equation with plane wave solutions

$$\delta\rho \propto \exp(i\boldsymbol{q}\cdot\boldsymbol{r} - i\omega t) \tag{5.6}$$

provided that

$$\omega^2 = \frac{\rho_0 e^2}{m\epsilon_0} + \frac{1}{m}\frac{\partial p}{\partial\rho}q^2. \tag{5.7}$$

These solutions are the plasma oscillations already referred to; at long wavelengths their frequencies correspond to energies of the order of 5 to 10 eV in good metals. These wave solutions must also arise in the microscopic theory.

To study the screening of the electrons we consider the effect on the charges of an external electric field $\boldsymbol{E}_e$ which varies slowly in space and time, with wave vector $\boldsymbol{q}$ and angular frequency ω. We simply have to add $\boldsymbol{E}_e$ to $\boldsymbol{E}$ in equation (5.1). The charge density induced by this external field is then found to be

$$e\delta\rho = \frac{(\rho_0 e^2/m)\,\mathrm{div}\,\boldsymbol{E}_e}{\omega^2 - \dfrac{\rho_0 e^2}{m\epsilon_0} - \dfrac{1}{m}\dfrac{dp}{d\rho}q^2}. \tag{5.8}$$

Considered as a function of ω in the complex plane, this diverges at the plasma frequencies. In the corresponding microscopic calculation, the induced charge, according to § 1.2, is dependent on the density–density correlation function. Thus this function should also have poles at the plasma frequencies. In quantum mechanics, the waves are quantized and the corresponding particles are called, by analogy with phonons and photons, plasmons. Actually in a full microscopic calculation, the plasmons can decay into particles. Thus the poles of the exact correlation function will be spread out into cuts.

The total electric field produced by the external field together with that due to the induced charge is

$$\boldsymbol{E}_t = \boldsymbol{E} + \boldsymbol{E}_e. \tag{5.9}$$

From equations (5.3) and (5.8) this is

$$\boldsymbol{E}_t = \frac{\left(\omega^2 - \dfrac{1}{m}\dfrac{dp}{d\rho}q^2\right)\boldsymbol{E}_e}{\omega^2 - \dfrac{\rho_0 e^2}{m\epsilon_0} - \dfrac{1}{m}\dfrac{\partial p}{\partial\rho}q^2} \tag{5.10}$$

and this is the field acting on any charge in the medium. Thus the effect of the medium is that of a dielectric with a frequency and wave vector dielectric constant $\epsilon(\boldsymbol{q}, \omega)$ given by

$$\epsilon(\boldsymbol{q}, \omega) = 1 - \frac{\rho_0 e^2/m\epsilon_0}{\omega^2 - \dfrac{1}{m}\dfrac{dp}{d\rho}q^2}. \tag{5.11}$$

For static fields ($\omega = 0$), the dielectric constant becomes infinite at long wavelengths. Thus the effect of the charged particles is to screen out the field at long wavelengths. Since the potential due to a stationary point particle of charge e is

$$\frac{e}{4\pi\epsilon_0 r} = \frac{e}{8\pi^2\epsilon_0}\int \frac{d^3q}{q^2}\exp(i\boldsymbol{q}\cdot\boldsymbol{r}),$$

the screened potential is

$$\frac{e}{8\pi^2\epsilon_0}\int \frac{d^3q\left(\dfrac{1}{m}\dfrac{\partial p}{\partial\rho}\right)\exp(i\boldsymbol{q}\cdot\boldsymbol{r})}{\dfrac{\rho_0 e^2}{m\epsilon_0}+\dfrac{1}{m}\dfrac{dp}{d\rho}q^2} = \frac{e}{4\pi\epsilon_0 r}\exp(-\kappa r),$$

where the inverse screening length κ is given by

$$\kappa^{-2} = (\rho_0 e^2/\epsilon_0)(dp/d\rho). \tag{5.12}$$

The size of the screening length for electrons in a metal can be estimated, provided that the density is not too high, using the kinetic energy of the electrons to provide an estimate of the pressure. This kinetic energy, on a free-electron model, is

$$E = \frac{2\mathscr{V}}{(2\pi)^3}\int_{|k|<k_F}\frac{k^2}{2m}d^3k = \frac{\mathscr{V}k_F^5}{10\pi^2 m} \tag{5.13}$$

Since

$$\rho_0 = k_F^3/3\pi^2, \tag{5.14}$$

$$p = -\partial E/\partial\mathscr{V} = (3\pi^2\rho_0)^{5/3}/15\pi^2 m,$$

$$\partial p/\partial\rho = (3\pi^2\rho_0)^{2/3}/3m,$$

$$\kappa^{-1} = 3\rho_0 e^2 m/\epsilon_0(3\pi^2\rho_0)^{2/3}. \tag{5.15}$$

For a good metal this yields a screening length of the order of 1 nm which is comparable with the distance between electrons.

This screening of a charge by all the other charges will also be effective in reducing the interaction between the original charges. Thus the effective interaction between the charges is a screened short-range one which, in suitable circumstances can be treated as a perturbation. These features of the macroscopic theory should also be consequences of the microscopic theory. We should find both the plasma oscillations (as poles in the density–density correlation function) and the screening of the interaction between charges.

§5.2 Microscopic theory

(a) *The Hartree approximation*

The macroscopic method takes account of the average effect of each charge upon the others. According to §3.7, this is the effect of the

Hartree approximation which we now proceed to use. To be specific we consider the free-electron model of a metal for which the Hamiltonian is

$$H = \sum_{k\sigma} \epsilon_k c_{k\sigma}^+ c_{k\sigma} + \tfrac{1}{2} \sum_{\substack{k,k',q \\ \sigma,\sigma'}} V(q) c_{k\sigma}^+ c_{k'+q\sigma'}^+ c_{k'\sigma'} c_{k+q\sigma}$$

$$- V(0)(N/\mathscr{V}) \sum_{k,\sigma} c_{k\sigma}^+ c_{k\sigma}, \tag{5.16}$$

where the last term is the interaction of the electrons with the uniform background of charge representing the lattice, and $\mathscr{V}$ is the volume. For the present model,

$$V(q) = \frac{e^2}{4\pi\epsilon_0 \mathscr{V}} \int \frac{d^3r}{r} \exp(iq \cdot r) = \frac{e^2}{\epsilon_0 q^2 \mathscr{V}}, \qquad V(0) \sim \mathscr{V}^{-1/3} \tag{5.17}$$

and

$$\epsilon_k = (k^2/2m) - \mu. \tag{5.18}$$

The last term in equation (5.16) and the $q = 0$ part of the second in the sum both depend only on N_{op}, the operator for the total number of electrons. In fact, together these two terms are

$$\mathscr{V}^{-1}[\tfrac{1}{2}V(0)(N_{\text{op}}^2 - N_{\text{op}}) - V(0)\langle N\rangle N_{\text{op}}].$$

If we use the canonical ensemble, N_{op} is fixed equal to $\langle N\rangle$ and this term is a constant which can be ignored. If, as we have been doing, we use a grand canonical ensemble and write

$$\delta N = N_{\text{op}} - \langle N\rangle$$

then the two terms together yield

$$(2\mathscr{V})^{-1}[-(\langle N\rangle + 2\langle N\rangle^2)V(0) + V(0)(\delta N)^2]. \tag{5.19}$$

The first of these is a constant which can be ignored. Since $(\delta N)^2$ is of the order of $\langle N\rangle$ the effect of the second term is negligibly small even compared with excitation energies. Hence, the two terms can be ignored and

$$H = \sum_{k,\sigma} \epsilon_k c_{k\sigma}^+ c_{k\sigma} + \tfrac{1}{2} \sum_{\substack{k,k',q \\ \sigma\sigma'}} V(q) c_{k\sigma}^+ c_{k'+q\sigma'}^+ c_{k'\sigma'} c_{k+q\sigma} \tag{5.20}$$

where we take

$$V(0) = 0. \tag{5.21}$$

From §§3.6 and 3.7 we see that the Hartree approximation for the self-energy depends on $V(0)$ and so vanishes in the present problem. To first order, then, the particles are free particles. (This is assumed in the macroscopic theory.) Before considering corrections to this result we turn to the response of electron gas to an electric field within the same approximation.

(b) *Random phase approximation*

As we have seen in §3.8 the extension of the Hartree approximation to find the linear response of the system to an external field is the random phase approximation (RPA). According to §3.8 a one-particle potential induces a change in the number density $\delta\rho(1)$ given (for fermions) by

$$\delta\rho(1) = G_1(1,4) \tag{5.22}$$

where

$$[(\partial/\partial\tau) + h_0(1) + v_0(1)]\,G_1(1,2) = -U_{\text{eff}}(1)G_0(1,2) \tag{5.23}$$

and

$$U_{\text{eff}}(1) = U(1) + \int d1'V(1,1')G_1(1',1'_+). \tag{5.24}$$

Here $U_{\text{eff}}(1)$ is the effective screened potential and $G_1(1,2)$ is the first-order correction to the single-particle Green's function. From (5.22) and (5.24) we find

$$U_{\text{eff}}(1) = U(1) + \int d1'V(1,1')\delta\rho(1'). \tag{5.25}$$

Hence the screened potential is the sum of the original potential and that due to the average charge distribution. (This is the analogue of equation (5.9)). Equation (5.23) can be integrated to yield

$$G_1(1,2) = \int G_0(1,1')U_{\text{eff}}(1')G_0(1',2)d1' \tag{5.26}$$

and hence

$$\delta\rho(1) = \int G_0(1,1')U_{\text{eff}}(1')G_0(1,1')d1'. \tag{5.27}$$

Equations (5.25) and (5.27) contain the results of the RPA and are the microscopic analogues of (5.1) to (5.4) with an external field, the relation between $U(1)$ and the electric field $E_e(1)$ being given by

$$eE_e = -\boldsymbol{\nabla}_1 U(1). \tag{5.28}$$

An alternative way of deriving equations (5.25) and (5.27) is to sum the diagrams of the RPA for $\delta\rho(1)$ shown in Fig. 5.1.

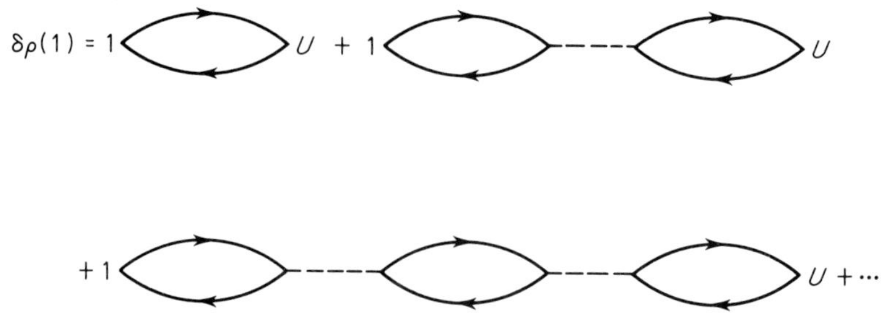

FIG. 5.1. The random phase approximation for $\delta\rho(1)$ linear in U.

Since the equations are linear we can again work with Fourier transforms in space and imaginary time. Then, taking into account the two spin states of an electron,

$$U_{\text{eff}}(q) = U(q) + V(\boldsymbol{q})\delta\rho(q), \tag{5.29}$$

$$\delta\rho(q) = 2\beta^{-1} \sum_k G_0(k) U_{\text{eff}}(q) G_0(k+q), \tag{5.30}$$

where

$$q \equiv (\boldsymbol{q}, \omega_\nu), \qquad k \equiv (\boldsymbol{k}, \zeta), \ \sum_k \equiv \sum_{\boldsymbol{k},\zeta} .$$

Hence

$$U_{\text{eff}}(q) = \frac{U(q)}{1 - 2\beta^{-1} V(\boldsymbol{q}) \sum_k G_0(k) G_0(k+q)} \equiv \frac{U(q)}{1 - V(\boldsymbol{q})\chi(q)} \tag{5.31}$$

or in terms of electric fields,

$$E_{\text{eff}}(q) = E_{\text{e}}(q)/[1 - V(\boldsymbol{q})\chi(q)]^{-1}. \tag{5.32}$$

This is the analogue of equation (5.10).

To extract the dielectric constant, we need to perform the analytic continuation from ω_ν to ω. For this purpose we perform the sum over ζ in $\chi(q)$;

$$\begin{aligned} \chi(q) &= 2\beta^{-1} \sum_{\boldsymbol{k},\zeta} \frac{1}{i\zeta - \epsilon_{\boldsymbol{k}}} \frac{1}{i(\zeta + \omega_\nu) - \epsilon_{\boldsymbol{k}+\boldsymbol{q}}} \\ &= 2 \sum_{\boldsymbol{k}} \frac{f(\epsilon_{\boldsymbol{k}}) - f(\epsilon_{\boldsymbol{k}+\boldsymbol{q}})}{i\omega_\nu + \epsilon_{\boldsymbol{k}} - \epsilon_{\boldsymbol{k}+\boldsymbol{q}}} \end{aligned} \tag{5.33}$$

In this expression we can continue analytically simply by replacing $i\omega_\nu$ by $\omega + i\delta$. Hence, the frequency and wave vector dependent dielectric constant is

$$\epsilon(\boldsymbol{q},\omega) = 1 - 2V(\boldsymbol{q}) \sum_k \frac{f(\epsilon_k) - f(\epsilon_{k+q})}{\omega + \epsilon_k - \epsilon_{k+q} + i\delta} = 1 - V(\boldsymbol{q})\chi(\boldsymbol{q},\omega). \quad (5.34$$

This is Lindhard's (1954) formula for the contribution of the electrons to the dielectric constant of a metal.

This result has been examined in some detail by Lindhard (1954) and by Pines and Nozieres (1966). Here we look at the result in the long-wavelength limit to compare with the macroscopic theory. In the following, we do not show δ explicitly but remember, if necessary, that ω has a small positive imaginary part. We exploit the antisymmetry of the numerator of the sum under the transformation

$$\boldsymbol{k} \to -\boldsymbol{k} - \boldsymbol{q}$$

to obtain

$$\sum_k \frac{f(\epsilon_k) - f(\epsilon_{k+q})}{\omega + \epsilon_k - \epsilon_{k+q}} = \sum_k \frac{[f(\epsilon_k) - f(\epsilon_{k+q})](\epsilon_{k+q} - \epsilon_k)}{\omega^2 - (\epsilon_k - \epsilon_{k+q})} \quad (5.35)$$

For small $\boldsymbol{q}$ this becomes

$$\begin{aligned}\omega^{-2} \sum_k [f(\epsilon_k) - f(\epsilon_{k+q})](\epsilon_{k+q} - \epsilon_k) \\ = 2\omega^{-2} \sum_k f(\epsilon_k)(\epsilon_{k+q} - \epsilon_k) \\ = Nq^2/2m\omega^2.\end{aligned}$$

Hence

$$\epsilon(0,\omega) = 1 - (\rho_0 e^2/m\omega^2\epsilon_0). \quad (5.36)$$

This agrees with the macroscopic result and for the plasma frequency in the long-wavelength limit (where $\epsilon(0,\omega)$ vanishes), it yields

$$\omega_{p0} = (\rho_0 e^2/m\epsilon_0)^{1/2} \quad (5.37)$$

The expansion in q can be continued to yield in next order

$$\omega_p(q) = \omega_{p0}\,[1 + \tfrac{3}{10}(qk_{\mathrm{F}}/m\omega_{p0})^2]. \quad (5.38)$$

This has the same form as the macroscopic result (from equation (5.7)) but the coefficient of the second term in the latter result is only 1/6.

One qualitative difference between the expression (5.34) and the corresponding macroscopic one is that $\epsilon(\boldsymbol{q},\omega)$ becomes complex when it is possible to satisfy

$$\omega = \epsilon_{k+q} - \epsilon_k. \quad (5.39)$$

This arises from the fact that a field of frequency ω and wave vector $\boldsymbol{q}$ can excite an electron and hole provided that energy is conserved according to equation (5.39). Energy can then be absorbed by the electrons from the field. In particular when $\omega_p(\boldsymbol{q})$ for a plasmon satisfies

$$\omega_p(\boldsymbol{q}) = \epsilon_{k+q} - \epsilon_k = (\boldsymbol{q} \cdot \boldsymbol{k}_F)/m + (q^2/2m) \qquad (5.40)$$

the plasmon can decay into an electron and hole and so have a finite lifetime. Since $\omega_p(0)$ is finite it is impossible to satisfy equation (5.40) as $q \to 0$. Thus the very-long-wavelength plasmons will have infinite lifetimes. We can estimate the wave vector q_c at which the lifetime first becomes finite by using approximations assuming $q \ll k_F$ in equation (5.40). Then we find that

$$q_c \approx m\omega_{p0}/k_F .$$

For metallic densities, this yields $q_c \sim k_F$ so that the approximation is not accurate. However, it does provide the order of magnitude of q_c.

We should also like to compare the result of RPA with the macroscopic result in the static limit. We have then

$$\epsilon(q, 0) = 1 - 2V(\boldsymbol{q}) \sum_k \frac{f(\epsilon_k) - f(\epsilon_{k+q})}{\epsilon_k - \epsilon_{k+q}} . \qquad (5.41)$$

At long wavelengths this yields

$$\epsilon(q, 0) = 1 - 2V(\boldsymbol{q}) \sum_k (\partial f/\partial \epsilon_k) = 1 + 2N(0)V(\boldsymbol{q})\mathscr{V}, \qquad (5.42)$$

where $N(0)$ is the density of states in one spin per unit volume and we have exploited the fact that for $k_B T \ll E_F$,

$$\partial f/\partial \epsilon_k \approx -\delta(\epsilon_k).$$

Since for the free-electron model,

$$\rho_0 = 4N(0)E_F/3,$$

we have

$$\epsilon(q, 0) = 1 + \frac{3\rho_0 m e^2}{\epsilon_0 k_F^2 q^2} . \qquad (5.43)$$

This agrees precisely with equation (5.10) in the static limit and so it yields the same screening length for a static charge. Thus we have seen that the RPA yields the same results as the macroscopic theory for the response of the electrons to an external field in the long-wavelength limit.

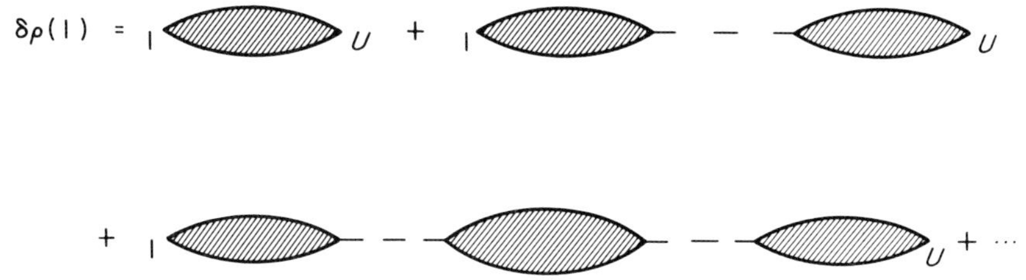

FIG. 5.2. The representation of $\delta\rho(1)$ in terms of polarization parts Q, the hatched areas. The polarization parts cannot be separated into two by cutting one potential line only.

Formally, it is not difficult to see how the results of this section can be generalized. We use the diagrammatic expansion for the charge $\delta\rho(1)$ induced by a small electric potential $U(1)$ where $\delta\rho(1)$ is given by equation (5.22), and define an irreducible polarization part $Q(q)$ as a part of the diagram connected to the remainder by two potential lines and which is such that it cannot be separated into two parts by cutting one potential line alone. The loops in Fig. 5.1 are irreducible polarization parts. If these polarization parts are denoted by shaded loops the expansion for $\delta\rho(1)$ can be written as illustrated in Fig. 5.2 and, in terms of Fourier transforms:

$$\delta\rho(q) = \frac{Q(q)U(q)}{1 - V(\boldsymbol{q})Q(q)}.$$

Equations (5.30) and (5.31) are of this form with $Q(q)$ approximated by its zeroth order term, the bare loop. The dielectric constant is, in general, given by

$$\epsilon(q) = 1 - V(\boldsymbol{q})Q(q).$$

Hence, the function Q does play the role of polarization in macroscopic theory. Improved approximations for Q can lead to improved results for the dielectric function. For further reading concerning such approximations the reader is referred to papers by Hubbard (1957) and Singwi *et al.* (1970).

(c) *The thermodynamic potential*

We have seen in §2.8 that the thermodynamic potential Ω can be expressed in terms of the one-particle Green's function, the most convenient expression to use being that given in equation (2.84). If we vary the strength of the Coulomb interaction by replacing e^2 by e'^2 $(= \lambda e^2)$, we can write equation (2.85) as

$$\Omega(e^2) - \Omega(0) = -\tfrac{1}{2} \int_0^{e^2} \frac{de'^2}{e'^2} \int d1d2\, V(1,2) G(1_+, 2_+; 1, 2), \tag{5.44}$$

where

$$\tau_{1+} = \tau_1 + \delta, \qquad \tau_{2+} = \tau_2 + \delta' \tag{5.45}$$

and δ, δ' prime are small positive quantities which are included to keep the right order of operators in the two-particle Green's function and which tend to zero at the end of the calculation. The expansion of the two-particle Green's function is illustrated by Fig. 3.10 with m and b equal to unity. If we introduce the space and time Fourier transforms of the Green's function and potential the expression can be written in terms of the polarization parts as

$$\begin{aligned}\Omega(e^2) - \Omega(0) &= -\tfrac{1}{2} \int_0^{e^2} \frac{de'^2}{e'^2} \frac{1}{\beta} \sum_q \frac{V(\boldsymbol{q}) Q(q)}{1 - V(q) Q(q)} \\ &= -\int_0^{e^2} \frac{de'^2}{e'^2} \frac{1}{2\beta} \sum_q \frac{V(q) Q(q)}{\epsilon(q)}.\end{aligned} \tag{5.46}$$

The expression in the sum in (5.46) depends on δ and δ' because, as a result of equation (5.45), the contribution from each diagram contains an extra factor,

$$\exp(i\zeta\delta + i\zeta'\delta'),$$

where ζ and ζ' are the energies which enter the vertices 1 and 2 respectively. For the lowest-order correction we replace $Q(q)$ by $\chi(q)$ defined by equation (5.32).

The term $\Omega(0)$ is the thermodynamic potential when there is no Coulomb interaction and is given by

$$\Omega(0) = 2 \sum_{k<k_F} \epsilon_k \tag{5.47}$$

$$\begin{aligned} &= \frac{2\mathscr{V}}{(2\pi)^3} \int_{k<k_F} d^3k \left(\frac{k^2}{2m} - \mu\right) \\ &= -\tfrac{2}{5} N\mu; \end{aligned} \tag{5.48}$$

in deriving this result we have used equation (5.14) for N and the fact that for the free-electron model the chemical potential lies at the Fermi surface with the result that

$$\mu = k_{\mathrm{F}}^2/2m. \tag{5.49}$$

The remainder of $\Omega(e^2)$ can be put in a commonly used form as follows. To find the lowest-order correction we replace Q by χ in equation (5.46) and rewrite it as,

$$\begin{aligned}\Omega(e^2) - \Omega(0) = &- \int_0^{e^2} \frac{de'^2}{e'^2} \frac{1}{2\beta} \sum_q V(q)\chi(q) \\ &- \int_0^{e^2} \frac{de'^2}{e'^2} \frac{1}{2\beta} \sum_q V(q)\chi(q) \left[\frac{1}{\epsilon(q)} - 1\right].\end{aligned} \tag{5.50}$$

Now as $\omega_\nu \to \infty$ both $\chi(q)$ as given by equation (5.35) and $[\epsilon(q)^{-1} - 1]$ behave like

$$\text{constant}/\omega_\nu^2 .$$

This makes the sum over q in the second term sufficiently convergent for one to be able to neglect δ and δ'. The first term in equation (5.50) has to be treated more carefully. We must keep δ, δ' in the definition of χ. Thus

$$\begin{aligned}\tfrac{1}{2}\chi(q) &= \beta^{-1} \sum_k G_0(k+q)G_0(k) \exp[i\omega_\nu\delta + i\zeta(\delta+\delta')] \\ &= \sum_k \frac{f(\epsilon_k)\exp(i\omega_\nu\delta)}{i\omega_\nu + \epsilon_k - \epsilon_{k+q}} - \frac{f(\epsilon_{k+q})\exp(-i\omega_\nu\delta')}{i\omega_\nu + \epsilon_k - \epsilon_{k+q}} \\ &= -\sum_k \left\{ \frac{[f(\epsilon_k) - f(\epsilon_{k+q})](\epsilon_{k+q} - \epsilon_k)\exp(i\omega_\nu\delta)}{(i\omega_\nu)^2 - (\epsilon_k - \epsilon_{k+q})^2} \right. \\ &\qquad \left. - \frac{f(\epsilon_{k+q})[\exp(-i\omega_\nu\delta') - \exp(i\omega_\nu\delta)]}{i\omega_\nu + \epsilon_k - \epsilon_{k+q}} \right\}.\end{aligned} \tag{5.55}$$

The first term when summed over ω_ν now converges sufficiently rapidly for δ to be put equal to zero. It is then $\frac{1}{2}\chi(q)$, as given before with $\delta = 0$, and can be recombined with the last term of equation (5.54). The last term of (5.55), which really derives from the need to change the order of the operators in the interaction Hamiltonian can be summed explicitly over ω_ν. One has

$$\beta^{-1} \sum_{k,\omega_\nu} \frac{f(\epsilon_{k+q})[\exp(-i\omega_\nu\delta') - \exp(i\omega_\nu\delta)]}{i\omega_\nu + \epsilon_k - \epsilon_{k+q}} \tag{5.56}$$

$$= -\sum_k \frac{1}{2\pi i} \oint \frac{d\omega\, f(\epsilon_{k+q})}{\omega + \epsilon_k - \epsilon_{k+q}} \left[\frac{\exp(-\omega\delta')}{\exp(-\beta\omega) - 1} + \frac{\exp(\omega\delta)}{\exp(\beta\omega) - 1}\right]$$

where the contour encloses the points

$$\omega = i\omega_\nu = i2\pi\nu/\beta.$$

As in Appendix A, the integral can be evaluated by deforming the contour to enclose the pole at $(\epsilon_{k+q} - \epsilon_k)$. The small positive quantities δ and δ' ensure that contributions from parts of the contour at infinity are zero. Then the expression (5.56) becomes

$$\begin{aligned}\sum_k f(\epsilon_{k+q}) \Big\{ &\{\exp[-\beta(\epsilon_k - \epsilon_{k+q})] - 1\}^{-1} \\ &+ \{\exp[\beta(\epsilon_k - \epsilon_{k+q})] - 1\}^{-1} \Big\} \\ = &- \sum_k f(\epsilon_{k+q}) \\ = &-\tfrac{1}{2}N.\end{aligned}$$

Hence

$$\Omega(e^2) - \Omega(0) = \sum_q \int_0^{e^2} \frac{de'^2}{e'^2} \frac{V(q)}{2} \left\{ -N - \frac{1}{\beta} \sum_{\omega_\nu} \frac{\chi(q)}{\epsilon(q)} \right\} \tag{5.57}$$

Now

$$\epsilon(q) = 1 - V(q)\chi(q)$$

and

$$V(q) = e'^2/\epsilon_0 q^2 \mathscr{V}.$$

Hence the integral over e^{12} can be performed to yield

$$\Omega(e^2) - \Omega(0) = \sum_q \left\{ -\frac{N}{2} V(\boldsymbol{q}) + \frac{1}{2\beta} \sum_{\omega_\nu} \ln[\epsilon(q, i\omega_\nu)] \right\}, \tag{5.58}$$

where $V(q)$ and ϵ are now expressed in terms of e^2. The ground state energy E_G can be obtained from this by letting $\beta \to 0$. Then the sum over ω_ν becomes an integral and we obtain

$$E_G(e^2) - E(0) = \sum_q \left\{ -\tfrac{1}{2} N V(\boldsymbol{q}) + (4\pi)^{-1} \int_{-\infty}^{\infty} d\omega \ln[\epsilon(\boldsymbol{q}, i\omega)] \right\}. \tag{5.59}$$

This result was first obtained by Gell-Mann and Brueckner (1957), who showed that it is correct in the limit of weak Coulomb interaction, that is when the mean potential energy is small compared with the mean kinetic energy. Conventionally this ratio is defined in terms of an electronic separation r_0 defined by

$$\tfrac{4}{3}\pi r_0^3 = \rho_0^{-1} = 3\pi^2/k_F^3. \tag{5.61}$$

Then a measure of the mean potential energy is e^2/r_0 and of the mean kinetic energy is $(mr_0^2)^{-1}$. The ratio is

$$e^2 m r_0 = r_0/a_0 = r_s, \tag{5.61}$$

where a_0 is the Bohr radius of the hydrogen atom and r_s is the interelectronic distance measured in units of the Bohr radius. A weak Coulomb interaction means small r_s and therefore high density of electrons. As $r_s \to 0$, Gell-Man and Brueckner have shown that, in Rydbergs,

$$E_G = \frac{2.21}{r_s^2} - \frac{0.916}{r_s} + 0.0622 \ln r_s - 0.096 + O(r_s). \tag{5.62}$$

The first two terms in this sum represent the kinetic energy and the potential energy or exchange energy in the unperturbed ground state. The remainder comes from the change in the wave function due to the interaction and is usually called the correlation energy.

It is possible to improve upon this result systematically. However, for real metals r_s lies in the range $1.8 \leqslant r_s \leqslant 5.6$. This is rather far removed from the range of validity of equation (5.62) and special approximations have been devised to estimate the correlation energy of real metals (Hubbard 1957, Singwi *et al.* 1970).

(d) *Improved self-energy and lifetime*

In the previous sections we have seen that an external field is screened by the Coulomb gas. In the same way we should expect the interaction between two electrons should be screened. This means that in order to treat correctly the interaction between two such electrons the polarization effects should be included. Thus the leading correction (for small r_s) to the self-energy is given not by the simple Hartree–Fock diagram but by the sum of diagrams shown in Fig. 5.3. The Hartree–Fock approximation itself is represented by the first term in the series.

In the limit of small r_s, it is possible to take as electron propagators those already calculated, that is, the unperturbed ones. Then the screened interaction, $V_s(q)$, represented by the wavy line satisfies

$$V_s(q) = V(\boldsymbol{q}) + V(\boldsymbol{q})\chi(q)V_s(q),$$

whence, as is to be expected,

$$V_s(q) = V(\boldsymbol{q})/\epsilon(q). \tag{5.63}$$

Hence

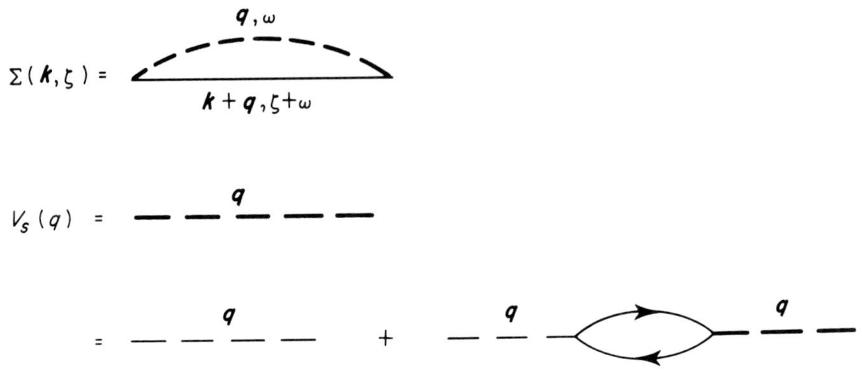

FIG. 5.3. Representation of an improved approximation for the self-energy. In the Hartree–Fock contribution the shielded potential $V_s(q)$, rather than the unshielded one $V(q)$ appears.

$$\Sigma(k, i\zeta) \equiv \Sigma(k) = \beta^{-1} \sum_q \frac{V(q)}{\epsilon(q)} \bar{G}(k-q) \exp[(i\zeta - i\omega_\nu)\delta]. \tag{5.64}$$

In this approximation Σ does depend on ζ and leads to a finite lifetime for an electron. In fact it contains within it many contributions to this lifetime as we can show explicitly by further evaluation.

We first perform the analytic continuation to $\Sigma(\boldsymbol{k}, \omega)$ by introducing the spectral form for $\bar{G}$ (equation 2.31). Then

$$\Sigma(k) = -\beta^{-1} \int dx \sum_q \frac{V(q)}{\epsilon(\boldsymbol{q}, i\omega_\nu)} \frac{A(\boldsymbol{k}-\boldsymbol{q}, x) \exp[i(\zeta - \omega_\nu)\delta]}{(i\zeta - i\omega_\nu - x)}$$

$$= -\frac{1}{2\pi i} \int dx \oint \frac{d\omega}{\exp(-\beta\omega) - 1}$$

$$\times \sum_q \frac{V(\boldsymbol{q})}{\epsilon(\boldsymbol{q}, \omega)} \frac{A(\boldsymbol{k}-\boldsymbol{q}, x) \exp[(i\zeta - \omega)\delta]}{(i\zeta - \omega - x)}.$$

Where the contour C, as usual, encloses all the points where

$$\omega = i\omega_\nu$$

(see Fig. 5.4). Now the convergence of the integrals is sufficiently rapid for the contour C to be deformed into C' which encloses

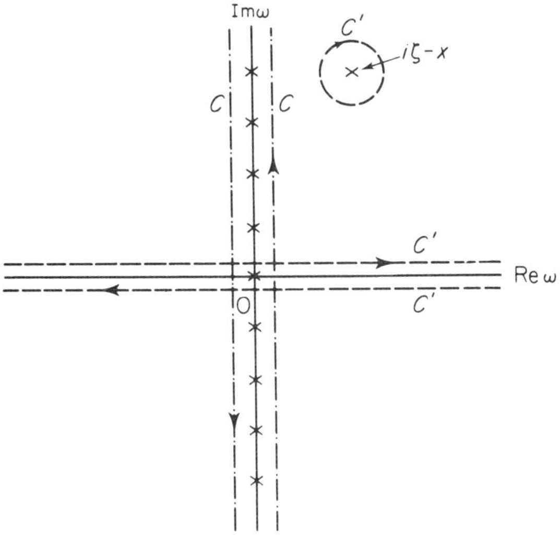

FIG. 5.4. The contours C and C' used in the evaluation of $\Sigma(\boldsymbol{k})$.

the real axis and the poles at $\omega = -x + i\zeta$, and the origin. Hence

$$\Sigma(\boldsymbol{k}, i\zeta) = \int_{-\infty}^{\infty} dx$$

$$\times \sum_{\boldsymbol{q}} V(\boldsymbol{q})A(\boldsymbol{k}-\boldsymbol{q}, x)\left\{-\frac{P}{2\pi i}\int_{-\infty}^{\infty}\frac{d\omega'}{\exp(-\beta\omega')-1}\cdot\frac{1}{(i\zeta-\omega'-x)}\right.$$

$$\left.\times\left[\frac{1}{\epsilon(\boldsymbol{q}, \omega'+i\delta)}-\frac{1}{\epsilon(\boldsymbol{q}, \omega'-i\delta)}\right]+\frac{1}{(\exp(\beta x)+1)\epsilon(\boldsymbol{q}, i\zeta-x)}\right\}.$$

This can now be analytically continued to ω simply by replacing $i\zeta$ by $\omega + i\delta$. If we note that $\epsilon(\boldsymbol{q}, \omega - i\delta)$ is the complex conjugate of $\epsilon(\boldsymbol{q}, \omega + i\delta)$, the result can be written

$$\Sigma(\boldsymbol{k}, \omega) = \int_{-\infty}^{\infty} dx \sum_{\boldsymbol{q}} V(\boldsymbol{q})A(\boldsymbol{k}-\boldsymbol{q}, x)$$

$$\times\left\{-\frac{1}{\pi}\int_{-\infty}^{\infty}\frac{d\omega'}{\exp(-\beta\omega')-1}\cdot\frac{1}{\omega-\omega'-x}\right.$$

$$\left.\times \operatorname{Im}\frac{1}{\epsilon(\boldsymbol{q}, \omega'+i\delta)}+\frac{1}{(\exp(\beta x)+1)\epsilon(\boldsymbol{q}, \omega-x)}\right\}. \tag{5.65}$$

Since $\epsilon(\boldsymbol{q}, 0)$ is real, the principal value does not affect the result and has been ignored.

To estimate the lifetime, we can use the previously obtained Green's function which is actually the perturbed one, on the right-hand side of equation (5.65). Hence

$$A(\boldsymbol{k}-\boldsymbol{q}, x) \approx \delta(x - \epsilon_{k-q}).$$

According to §3.5, the lifetime τ_k of an electron of momentum $\boldsymbol{k}$ is given by

$$\tau_k^{-1} = -2 \operatorname{Im} \Sigma(\boldsymbol{k}, \epsilon_k + i\delta).$$

Therefore, in this case,

$$\frac{1}{\tau_k} = 2 \sum_q V(\boldsymbol{q}) \left\{ \int_{-\infty}^{\infty} \frac{d\omega'}{\exp(-\beta\omega') - 1} \delta(\epsilon_k - \epsilon_{k-q} - \omega') \times \operatorname{Im} \frac{1}{\epsilon(\boldsymbol{q}, \omega' + i\delta)} + f(\epsilon_{k-q}) \operatorname{Im} \frac{1}{\epsilon(\boldsymbol{q}, \epsilon_k - \epsilon_{k-q} + i\delta)} \right\} \tag{5.66}$$

$$= 2 \sum_q V(\boldsymbol{q}) [n(\epsilon_{k-q} - \epsilon_k) + f(\epsilon_{k-q})] \operatorname{Im} \frac{1}{\epsilon(\boldsymbol{q}, \epsilon_k - \epsilon_{k-q} + i\delta)}$$

where $n(x)$ is the Boson distribution function

$$n(x) = [\exp(\beta x) - 1]^{-1}.$$

At the absolute zero of temperature and for electrons above the Fermi surface ($k > k_F$) the factor in square brackets in (5.66) is non-zero only for

$$\epsilon_k > \epsilon_{k-q} > 0.$$

If ϵ_k is small, we can write

$$-\operatorname{Im} \frac{1}{\epsilon(\boldsymbol{q}, \epsilon_k - \epsilon_{k-q} + i\delta)} = \frac{\operatorname{Im} \epsilon(q, \epsilon_k - \epsilon_{k-q} + i\delta)}{|\epsilon(\boldsymbol{q}, \epsilon_k - \epsilon_{k-q} + i\delta)|^2} \tag{5.67}$$

and neglect the weak dependence of the dielectric function on frequency. The numerator, according to equation (5.34), is (at absolute zero)

$$2\pi V(\boldsymbol{q}) \sum_{k'} [\theta(\epsilon_{k'}) - \theta(\epsilon_{k'+q})] \, \delta(\omega + \epsilon_{k'} - \epsilon_{k'+q}). \tag{5.68}$$

Inserted into equation (5.66), this yields

$$\frac{1}{\tau_k} = 4\pi \sum_{\substack{k', q \\ k > |k-q| > k_F}} \left[\frac{V(\boldsymbol{q})}{\epsilon(q, 0)}\right]^2 \{\theta(\epsilon_{k'})[1 - \theta(\epsilon_{k'+q})] - \theta(\epsilon_{k'+q})[1 - \theta(\epsilon_{k'})]\} \times \delta(\epsilon_k - \epsilon_{k-q} + \epsilon_{k'} - \epsilon_{k'+q}). \tag{5.69}$$

From the delta-function and the restrictions on ϵ_k and ϵ_{k-q} it follows that $\epsilon_{k'+q} - \epsilon_{k'}$ is positive. Hence the second term in the curly bracket is zero. It is clear from the delta- and theta-functions that the expression for the lifetime results from the scattering of an electron of momentum $\boldsymbol{k}$ to $\boldsymbol{k} - \boldsymbol{q}$ with the creation of an electron–hole pair. The result (5.69) can also be obtained using the Golden Rule or Born approximation with the *screened* scattering potential.

Quinn and Ferrell (1958) have evaluated the expression (5.69) explicitly for small ϵ and they find that

$$\frac{1}{\tau_k} = \frac{\sqrt{3}\pi^2}{128}\,\omega_p\left(\frac{\epsilon_k}{E_{\mathrm{F}}}\right)^2. \tag{5.70}$$

The important point about the result is that for an electron near the Fermi surface ($|\epsilon| \ll E_{\mathrm{F}}$), the lifetime tends to infinity like ϵ^{-2}. Hence the level width of a state of energy ϵ is ϵ^2 which becomes much smaller than ϵ as $\epsilon \to 0$. Such quasi-particle states are therefore, well defined. This idea has been exploited by Landau to provide a very general basis for discussing Fermi systems, including electrons in metals (cf. Chapter 6).

For higher-energy electrons other processes for the redistribution of the energy, besides pair creation, become important. For example for very fast electrons with $k \gg k_{\mathrm{F}}$ there is a contribution to (5.66) from small $|q|$ values which still leave $\epsilon_k - \epsilon_{k-q} (\approx \boldsymbol{q}\cdot\boldsymbol{v})$ quite large. For given $|q|$ the energy difference is greatest when $\boldsymbol{q}$ is parallel to $\boldsymbol{v}$ so the electron is scattered through only a small angle. In this case we can approximate the dielectric function by equation (5.36). Then

$$\operatorname{Im}\frac{1}{\epsilon(\boldsymbol{q}, \boldsymbol{v}\cdot\boldsymbol{q} + i\delta)} \approx \operatorname{Im}\frac{(\boldsymbol{v}\cdot\boldsymbol{q})^2}{(\boldsymbol{v}\cdot\boldsymbol{q} + i\delta)^2 - \omega_p^2}$$

$$= \tfrac{1}{2}\pi\omega_p\,[\delta(\boldsymbol{v}\cdot\boldsymbol{q} - \omega_p) + \delta(\boldsymbol{v}\cdot\boldsymbol{q} + \omega_p)]. \tag{5.71}$$

Since $\boldsymbol{v}\cdot\boldsymbol{q}$ is positive (to make $\epsilon_{k-q} < \epsilon_k$), the second term is zero and the contribution to the lifetime from such values of q is

$$\approx \tfrac{1}{2}\pi\omega_p \sum_q V(\boldsymbol{q})\,\delta(\boldsymbol{v}\cdot\boldsymbol{q} - \omega_p). \tag{5.72}$$

The delta-function shows that the energy lost by the electron goes into a plasmon.

Expression (5.66), even for high-energy electrons, will also include the pair creation processes we have already discussed. It will also include values of q which do not correspond to scattering by long-lived plasmons or quasi-particles. All such processes are involved in the stopping of high energy electrons by an electron gas and are discussed in more detail by Pines and Nozières (1966, Section 4.4).

References

General

Pines, D. and Nozières, P. (1966). "The Theory of Quantum Liquids", Vol. 1. Benjamin, New York.

Pines, D. (1961). "The Many-Body Problem". Benjamin, New York.

Special

Gell-Mann, M. and Brueckner, K. A. (1957). *Phys. Rev.* **106**, 364.

Hubbard, J. (1957). *Proc. Roy. Soc. (London)* **A243**, 336.

Lindhard, J. (1954). *Kgl Danske Videnskab. Selskab., Mat-fgs. Medd.* **28**, 8.

Quinn, J. J. and Ferrell, R. A. (1958). *Phys. Rev.* **112**, 812.

Singwi, K. S., Sjolander, A., Tosi, M. P. and Land, R. H. (1970). *Phys. Rev.* **B1**, 1044.

Problems

1. Evaluate the frequency and wavelength dependent magnetic spin susceptibility $m(\boldsymbol{q}, \omega)$ in the random phase approximation by following the argument of §5.2(b). The magnetic moment density operator is given in Table A1. (Appendix A) and the interaction term is

$$-\frac{1}{2}g \int \psi^+(\boldsymbol{r})\sigma_z H(\boldsymbol{r}, t)\psi(\boldsymbol{r})d^3r$$

when the external magnetic field $\boldsymbol{H}(\boldsymbol{r}, t)$ is in the z-direction. Show that the contributions of bubbles which contain exactly one vertex at which σ_z acts are zero. Hence show that

$$m(\boldsymbol{q}, \omega) = -\tfrac{1}{4}g^2\chi(\boldsymbol{q}, \omega).$$

2. Since the screening of a charge by electrons depends on their density of states at the Fermi surface, one would not expect it to be altered very much by a dilute concentration of impurities. Calculate the effect of impurities (including only s-wave scattering) on the contribution of a bubble, $\chi(\boldsymbol{q}, \omega)$, by using the method of §4.3. Show that the vertex correction cancels the self-energy and that $\chi(\boldsymbol{q}, \omega)$ is unchanged. (The main difference between the transport problem and this one is that the factor $\cos\theta$ occuring in the former is replaced by unity.)

Chapter 6

Landau's Theory of Normal Fermi Liquids

§6.1 The neutral liquid

(a) *Introduction*

One of the remarkable facts about electrons in metals is that, despite the Coulomb force which acts between them and which is both long-range and strong, they behave as weakly interacting particles. In the last chapter we showed that the long-range of the interaction gives rise to collective high-frequency modes of the electrons, the plasmons. Because of their high frequencies these modes play little part in the thermal and transport properties of metals. Once these modes are accounted for there remains a residual short-range interaction between the electrons. At normal electron densities in metals this interaction is still comparable with the kinetic energy and some explanation of the apparently independent behaviour of the charges is still required. Such an explanation is provided by Landau's theory of normal Fermi liquids (Landau, 1957a, b, 1959).

We saw in the last chapter that, in the high-density limit, the lifetime of low-lying excitations of energy ϵ (measured from the Fermi surface) is proportional to ϵ^2. The level widths of these states are, therefore, much less than their excitation energies and they are approximate eigenstates of the Hamiltonian. They have a well defined momentum and so are good quasi-particles. The result for the lifetime depends only on the short range of the residual interaction and on the amount of phase space available for the scattering of one electron through the creation of an electron–hole pair. It therefore has a much wider range of validity than appears from the proof. Even when the interaction is strong then, the low lying excitations are good quasi-particles. Landau's theory of a Fermi liquid exploits this property as well as the degeneracy of the fermions. The aim is to relate the thermodynamic properties of the liquid and also its response to fields varying slowly in space and time to a few parameters which are ultimately determined empirically. Landau's theory was advanced originally in a form suitable for a neutral Fermi liquid with only short-range forces such as normal liquid ^{3}He, although Landau had in mind the application to electrons in metals. Silin (1957) extended the theory to apply to charged Fermi liquids.

In this section we provide an account of the theory for neutral liquids following an approach due to Leggett (1965) of considering molecular fields. We then turn to the justification of this theory from microscopic theory. Finally, we consider the effect of the Coulomb force.

The assumptions of Landau's theory are:

(i) There is a 1–1 correspondence between the low lying excited states of the interacting system and the non-interacting one from which it derives. In particular, the same quantum numbers can be used to label the corresponding states. For a translationally invariant system, the momentum $\boldsymbol{k}$ and spin $\boldsymbol{\sigma}$ are used to label the states.

(ii) The low lying states of the system are quasi-particle states which can be specified by stating the number of quasi-particles present in each state. This means in particular that the energy is a function only of the occupation numbers of the quasi-particle states.

(iii) At low temperatures and under weak perturbations the changes in occupation of the quasi-particle states are so small that cubic terms in these quantities make a negligible contribution to the energy.

(iv) In fields which vary slowly in space and time, it is possible to define a local instantaneous energy which depends only on the local, instantaneous occupation numbers.

These assumptions define a *normal* Fermi liquid and fail for superfluid systems.

We shall consider only translationally invariant systems at or close to the absolute zero of temperature. This means that at the present time and in this form, the theory applies only to liquid ^{3}He and to that only at temperatures below about 50 mK and above its superfluid transition at 2.6 mK. However, much of what we say, when modified by taking charge into account, applies to metals if momentum is interpreted as crystal momentum.

For the translationally invariant system the state is given completely by the occupation numbers $n_{\mathbf{k},\sigma}$ of the states of momentum $\mathbf{k}$ and spin σ. The energy E is therefore a function of $\delta n_{\mathbf{k},\sigma}$, the deviations in the values of the occupation numbers from the ground state ones, i.e.

$$E \equiv E[\delta n_{\mathbf{k},\sigma}]. \tag{6.1}$$

According to the second assumption, this can be expanded as

$$E = \sum_{\mathbf{k}\sigma} \epsilon_{\mathbf{k}}\, \delta n_{\mathbf{k},\sigma} + \tfrac{1}{2} \sum_{\mathbf{k},\mathbf{k}',\sigma,\sigma'} f(\mathbf{k},\mathbf{k}',\sigma,\sigma')\delta n_{\mathbf{k}\sigma}\, \delta n_{\mathbf{k}'\sigma'}, \tag{6.2}$$

where f is a new function introduced by Landau to describe the Fermi liquid. For the homogeneous isotropic system $f(\mathbf{k},\mathbf{k}',\sigma,\sigma')$ depends only on the vector difference $(\mathbf{k}-\mathbf{k}')$.

The excitation energy for a quasi-particle in the state $\mathbf{k},\sigma$ is simply the energy required to increase $\delta n_{\mathbf{k}\sigma}$ by unity. From equation (6.2) this is

$$\tilde{\epsilon}_{\mathbf{k}\sigma} = \partial E/\partial(\delta n_{\mathbf{k}\sigma}) = \epsilon_{\mathbf{k}} + \sum_{\mathbf{k}',\sigma'} f(\mathbf{k},\mathbf{k}',\sigma,\sigma')\delta n_{\mathbf{k}'\sigma'} \tag{6.3}$$

and depends on the distribution of other quasi-particles. In equilibrium at the absolute zero of temperature all $\delta n_{\mathbf{k}\sigma}$ are zero and

$$\tilde{\epsilon}_{\mathbf{k}} = \epsilon_{\mathbf{k}}. \tag{6.4}$$

If energies are measured from the chemical potential, the Fermi surface is defined by

$$\epsilon_{\mathbf{k}} = 0. \tag{6.5}$$

For an isotropic system ϵ depends on k only and the Fermi wave vector, k_{F}, is defined by

$$\epsilon_{k_{\mathrm{F}}} = 0.$$

It is assumed that the interaction does not change the ordering of the levels so that

$$\begin{aligned} \epsilon_k &> 0, \qquad k > k_{\mathrm{F}}, \\ \epsilon_k &< 0, \qquad k < k_{\mathrm{F}}. \end{aligned} \tag{6.6}$$

From assumption (1) the number of states with $k < k_{\mathrm{F}}$ must be equal to the mean density n of particles. Hence

$$n = [2/(2\pi)^3] \int d^3k\theta(k_F - k) = k_F^3/3\pi^2. \tag{6.7}$$

For the phenomena to which the theory is applicable, only changes of occupation number near the Fermi surface are important. This means that $|k|$ and $|k'|$ can be fixed at k_F in the function f and this function depends only on the angle between $\boldsymbol{k}$ and $\boldsymbol{k}'$ as well as spin. Also we can expand ϵ_k in powers of $(k - k_F)$ keeping only the first-order term. Thus we write

$$\epsilon_k = (k_F/m^*)(k - k_F), \tag{6.8}$$

where this equation defines an effective mass m^*. The properties of the system are then described by the empirical quantities m^* and $f(\boldsymbol{k}_F, \boldsymbol{k}_F', \sigma, \sigma')$. For a translationally invariant system, m^* is not independent of f, as we see below.

The occupation numbers $\delta n_{k\sigma}$ are subject to both quantum mechanical and thermal fluctuations about their mean values. However, the deviations from the mean values $\langle \delta n_{k\sigma} \rangle$ are small compared with $\langle \delta n_{k\sigma} \rangle$. In the second term of the energy (6.2) such deviations occur quadratically and would give rise to contributions of order $(\delta n)^3$, which we have already neglected. We can therefore neglect terms quadratic in $\delta n_{k\sigma} - \langle \delta n_{k\sigma} \rangle$ in E and write

$$E = \epsilon_k \delta n_{k\sigma} + \sum_{k, k', \sigma, \sigma'} f(\boldsymbol{k}, \boldsymbol{k}', \sigma, \sigma') \langle \delta n_{k'\sigma} \rangle \delta n_{k\sigma}$$

$$- \tfrac{1}{2} \sum f(\boldsymbol{k}, \boldsymbol{k}', \sigma, \sigma') \langle \delta n_{k\sigma} \rangle \langle \delta n_{k'\sigma'} \rangle$$

$$= \sum_{k, \sigma} (\epsilon_k + U_{k\sigma}) \delta n_{k\sigma} - \tfrac{1}{2} \sum_{k\sigma} U_{k\sigma} \langle \delta n_{k\sigma} \rangle, \tag{6.9}$$

where

$$U_{k\sigma} = \sum_{k', \sigma'} f(\boldsymbol{k}, \boldsymbol{k}', \sigma, \sigma') \langle \delta n_{k'\sigma'} \rangle. \tag{6.10}$$

In this, $U_{k\sigma}$ is an average molecular field acting on one quasi-particle and due to all of them. We have here a generalization of the Hartree picture (cf. §3.7) with the self-consistent or molecular field defined by (6.10).

Equation (6.9) shows that the quasi-particles behave like ordinary particles with energies $\epsilon_k + U_{k\sigma}$. At a finite temperature T the occupation numbers become

$$\delta n_{k\sigma} = \{\exp[-\beta(\epsilon_k + U_{k\sigma})] + 1\}^{-1} - \theta(k_F - k).$$

(Note that in our notation, for $k < k_F$, $\delta n_{k\sigma}$ is minus the number of holes present.) It follows that $U_{k\sigma}$ is independent of angle and spin and is, therefore, a constant. Also since the number of electron-like quasi-particles is equal to the number of holes, $\Sigma\langle \delta n_{k\sigma}\rangle$ is zero and so, from equation (6.10), $U_{k\sigma}$ is zero. Hence

$$\delta n_{k\sigma} = [\exp(-\beta\epsilon_k) + 1]^{-1} - \theta(k_F - k). \tag{6.11}$$

The free energy then has the same form as for independent fermions and leads to a specific heat per unit volume,

$$C_V = \tfrac{2}{3}\pi^2 N(0) k_B^2 T, \tag{6.12}$$

where $N(0)$ is the density of states for one spin at the Fermi surface, that is,

$$N(0) = m^* k_F / 2\pi^2. \tag{6.13}$$

(b) *Effective mass*

For a translationally invariant system there is a relation between m^* and f which can be obtained as follows. Suppose the system is given a velocity $\boldsymbol{v}$ as a whole. The interactions are unchanged and the system has total momentum

$$\boldsymbol{p} = Nm\boldsymbol{v}, \tag{6.14}$$

where m is the mass of one of the fermions when free.

The change of momentum gives rise to a change in the occupation numbers of the quasi-particles which must be consistent with the change in momentum. This requirement of consistency leads to the relation between m^* and the function f.

Because the interaction is unchanged the effect on the energy is to add to it a term $-\boldsymbol{v}.\boldsymbol{p}$. Since we also have

$$\boldsymbol{p} = \sum_{\boldsymbol{k},\sigma} \boldsymbol{k}\,\delta n_{\boldsymbol{k},\sigma}, \tag{6.15}$$

the energy is now

$$E(\boldsymbol{v}) = \sum_{\boldsymbol{k},\sigma} (\epsilon_k - \boldsymbol{v}.\boldsymbol{k} + U_{k\sigma})\delta n_{k\sigma}. \tag{6.16}$$

Hence the distribution function is

$$\delta n_{k\sigma} = \{\exp[-\beta(\epsilon_k - \boldsymbol{v}.\boldsymbol{k} + U_{k\sigma})] + 1\}^{-1} - \theta(k_F - k)$$

and (6.17)

$$U_{k\sigma} = \sum_{\boldsymbol{k}'\sigma'} f(\boldsymbol{k}, \boldsymbol{k}', \sigma, \sigma')\delta n_{\boldsymbol{k}'\sigma'}$$

$$= \sum_{k'\sigma'} f(k, k', \sigma, \sigma')(U_{k',\sigma'} - v\cdot k')\frac{\partial n_{k'\sigma'}}{\partial \epsilon_{k'}}, \tag{6.18}$$

where

$$n_{k\sigma} = \delta n_{k\sigma} + \theta(k_F - k).$$

Since f depends only on the angle between k and k', the solution of (6.18) will be of the form

$$U_{k\sigma} = av\cdot k \tag{6.19}$$

where a is a constant which satisfies

$$a = (a-1)\sum_{k'\sigma'} f(k, k', \sigma, \sigma')\cos\theta_{kk'}\frac{\partial n_{k'\sigma'}}{\partial \epsilon_{k'}}.$$

Neglecting terms of relative order $k_B T/E_F$, we can replace

$$\partial n_{k'\sigma'}/\partial \epsilon_{k'} \quad \text{by} \quad -\delta(\epsilon_{k'}).$$

Hence

$$a = 2(1-a)N(0)\int f(k, k', \sigma, \sigma')\cos\theta d\Omega/4\pi, \tag{6.20}$$

where θ is the angle between k and k'.

It is now common practice to expand f in terms of spherical harmonics. Since there is no particular axis for spin f will depend on $\boldsymbol{\sigma}, \boldsymbol{\sigma}'$ only through the combination (for spin $\frac{1}{2}$ particles) $\boldsymbol{\sigma}\cdot\boldsymbol{\sigma}'$. Hence we can write

$$2N(0)f(k, k', \sigma, \sigma') = \sum_l (F_l + Z_l\,\boldsymbol{\sigma}\cdot\boldsymbol{\sigma}')P_l(\cos\theta). \tag{6.21}$$

From equations (6.20) and (6.21) one finds

$$(1-a) = (1+\tfrac{1}{3}F_1)^{-1} \tag{6.22}$$

and

$$\delta n_{k\sigma} = \exp[-\beta(\epsilon_k - v_{\text{eff}}\cdot k)]^{-1} - \theta(k_F - k), \tag{6.23}$$

where

$$v_{\text{eff}} = (1-a)v \tag{6.24}$$

is an effective velocity for each quasi-particle. The total momentum is now given by expressions (6.14) and (6.15). Hence

$$Nmv = \sum_{k,\sigma}\delta n_{k\sigma} = -\sum_{k,\sigma} k\frac{\partial n_{k\sigma}}{\partial \epsilon_k} v_{\text{eff}}\cdot k$$

$$= \tfrac{2}{3}N(0)k_F^2 v_{\text{eff}}. \tag{6.25}$$

From equations (6.7), (6.24) and (6.25) it follows that

$$m^*/m = 1 + \tfrac{1}{3}F_1. \tag{6.26}$$

This is the requirement for consistency.

This result applies to liquid ^{3}He for which, from specific heat measurements (Wheatley 1975), it is found that F_1 varies from 6.3 at saturated vapour pressure to approximately 15 on the melting curve. Because of the crystal lattice, the result is not valid for electrons in metals, even for a spherical Fermi surface.

An alternative way to look at the result is that under the translation each quasi-particle has its momentum increased by $m^* v_{\text{eff}}$ which for momentum balance must be the same as mv. This yields equation (6.28). The difference between $m^* v_{\text{eff}}$ and $m^* v$ arises from the backflow of the other quasi-particles induced, through the interaction, by the motion of the first particle.

(c) *The velocity of sound*

According to hydrodynamic theory, the velocity of sound, c_1, is given by

$$c_1^2 = \frac{1}{m}\frac{\partial P}{\partial \rho}$$

where ρ is the number density and P the pressure. On the other hand, from thermodynamics it is known that

$$\partial P/\partial \rho = N\,\partial\mu/\partial N.$$

Hence

$$c_1^2 = \frac{N}{m}\frac{\partial \mu}{\partial N}. \tag{6.27}$$

We shall use Fermi liquid theory to calculate $\partial N/\partial\mu$ and so find the velocity of sound.

Now a change in μ will produce a change in the molecular field $U_{k\sigma}$ which will be spherically symmetric and therefore a constant. Also since ϵ_k is measured from the chemical potential

$$\delta\epsilon_k = -\delta\mu.$$

Hence

$$\delta(\delta n_{k\sigma}) = \delta\{\exp[-\beta(\epsilon_k + U)] + 1\}^{-1}$$

$$= \frac{\partial n_{k\sigma}}{\partial \epsilon_k}\left[-1 + \frac{\partial U}{\partial \mu}\right]\delta\mu. \tag{6.28}$$

It follows from this and (6.10) that the equation for $\partial U/\partial\mu$ is

$$\frac{\partial U}{\partial \mu} = \sum_{k',\sigma'} f(k, k', \sigma, \sigma')\frac{\partial n_{k\sigma}}{\partial \epsilon_k}\left[-1 + \frac{\partial U}{\partial \mu}\right]$$

$$= F_0\left(1 - \frac{\partial U}{\partial \mu}\right).$$

Hence

$$(1 - \partial U/\partial\mu) = (1 + F_0)^{-1} \tag{6.29}$$

The change in the number of particles produced by the change in the chemical potential is

$$\begin{aligned}\delta N &= \Sigma(\delta n_{k\sigma}) \\ &= \Sigma \frac{\partial n_{k\sigma}}{\partial\epsilon_k}\left[-1 + \frac{\partial U}{\partial\mu}\right]\delta\mu \\ &= 2N(0)/(1 + F_0)\delta\mu.\end{aligned}$$

Since μ is the only dependent variable undergoing a change we have that

$$\partial\mu/\partial N = (\partial N/\partial\mu)^{-1}$$

and

$$c_1^2 = N(1 + F_0)/2mN(0). \tag{6.30}$$

With equations (6.7) and (6.13) this yields

$$\begin{aligned}c_1^2 &= (1 + F_0)k_F^2/3m^*m \\ &= \frac{1 + F_0}{1 + \frac{1}{3}F_1}\frac{k_F^2}{3m^2}.\end{aligned} \tag{6.31}$$

The last factor is the result for the free Fermi gas; the first factor contains the Fermi liquid effects. The values of F_0 for liquid ^3He derived from experiment vary from 10.8 at s.v.p. to approximately 100 near the melting curve. The prefactor, therefore, varies from about 3.8 to 17. In this case the Fermi liquid corrections are large and one is a long way from the free-fermion model.

(d) *Effects of external fields*

If the system departs from equilibrium in a way which varies slowly in space and time, we assume that there exists a local energy density which depends on local values of the distribution functions $\delta n_k(r, t)$. Thus

$$\begin{aligned}E &= \int d^3rE(r) \\ &= \int d^3r\left\{\sum_{k\sigma}\epsilon_k\delta n_{k\sigma}(r) + \sum_{\substack{k,k' \\ \sigma,\sigma'}} f(k, k', \sigma, \sigma')\delta n_{k'\sigma'}(r)\delta n_{k\sigma}(r)\right\} \\ &= \int d^3r\left\{\sum_{k,\sigma}\epsilon_k\delta n_{k\sigma}(r) + \sum_{k,\sigma} U_{k\sigma}(r)\delta n_{k\sigma}(r)\right\},\end{aligned} \tag{6.32}$$

where

$$U_{k\sigma}(r) = \sum_{k',\sigma'} f(k, k', \sigma, \sigma')\langle\delta n_{k'\sigma'}(r)\rangle. \tag{6.33}$$

This assumes that the range of the interaction is much less than the distance over which $\delta n_{k\sigma}(r)$ varies significantly and that the time for the interaction to take place is much less than the time over which $\delta n_{k\sigma}(r)$ varies.

To find the effects of slowly varying fields we treat E as a Hamiltonian for the quasi-particles and add to it the interaction with the external fields. We need to decide what operators to choose for the number densities $\delta n_{k\sigma}(r)$. The total number density and number current density operators are

$$\rho(r) = \psi^+(r)\psi(r)$$

$$= \mathscr{V}^{-1} \sum_{k,q,\sigma} c^+_{k-q/2\sigma} c_{k+q/2\sigma} \exp(iq \cdot r)$$

and

$$J(r) = (-i/2m)[\psi^+\nabla\psi - (\nabla\psi^+)\psi]$$

$$= \mathscr{V}^{-1} \sum_{k,q,\sigma} (k/m) c^+_{k-q/2\sigma} c_{k+q/2\sigma} \exp(iq \cdot r).$$

Hence, we can take

$$n_{k\sigma}(r) = \mathscr{V}^{-1} \sum_{q} c^+_{k-q/2\sigma} c_{k+q/2\sigma} \exp(iq \cdot r)$$

and

$$H = \sum_{k,\sigma} \epsilon_k [c^+_{k\sigma} c_{k\sigma} - \theta(k_F - k)] + \sum_{k,\sigma} U_{k\sigma}(q) c^+_{k-q/2\sigma} c_{k+q/2\sigma}, \tag{6.34}$$

where

$$U_{k\sigma}(q) = \int d^3r U_{k\sigma}(r) \exp(iq \cdot r)$$

$$= \sum_{k',\sigma'} f(k, k', \sigma, \sigma') \langle c^+_{k'+q/2\sigma'} c_{k'-q/2\sigma'}\rangle. \tag{6.35}$$

Thus we have a molecular field problem which is a generalization of the Hartree solution and in which the molecular field $U_{k\sigma}$ is related to the molecular parameters $\langle c^+c\rangle$ through equation (6.35).

Suppose now that the liquid is acted upon by an external field which introduces an extra, possibly time dependent, term

$$\sum_{k,q,\sigma} \phi_{k\sigma}(q, t) c^+_{k-q/2\sigma} c_{k+q/2\sigma} \tag{6.45}$$

into the Hamiltonian. The quasi-particles are then acted upon altogether by an effective field

$$\phi_{k\sigma e}(\boldsymbol{q}, t) = \phi_{k\sigma}(\boldsymbol{q}, t) + U_{k,\sigma}(\boldsymbol{q}, t), \tag{6.46}$$

where the molecular field will, in general, also depend on t. The total Hamiltonian is consequently

$$H = \sum_{k\sigma} \epsilon_k [c_{k\sigma}^+ c_{k\sigma} - \theta(k_F - k)] + \sum_{k} \phi_{k\sigma e}(\boldsymbol{q}, t) c_{k-q/2\sigma}^+ c_{k+q/2\sigma}. \tag{6.47}$$

The temperature Green's function for the unperturbed quasi-particles is now given by

$$G_0(\boldsymbol{k}, \omega_\nu)^{-1} = i\omega_\nu - \epsilon. \tag{6.48}$$

The effect of the field can be expressed in terms of

$$\begin{aligned}\langle n_{kq\sigma}(t)\rangle &= \langle c_{k'+q/2\sigma}^+(t) c_{k'-q/2\sigma}(t)\rangle \\ &= G_\sigma(\boldsymbol{k}' - \boldsymbol{q}/2, t, \boldsymbol{k}' + \boldsymbol{q}/2, t_+).\end{aligned}$$

We are concerned with the linear response of the liquid for which one time Fourier component of ϕ leads to a corresponding Fourier component of $\langle n\rangle$. The latter can be obtained by analytic continuation from (cf. §§3.2 and 5.2),

$$\begin{aligned}\langle n_{kq\sigma}(i\Omega_\nu)\rangle &= \beta^{-1} \sum_{\omega_\nu} G_0(\boldsymbol{k} - \boldsymbol{q}/2, \omega_\nu) G_0(\boldsymbol{k} + \boldsymbol{q}/2, \omega_\nu + \Omega_\nu) \phi_{k,\sigma}(\boldsymbol{q}, i\omega_\nu) \\ &= \frac{[f(\epsilon_{k-q/2}) - f(\epsilon_{k+q/2})]\phi_{k\sigma e}}{i\Omega_\nu + \epsilon_{k-q/2} - \epsilon_{k+q/2}},\end{aligned} \tag{6.49}$$

where $f(x)$ is again the Fermi distribution function.

As we have explained, this theory is expected to be valid only in the wavelength limit $q \to 0$. Thus we can write

$$n_{k\sigma}(i\Omega_\nu) = \frac{-\boldsymbol{q}\cdot\boldsymbol{v}\,\partial f/\partial\epsilon}{i\Omega_\nu - \boldsymbol{q}\cdot\boldsymbol{v}} \phi_{k\sigma e}(\boldsymbol{q}, i\Omega_\nu) \tag{6.50}$$

and

$$\phi_{k\sigma e}(\boldsymbol{q}, i\Omega_\nu) = \phi_{k,\sigma}(q) + \sum_{k',\sigma'} \frac{f(\boldsymbol{k}, \boldsymbol{k}', \sigma, \sigma')\boldsymbol{q}\cdot\boldsymbol{v}'\,\partial f/\partial\epsilon'\,\phi_{k'\sigma' e}(q)}{\boldsymbol{v}'\cdot\boldsymbol{q} - i\Omega_\nu}. \tag{6.51}$$

We have also written

$$\boldsymbol{v} = \partial\epsilon_k/\partial\boldsymbol{k} = \boldsymbol{k}/m^*, \qquad q = (\boldsymbol{q}, i\Omega_\nu). \tag{6.52}$$

The complete solution for the response is now given by equations (6.50) and (6.51) both of which can be continued analytically by replacing $i\Omega_\nu$ by $\Omega + i\delta$. Hence

$$n_{k\sigma}(\boldsymbol{q}, \Omega) = \frac{\boldsymbol{q}\cdot\boldsymbol{v}\,\partial f/\partial\epsilon}{\boldsymbol{q}\cdot\boldsymbol{v} - \Omega - i\delta}\phi_{k\sigma e}(\boldsymbol{q}, \Omega) \tag{6.53}$$

$$\phi_{k\sigma e}(\boldsymbol{q}, \Omega) = \phi_{k\sigma}(\boldsymbol{q}, \Omega) + \sum_{k'\sigma'} \frac{f(\boldsymbol{k}, \boldsymbol{k}', \sigma, \sigma')\boldsymbol{v}'\cdot\boldsymbol{q}\,\partial f/\partial\epsilon'}{\boldsymbol{v}'\cdot\boldsymbol{q} - \Omega - i\delta}\phi_{k'\sigma' e}(\boldsymbol{q}, \Omega). \tag{6.54}$$

(These results can also be obtained directly if we use the retarded Green's functions. We have derived them using the temperature Green's functions in the first instance for comparison with the microscopic theory given in succeeding sections.) Equation (6.54) is an implicit equation for the total effective field $\phi_{k\sigma e}(\boldsymbol{q}, \Omega)$.

Note that we have calculated only the response of the quasi-particles to the external field. There could be a response of the remaining excitations to the external field. However these will be of high energy, and for long-wavelength low-energy fields they will contribute at most a constant response to the fields.

We now apply these results to a number of specific problems.

(e) *Spin susceptibility*

We consider the response to a static magnetic field $\boldsymbol{H}(\boldsymbol{r})$. The extra term in the Hamiltonian is

$$\tfrac{1}{2}\int d^3r\,\psi^+ g\,\boldsymbol{\sigma}\cdot\boldsymbol{H}(\boldsymbol{r})\psi(\boldsymbol{r}).$$

where g is the gyromagnetic ratio for the particles. This term can be rewritten

$$-\tfrac{1}{2}g\sum_{k,q} H(\boldsymbol{q})(c^+_{k-q/2\uparrow}c_{k+q/2\uparrow} - c^+_{k-q/2\downarrow}c_{k+q/2\downarrow}), \tag{6.55}$$

where the field is taken to be always in the z-direction. In this case,

$$\phi_{k\uparrow}(\boldsymbol{q}) = -\tfrac{1}{2}gH(\boldsymbol{q}), \qquad \phi_{k\downarrow}(\boldsymbol{q}) = \tfrac{1}{2}gH(\boldsymbol{q}). \tag{6.56}$$

and ϕ is antisymmetric under a change of spin. Similarly, the effective field will also be antisymmetric under a change of spin. The only part of $f(\boldsymbol{k}, \boldsymbol{k}', \sigma, \sigma')$ which contributes is, therefore, that proportional to $\sigma_z\sigma_z'$. We therefore have

$$-\phi_{k\downarrow e}(q) = \phi_{k\uparrow e}(q) = \phi_{k\uparrow}(q) - \sum_{l} \frac{Z_l}{4} \int \frac{d\Omega}{4\pi} P_l(\cos\theta)\phi_{k'\uparrow e}(q)$$

$$= \phi_{k\uparrow}(q) - \frac{Z_0}{4}\phi_{k\uparrow e}(q).$$

Hence

$$\phi_{k\sigma e}(q) = \phi_{k\sigma}(q)/(1 + \tfrac{1}{4}Z_0), \tag{6.57}$$

and the molecular field reduces the effective field by a factor $(1 + \frac{1}{4}Z_0)$.

The total magnetic moment is

$$\tfrac{1}{2}g \sum_{k} (n_{k\uparrow} - n_{k\downarrow}) = -\frac{\frac{1}{2}g^2 H}{1 + \frac{1}{4}Z_0} \sum_{k} \frac{\partial f}{\partial \epsilon} = \frac{\frac{1}{2}N(0)g^2 H}{1 + \frac{1}{4}Z_0}. \tag{6.58}$$

Because $N(0)$ is proportional to the effective mass we find that the magnetic susceptibility χ is given by

$$\chi/\chi_0 = (1 + \tfrac{1}{3}F_1)/(1 + \tfrac{1}{4}Z_0), \tag{6.59}$$

where χ_0 is the calculated susceptibility when Fermi liquid effects are completely ignored. For liquid ^{3}He the denominator varies from about 0.33 at saturated vapour pressure to 0.26 on the melting curve. Hence χ is enhanced by a factor of from 3 to 4 above $\frac{1}{2}N(0)g^2$, the value one would obtain assuming free fermions and an effective mass given by the specific heat.

If $(1 + \frac{1}{4}Z_0)$ were to decrease to zero and become negative the liquid would become ferromagnetic. The small positive value of $(1 + \frac{1}{4}Z_0)$ implies that it is almost ferromagnetic. Spin fluctuations of the liquid are easy to excite and play an important part in the general properties of liquid ^{3}He. (For further details the reader is referred to papers by Doniach and Engelsberg 1966, Rice 1967, Pethick 1969 and Emery 1968.)

The results for m^* and $\partial N/\partial\mu$ can be obtained in a similar way with appropriate choices for ϕ.

(f) *Collective oscillations*

Frequencies of collective motions show up as poles in the response function of the system. In the present case the response function is given by the expression (6.53) for $n_{k\sigma}(\Omega)$. Now this expression has poles where $\Omega = \boldsymbol{q}\cdot\boldsymbol{v}$. In fact $\boldsymbol{q}\cdot\boldsymbol{v}$ arose from the approximation

$$\epsilon_{k+q/2} - \epsilon_{k-q/2} = \boldsymbol{q}\cdot\boldsymbol{v},$$

and is the excitation energy of a particle-hole pair. However $n_k(q, \Omega)$ will also have poles where $\phi_{kqe}(q, \Omega)$ has poles. Since $\phi_{k\sigma}$ represents the external field it will, in general, not have poles for the same values of q, Ω. Hence near poles of $\phi_{k\sigma e}$, $\phi_{k\sigma}$ can be ignored in equation (6.54). The poles of $\phi_{k\sigma e}$ are therefore the values of Ω for which the homogeneous equation,

$$\phi_{k\sigma e}(q, \Omega) = \sum_{k', \sigma'} \frac{f(k, k', \sigma, \sigma')v' . q}{v' . q - \Omega - i\delta} \frac{\partial f}{\partial \epsilon'} \phi_{k'\sigma' e}(q, \Omega), \quad (6.60)$$

has solutions.

For a general form of the scattering function f, equation (6.60) is rich in possible solutions. In general, there will be one solution in which $\phi_{k\sigma e}$ depends only on the angle between k and q and not on σ. (For other solutions ϕ can depend on $\boldsymbol{\sigma}$ as well as on the azimuthal angle of k.) The corresponding mode of oscillation is called zero sound and is the only one of the modes so far observed. Even in this case, the solution is, for a general f, quite complicated. If we suppose that f is dominated by the $l = 0$ term in the Legendre expansion (6.21), then $\phi_{k\sigma e}$ is independent of angle and equation (6.60) reduces to

$$1 = -\frac{F_0}{2} \int \frac{d\mu\, v_F q \mu}{v_F q \mu - \Omega - i\delta}. \quad (6.61)$$

If we write c_0 for the zero sound velocity Ω/q and s_0 for the ratio c_0/v_F, equation (6.61) yields the implicit equation for s_0,

$$\frac{s_0}{2} \ln \left| \frac{1 + s_0}{1 - s_0} \right| = 1 + F_0^{-1}. \quad (6.62)$$

The corresponding distribution function is

$$n_{k\sigma}(q) = \frac{\cos\theta (\partial f/\partial \epsilon)}{\cos\theta - s_0} \phi. \quad (6.63)$$

For $s_0 \gg 1$ (corresponding to $F_0 \gg 3$), this is proportional to $\cos\theta$ and has the same form as for an ordinary sound wave. For $s_0 \approx 1$ (corresponding to $|F_0| \ll 1$) the distribution is peaked in the direction of propagation of the wave.

For liquid ^{3}He, it seems that f is dominated by the $l = 0$ and $l = 1$ terms. In this case the equation for s becomes

$$\frac{s_0}{2} \ln \left| \frac{1 + s_0}{1 - s_0} \right| = [F_0 + s_0^2 F_1 (1 + \tfrac{1}{3} F_1)]^{-1}. \quad (6.64)$$

For F_0 large, one finds from equations (6.31) and (6.64) a relationship between the first and second sound velocities

$$(c_0^2 - c_1^2)/c_1^2 = 4/[5(1 + F_0)]. \tag{6.65}$$

Evidence for zero sound comes from measurements of the velocity and attenuation of sound as a function of temperature at two frequencies by Abel *et al.* (1966). At low temperatures the attenuation is independent of frequency but proportional to T^2. At higher temperatures the attenuation is proportional to ω^2/T^2. These measurements are consistent with the interpretation that at low temperatures, where the mean free path is large, the sound is zero sound while at higher temperatures it is ordinary sound. Near the attentuation maximum the velocity drops from 194 ms^{-1} at low temperatures to 188 ms^{-1} at higher temperatures. From equation (6.65) this leads to $F_0 \approx 11$ at 0.32 atm, the pressure at which the experiment was conducted. This is consistent with the value of F_0 obtained from other measurements.

§6.2 Microscopic basis of Fermi liquid theory

We can show that provided the potential is not long-range and that perturbation theory converges, one can use perturbation theory to derive the results of §6.1. Thus a normal neutral Fermi liquid is one for which perturbation theory converges. The perturbation expansion naturally provides the required 1–1 correspondence between the states of the unperturbed liquid and the perturbed ones.

We have seen that the exact Green's function can be written in the form

$$G(\boldsymbol{k}, \omega) = [\omega - \epsilon_{k0} - \Sigma(k, \omega)]^{-1}, \tag{6.66}$$

where Σ is the self-energy and we have written

$$\epsilon_{k0} = k^2/2m - \mu \tag{6.67}$$

for the energies of the free fermions measured from the chemical potential. The Fermi wave vector satisfies

$$\epsilon_{k_F 0} + \Sigma(k_F, 0) = 0. \tag{6.68}$$

The discussion of §5.2(d) and the phase space argument of §6.1 show that near $k = k_F$ and $\omega = 0$, $\Sigma(k, \omega)$ is a weak function of its arguments. For these values of k and ω it is possible to expand ϵ_{k0} and $\Sigma(k, \omega)$ in powers of $(k - k_F)$ and ω. Then

$$\epsilon_{k0} + \Sigma(k, \omega) = (k - k_F)\left[\frac{k_F}{m} + \frac{\partial\Sigma(k_F, 0)}{\partial k}\right] + \omega\,\frac{\partial\Sigma(k_F, 0)}{\partial\omega} + \dots \tag{6.69}$$

Further, because of the phase space limitations discussed in §6.1, the imaginary part of Σ will be quadratic in $(k - k_F)$ and ω. The terms displayed in (6.69) are therefore all real and, if we neglect quadratic ones, we are led to (cf. §3.5)

$$G(\boldsymbol{k}, \omega) = [Z(\omega - \epsilon_k)]^{-1} \equiv a(\omega - \epsilon_k)^{-1}, \tag{6.70}$$

where

$$a^{-1} = Z = 1 - \partial\Sigma(k_F, 0)/\partial\omega, \tag{6.71}$$

$$\epsilon_k = (k - k_F)/m^* \tag{6.72}$$

with

$$m^{*-1} = [m^{-1} + k_F^{-1}\,\partial\Sigma(k_F, 0)/\partial k]Z^{-1}. \tag{6.73}$$

Equation (6.70) gives the correct behaviour of G close to its poles near the origin. In general we can write

$$G(\boldsymbol{k}, \omega) = a(\omega - \epsilon_k)^{-1} + G'(\boldsymbol{k}, \omega), \tag{6.74}$$

where the pole near the origin is separated out. It is this pole which represents the quasi-particles.

Note that because of the factor a, due to the wave function renormalization, the first term of G is not quite the same as that used in the previous section. This is because the particle operators c_k used in the previous section refer to quasi-particles and not particles. It is possible to redefine G in terms of new operators (Nozières 1964) to make the two results agree but this is unnecessary. As we shall now see, the important point is to examine whether the observable results of the exact and Landau theories agree.

Now all the results of Landau's theory are contained in equations (6.53) and (6.54) for the linear response of the Fermi liquid to slowly varying external fields. We therefore aim to show that these results can be obtained from perturbation theory to all orders in the interaction. We shall use temperature perturbation theory and shall aim to derive equations (6.50) and (6.51), in the first instance, for small values of $\boldsymbol{q}$ and Ω_ν. According to the rules we have constructed the response is given by the diagram of Fig. 6.1, where the square represents all ways in which a particle–hole pair of total momentum $\boldsymbol{q}$ and "energy" Ω can scatter to another pair with the same total momentum and "energy".

Since q and Ω are small, why not obtain the response by setting q and Ω both zero in these diagrams? This would yield a result of the form

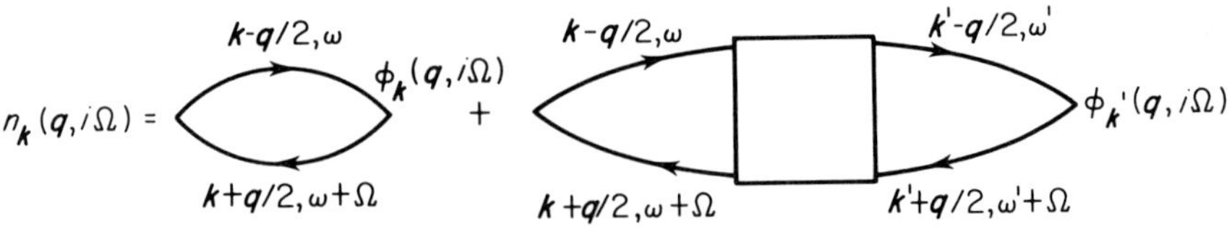

FIG. 6.1. The first-order contributions to $n_k(q)$ arising from an external field $\phi_k(q)$. The square represents all ways in which an electron–hole pair can scatter to another electron–hole pair with the same total momentum and energy.

$$n_k(q, i\Omega) = \sum_{k'} K(k, k')\phi_{k'}(q, i\Omega), \tag{6.75}$$

which actually represents the static results ($\Omega = 0$) of equations (6.50) and (6.51). However, there is a difficulty about doing this which becomes apparent if we look at the first diagram of Fig. 6.1. This contribution yields

$$n_{k\sigma}(q, i\Omega) = \beta^{-1} \sum_{\omega_\nu} G(k-q/2, \omega_\nu)G_0(k+q/2, \omega_\nu+\Omega_\nu)\phi_{k,\sigma}(q, i\Omega_\nu).$$

If only the quasi-particle parts of the Green's functions are retained, we obtain, for small q and Ω,

$$n_{k\sigma}(q, i\Omega) = \frac{a^2 q \cdot v\, \partial f/\partial \epsilon}{q \cdot v - i\Omega_\nu}\phi_{k\sigma}(q, i\Omega). \tag{6.76}$$

Like the Fermi liquid result this has an ill defined limit as q and Ω tend to zero. To make the limit definite it is necessary to stipulate how the ratio q/Ω behaves as $q \to 0$.

Now this ambiguity arises because in the limit of q, $\Omega \to 0$, the quasi-particle poles of the Green's functions coincide. Such an ambiguity will arise wherever poles coincide in the limit q, $\Omega \to 0$. For short-range potentials this will arise only in the propagation of particle–hole pairs. We therefore separate out explicitly propagators of electron–hole pairs and write the diagrams of Fig. 6.1 as shown in Fig. 6.2.

Here the circles denote all diagrams for the scattering of particle–hole pairs which cannot be cut into two parts by cutting a pair of particle–hole lines only. The contributions from the circles have unambiguous limits as q, Ω tend to zero. They can, therefore, be written as $\Gamma_0(k, \omega, \sigma; k'', \omega'', \sigma'')$ where k, ω, σ are the momentum "energy" and the spin of the particle of one of the pairs and k'', ω'', σ'', the momentum, energy and spin of the other pair (see Fig. 6.2).

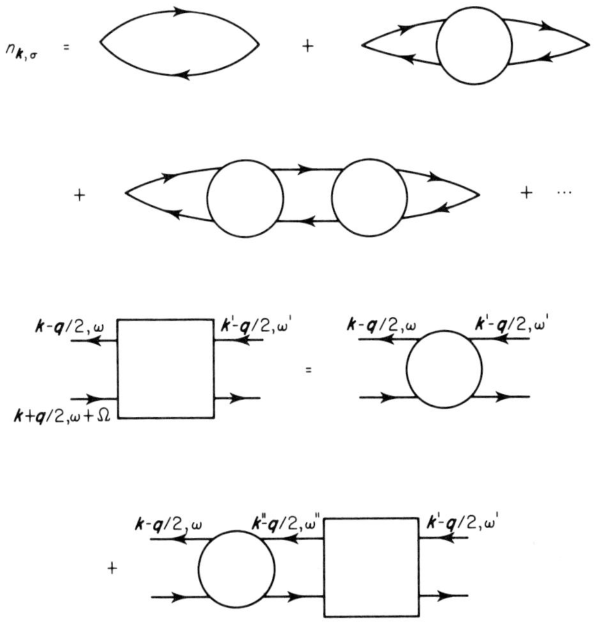

FIG. 6.2. The separation of electron–hole pairs in the diagrams of Fig. 6.1. Here the circles represent contributions which cannot be separated into two parts by cutting one electron and one hole line only.

If we write the contribution from the square of Fig. 6.2 as $\Gamma(\boldsymbol{k}, \omega, \boldsymbol{\sigma}; \boldsymbol{k}'', \omega'', \boldsymbol{\sigma}'', \boldsymbol{q}, i\Omega)$ the diagrams represent the equations

$$n_{k\sigma}(\boldsymbol{q}, i\Omega) = \beta^{-1} \sum_{\omega} G(\boldsymbol{k} - \boldsymbol{q}/2, \omega) G(\boldsymbol{k} + \boldsymbol{q}/2, \omega + \Omega) \phi_{k\sigma}(\boldsymbol{q}, i\Omega)$$

$$+ \beta^{-2} \sum_{\omega, \boldsymbol{k}', \omega', \sigma'} G(\boldsymbol{k} - \boldsymbol{q}/2, \omega) G(\boldsymbol{k} + \boldsymbol{q}/2, \omega + \Omega)$$

$$\times \Gamma(\boldsymbol{k}, \omega, \boldsymbol{\sigma}; \boldsymbol{k}', \omega', \sigma'; \boldsymbol{q}, i\Omega) G(\boldsymbol{k}' - \boldsymbol{q}/2, \omega')$$

$$\times G(\boldsymbol{k}' + \boldsymbol{q}/2, \omega' + \Omega) \phi_{k'\sigma'}(\boldsymbol{q}, i\Omega), \tag{6.77}$$

and

$$\Gamma(\boldsymbol{k}, \omega, \boldsymbol{\sigma}; \boldsymbol{k}', \omega' \boldsymbol{\sigma}'; \boldsymbol{q}, i\Omega) = \Gamma_0(\boldsymbol{k}, \omega, \boldsymbol{\sigma}; \boldsymbol{k}', \omega', \sigma')$$

$$+ \beta^{-1} \sum_{\boldsymbol{k}'', \omega'' \sigma''} \Gamma_0(\boldsymbol{k}, \omega, \boldsymbol{\sigma}; \boldsymbol{k}'', \omega'', \boldsymbol{\sigma}'') G(\boldsymbol{k}'' - \boldsymbol{q}/2, \omega'')$$

$$\times G(\boldsymbol{k}'' + \boldsymbol{q}/2, \omega'' + \Omega) \Gamma(\boldsymbol{k}'', \omega'', \boldsymbol{\sigma}''; \boldsymbol{k}', \omega', \boldsymbol{\sigma}'; \boldsymbol{q}, i\Omega). \tag{6.78}$$

Now, at a finite temperature real processes can take place and these complicate the discussion beyond what is appropriate for this book. We therefore specialize the discussion to zero temperature and refer the interested reader to the paper by Leggett (1965) for the discussion of the problem at finite temperatures. Since

$$\omega_\nu = (2n+1)\pi/\beta$$

we accomplish the limit $T \to 0$ by making ω a continuous variable and replacing

$$\beta^{-1} \sum_{\omega_\nu} \text{ by } (2\pi)^{-1} \int_{-\infty}^{\infty} d\omega$$

in equations (6.77) and (6.78). The continuation from $n(i\Omega)$ to $n(\Omega)$ can then be carried out by replacing $i\Omega$ by $\Omega + i\delta$.

Consider now the quasi-particle contribution of the product of Green's functions

$$G(\boldsymbol{k}-\boldsymbol{q}/2, \omega)G(\boldsymbol{k}+\boldsymbol{q}/2, \omega - i\Omega)$$

which occurs in equation (6.77) and (6.78). This contribution has the double pole at $i\omega = \epsilon$ when q, Ω become zero, and is greatest when $\omega = \epsilon = 0$. Consequently put $\omega = 0$ and $k = k_{\mathrm{F}}$ in the slowly varying function it multiplies. The quasi-particle contribution to the integral of the product over ω is therefore,

$$\begin{aligned} &\int_{-\infty}^{\infty} \frac{d\omega}{2\pi} a^2 [(i\omega - \epsilon_{k-q/2})(i\omega + \Omega - \epsilon_{k+q/2})]^{-1} \\ &= a^2[\theta(\epsilon_{k+q/2}) - \theta(\epsilon_{k-q/2})]/[\Omega + \epsilon_{k-q/2} - \epsilon_{k+q/2}] \\ &\approx [a^2 \boldsymbol{q} \cdot \boldsymbol{v} \delta(\epsilon_k)]/(\Omega - \boldsymbol{q} \cdot \boldsymbol{v}) \end{aligned} \tag{6.79}$$

for q small. Hence we can write

$$G(\boldsymbol{k}-\boldsymbol{q}/2, \omega)G(\boldsymbol{k}+\boldsymbol{q}/2, \omega - i\Omega) = Q(\boldsymbol{q}, \Omega, \boldsymbol{k}, \omega) + g(\boldsymbol{k}, \omega) \tag{6.80}$$

where $g(\boldsymbol{k}, \omega)$ is the remaining contribution to the product and

$$Q(\boldsymbol{q}, \Omega, \boldsymbol{k}, \omega) = [a^2 \boldsymbol{q} \cdot \boldsymbol{v} \delta(\epsilon_k)\delta(\omega)]/(\Omega - \boldsymbol{q} \cdot \boldsymbol{v}). \tag{6.81}$$

Since the function g is not singular as q, Ω tend to zero these quantities have been put equal to zero in it.

The discussion now becomes very formal. The quantities $\boldsymbol{q}$ and

Ω are parameters in equations (6.77) and (6.78). We shall, therefore, not indicate the dependence of the various quantities on them but take this for granted. Now the equations can be conveniently written using a matrix notation with the variables $\boldsymbol{k}$, ω, $\boldsymbol{\sigma}$ labelling the rows and columns. Thus Γ is the matrix with components

$$\Gamma(\boldsymbol{k}, \omega, \boldsymbol{\sigma}; \boldsymbol{k}', \omega', \boldsymbol{\sigma}')$$

The product of two matrices AB has the components

$$(AB)(\boldsymbol{k}, \omega, \boldsymbol{\sigma}; \boldsymbol{k}', \omega', \boldsymbol{\sigma}')$$

$$= \frac{1}{2\pi}\int \mathrm{d}\omega'' \sum_{\boldsymbol{k}''\sigma''} A(\boldsymbol{k}, \omega, \boldsymbol{\sigma}; \boldsymbol{k}'', \omega'', \boldsymbol{\sigma}'')B(\boldsymbol{k}'', \omega'', \boldsymbol{\sigma}''; \boldsymbol{k}', \omega', \boldsymbol{\sigma}').$$

The functions $g(\boldsymbol{k}, \omega)$, $Q(\boldsymbol{k}, \omega)$ are taken to be the elements of diagonal matrices g and Q. Then equation (6.78) can be written in this notation as

$$\Gamma = \Gamma_0 + \Gamma_0(g + Q)\Gamma. \tag{6.82}$$

To rewrite (6.77), we recognize that the physical quantities are all sums of the occupation numbers of the form

$$P = \sum_{\boldsymbol{k},\sigma} p_{\boldsymbol{k}\sigma} n_{\boldsymbol{k}\sigma}. \tag{6.83}$$

Thus, if we regard $p_{\boldsymbol{k}\sigma}$ and $\phi_{\boldsymbol{k},\sigma}$ as elements of vectors (in the matrix space), independent of ω, equation (6.77) can be replaced by the equation for P,

$$P = p^{+}(Q + g)[1 + \Gamma(Q + g)]\phi. \tag{6.84}$$

If we do wish to obtain the occupation numbers we can use

$$n_{\boldsymbol{k}\sigma} = \partial P/\partial p_{\boldsymbol{k}\sigma}. \tag{6.85}$$

Now we pick out the non-singular part of Γ by writing Γ_1 for the matrix which satisfies

$$\Gamma_1 = \Gamma_0 + \Gamma_0 g \Gamma_1. \tag{6.86}$$

Since

$$\lim_{q\to 0} Q = 0$$

where Ω is finite, Γ_1 satisfies the same equation as $\lim_{\Omega\to 0}\lim_{q\to 0}\Gamma$ (where the order of limits is important) and can be identified with this quantity.

In the following, we assume that the inverse of the matrices we need exist; if they do not there must be physical effects which we have not considered. Then, from equation (6.82)

$$\Gamma = (1-\Gamma_0 g)^{-1}[\Gamma_0 + \Gamma_0 Q\Gamma] = \Gamma_1 + \Gamma_1 Q\Gamma, \qquad (6.87)$$

where in the second step we have used equation (6.86). Straightforward manipulation of the matrices enables us to use equation (6.88) to write (6.84) in the forms

$$\begin{aligned} P &= p^+g(1+\Gamma_1 g)\phi + p^+(1+g\Gamma_1)(1-Q\Gamma_1)^{-1}Q(1+\Gamma_1 g)\phi \\ &= p^+g(1+\Gamma_1 g)\phi + p^+(1+g\Gamma_1)Q(1-\Gamma_1 Q)^{-1}(1+\Gamma_1 g)\phi. \end{aligned} \qquad (6.88)$$

In the last step we have used the identity

$$Q(1-\Gamma_1 Q)^{-1} = (1-Q\Gamma_1)^{-1}Q$$

which is obvious from the expansion of the two sides as well as from the symmetry in the problem.

The first term in equation (6.88) is non-singular and contains the non-quasi-particle contribution to P. It provides the total response in the limit when q tends to zero before Ω, i.e.

$$\lim_{\Omega\to 0}\lim_{q\to 0} P = p^+g(1+\Gamma_1 g)\phi. \qquad (6.89)$$

Now in cases where either the interaction operator, or the operator the response of which is being calculated commutes with the Hamiltonian (in the limit $q \to 0$), this term is quite generally zero. Indeed, according to equation (2.12) specialized to zero temperature, the exact response is

$$P(\Omega) = -\sum_n \left(\frac{P_{0n}\phi_{n0}}{E_n - E_0 - \Omega} + \frac{P_{n0}\phi_{0n}}{E_n - E_0 + \Omega}\right), \qquad (6.90)$$

where the labels n refer to the exact eigenstates of the Hamiltonian with corresponding energies E_n. The label "0" refers to the ground state. The operators P and ϕ are defined by equation (6.83) and

$$\phi = \Sigma\phi_{k\sigma}n_{k\sigma} \qquad (6.91)$$

respectively, and their matrix elements are denoted by P_{0n}, ϕ_{0n}. Now if ϕ (or P) commutes with the Hamiltonian, the ground state of the normal Fermi liquid is an eigenstate of ϕ (or P). Hence only the term $n = 0$ contributes to the sum and $P(\Omega)$ is zero.

Let us now turn to the second term of equation (6.88).

Because of the delta-function factors in Q [see equation (6.81)] the vectors

$$(1+\Gamma_1 g)\phi \quad \text{and} \quad p^+(1+g\Gamma_1)$$

have to be evaluated at $\omega = 0$, $k = k_F$. Written out in full the first of these is

$$\phi_{k\sigma} + (2\pi)^{-1} \int d\omega' \sum_{k'\sigma'} \Gamma_1(k, \omega, \boldsymbol{\sigma}; k', \omega', \sigma') g(k', \omega', \sigma') \phi_{k',\sigma'} \Big|_{\substack{\omega=0 \\ k=k_F}} \tag{6.92}$$

and depends only on the direction of $\boldsymbol{k}$ and on $\boldsymbol{\sigma}$. For the functions $\phi_{k\sigma}$ of importance, $(1, \boldsymbol{\sigma}, \boldsymbol{k}, \boldsymbol{\sigma}.\boldsymbol{k})$ and for isotropic systems, this means that (6.92) must be proportional to $\phi_{k,\sigma}$. Hence

$$(1 + \Gamma_1 g)\phi = R_\phi \phi, \tag{6.93}$$

where R_ϕ is a number which depends on ϕ and gives the renormalization of the interaction with the external field. Similarly

$$p^+(1 + g\Gamma_1) = R_p p^+.$$

Hence we can write for the required response

$$P = R_p R_\phi p^+ Q(1 - \Gamma_1 Q)^{-1} \phi. \tag{6.94}$$

To put this into a form which resembles that obtained from Fermi liquid theory, we rewrite this as

$$P = R_p R_\phi p^+ Q \phi_e$$

where

$$(1 - \Gamma_1 Q)\phi_e = \phi.$$

Written out explicitly these equations are

$$P = a^2 R_p R_\phi \sum_{k,\sigma} p_{k\sigma} \frac{\boldsymbol{q}.\boldsymbol{v}\delta(\epsilon_k)}{\Omega - \boldsymbol{q}.\boldsymbol{v}} \phi_{k\sigma} \tag{6.95}$$

and

$$\phi_{k\sigma e} = \phi_{k\sigma} + \sum_{\sigma',k'} \frac{\boldsymbol{q}.\boldsymbol{v}'\delta(\epsilon_{k'})}{\Omega - \boldsymbol{q}.\boldsymbol{v}'} a^2 \Gamma_1(k, 0, \boldsymbol{\sigma}; k', 0, \sigma') \phi_{k'\sigma' e} \tag{6.96}$$

If we identify

$$a^2 \Gamma_1(k, 0, \boldsymbol{\sigma}; k', 0, \boldsymbol{\sigma}') = f(k, k', \boldsymbol{\sigma}, \boldsymbol{\sigma}'), \tag{6.97}$$

equation (6.96) is precisely (6.52) at $T = 0$ and after the analytic continuation from $\phi(i\Omega)$ to $\phi(\Omega)$. Also equation (6.95) is the result one obtains for P from equation (6.51) except for the factor $a^2 R_p R_\phi$ which takes into account the possible differences between the coupling of the particles and quasi-particles to the external field. In fact, when the coupling is through quantities which are conserved, the quasi-particles and particles will couple in the same way and the renormalization will be unity. We prove that this is so.

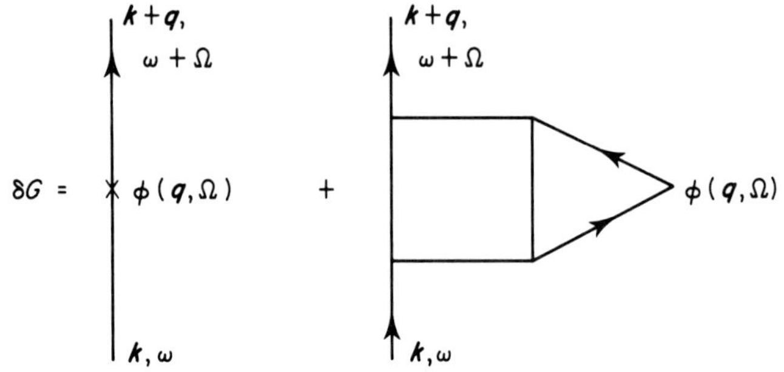

FIG. 6.3. Diagrams representing the first-order change in the one-particle Green's function due to the external potential ϕ.

Consider all the diagrams of Fig. 6.3. These represent the first-order change in G due to the introduction of the external field $\phi(\boldsymbol{q}, \Omega)$. If we denote this change by δG we can write

$$\delta G_\sigma(\boldsymbol{k}, \omega; \boldsymbol{k}+\boldsymbol{q}, \omega+\Omega) = G(k)\,G(k+q)$$

$$\times\ \phi_{k\sigma} + (2\pi)^{-1}\int d\omega' \sum_{k'} \Gamma(\boldsymbol{k}, \omega, \boldsymbol{\sigma}; \boldsymbol{k}', \omega', \boldsymbol{\sigma}')\,G(k')G(k'+q)\phi_{k'\sigma'}. \tag{6.98}$$

We now divide both sides by $G(k)\,G(k+q)$ and let q, Ω tend to zero in such a way that q/Ω tends to zero, the equation becomes

$$-\delta G^{-1}(k) = \phi_{k\sigma} + (2\pi)^{-1}\int d\omega' \sum_{k'} \Gamma_1(\boldsymbol{k}, \omega, \boldsymbol{\sigma}; \boldsymbol{k}', \omega', \boldsymbol{\sigma}')g(k')\phi_{k'\sigma'}, \tag{6.99}$$

where we have used the identity

$$-\delta G^{-1}(k) = G^{-1}(k)\delta G(k)G^{-1}(k). \tag{6.100}$$

The right-hand side of equation (6.99) is the $\boldsymbol{k}$, ω, $\boldsymbol{\sigma}$ component of the vector $(1+\Gamma g)\phi$. When evaluated at $\omega = 0$, $k = k_{\mathrm{F}}$, it yields the quantity required in (6.93). Hence

$$R_\phi \phi_{k\sigma} = -\delta G^{-1}(k, \omega)\Big|_{\substack{\omega=0\\ k=k_{\mathrm{F}}}}. \tag{6.101}$$

Now the three principal quantities the theory deals with are number, spin and momentum for which $\phi_{k\sigma}$ is proportional to 1, $\boldsymbol{\sigma}$ and $\boldsymbol{k}$ respectively. All of these are conserved quantities which can be used to label the states and the creation operators $c_{k\sigma}^{+}$ (as is clear from the notation). Hence the effect of $\phi_{k\sigma}$ in any of these cases is to add energy $\phi_{k\sigma}$ to any state created by $c^+_{k\sigma}$. In the presence of the field ϕ, $G_\sigma(\boldsymbol{k}, \omega)$ is changed to $G_\sigma(\boldsymbol{k}, \omega - \phi_{k\sigma})$.

More formally, the one-particle Green's function, even in the presence of ϕ, is given exactly by equation (2.12) which at absolute zero becomes

$$G_\sigma(\boldsymbol{k}, \omega, \phi) = \sum_n \left[\frac{|c_{k\sigma 0n}|^2}{\tilde{E}_n - \tilde{E}_0 - \omega} - \frac{|c_{k\sigma n0}|^2}{\tilde{E}_n - \tilde{E}_0 + \omega} \right]. \quad (6.102)$$

But for the states for which $c_{k\sigma 0n}$ is non-zero,

$$\tilde{E}_n - \tilde{E}_0 = \phi_{k\sigma}.$$

while for those states for which $c_{k\sigma n0}$ is non-zero

$$\tilde{E}_n - \tilde{E}_0 = -\phi_{k\sigma}.$$

Hence

$$G_\sigma(\boldsymbol{k}, \omega, \phi) = G_\sigma(\boldsymbol{k}, \omega - \phi_{k\sigma})$$

and

$$\delta \bar{G}_\sigma(\boldsymbol{k}, \omega) = -\frac{\partial G_\sigma^{-1}(\boldsymbol{k}, \omega)}{\partial \omega} \phi_{k\sigma}. \quad (6.103)$$

But

$$G_\sigma^{-1}(\boldsymbol{k}, \omega) = \omega - \epsilon_k - \Sigma(\boldsymbol{k}, \omega)$$

$$R_\phi = \left. \frac{\partial G_\sigma^{-1}(\boldsymbol{k}, \omega)}{\partial \omega} \right|_{\substack{\omega = 0 \\ k = k_F}}$$

$$= [1 - \partial \Sigma(k_F, 0)/\partial \omega] = a^{-1}. \quad (6.104)$$

In the last step we have used equation (6.71). Thus if p and ϕ belong to this group of conserved quantities the total renormalization is unity and one has Landau's theory exactly. Because this result stems from the behaviour of conserved quantities one would expect it to be related to the Ward identities discussed in §3.9. The proof based on these identities is given by Nozieres (1964). Leggett (1965) has given a method of calculating R_ϕ for non-conserved quantities.

§6.3 The charged Fermi liquid

We have already seen in Chapter 5 that the long range of the Coulomb force is important in determining the effective interaction of charged particles. This long range invalidates the treatment of the previous section because the assumption that Γ_0 and, hence Γ_1, is well behaved as $q \to 0$ is no longer valid. For example the contribution to Γ_0 from the diagram shown in Fig. 6.4 is simply $V(q)$ which, for the Coulomb

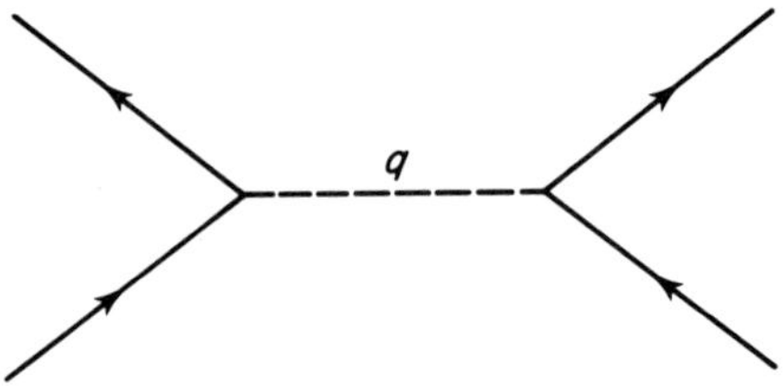

FIG. 6.4. Electron–hole scattering via the Coulomb potential.

interaction, becomes infinite as $q \to 0$. This is in fact the contribution to Γ_0 which does diverge. As we have seen in Chapter 5 diagrams involving this particle–hole scattering are important for yielding the proper long-range behaviour of the Coulomb gas.

Silin (1959) proposed that the way to deal with the effect of the long-range force is to separate out the effect of the average change in the charge distribution. This average gives rise to the macroscopic electric field $\boldsymbol{E}$ which was discussed in §5.1. The remaining interaction arises from fluctuations in the charge density and these are screened and short-range in nature. They will, therefore, give rise to internal fields which can be described by a function $f(\boldsymbol{k}, \boldsymbol{k}')$ as in neutral Fermi liquid theory. There is, then, a total effective field which comprises the external field, the electric field due to the average charge and the short-range field. Couched in terms of quasi-particles the effective Hamiltonian is now

$$H = \sum_{k\sigma} \epsilon_k c_{k\sigma}^+ c_{k\sigma} + \sum_{k,q,\sigma} (U_{k\sigma} + \psi(\boldsymbol{q}) + \phi_{k,q,\sigma}) c_{k\sigma}^+ c_{k+q\sigma}, \tag{6.105}$$

where ϕ is the external potential, and

$$\psi(\boldsymbol{q}) = \sum_{k,\sigma} V(q) \langle c_{k+q\sigma}^+ c_{k\sigma} \rangle, \tag{6.106}$$

where $V(q)$ is the bare Coulomb interaction,

$$V(q) = e^2/\mathscr{V}\epsilon_0 q^2; \tag{6.107}$$

the short-range molecular field $U_{k\sigma}$ is given by

$$U_{k\sigma} = \sum_{k'\sigma'} f(\boldsymbol{k}, \boldsymbol{k}', \boldsymbol{\sigma}, \boldsymbol{\sigma}') \langle c_{k'+q\sigma'}^+ c_{k'\sigma'} \rangle. \tag{6.108}$$

Significant differences from the neutral theory are to be expected for the response to longitudinal fields and for the collective modes.

To see how these arise, let us look at the charge produced by a sinusoidal potential $\phi_q(\Omega)$. As before, the resulting charge density (at absolute zero) is

$$\rho_q = e \sum_{k,\sigma} n_{k\sigma} = 2 \sum_k \frac{\boldsymbol{q}\cdot\boldsymbol{v}\,\delta(\epsilon)}{\Omega - \boldsymbol{q}\cdot\boldsymbol{v}} \phi_{ke}(\boldsymbol{q}, \Omega), \tag{6.109}$$

where the effective potential is now

$$\phi_{ke}(\boldsymbol{q}, \Omega) = \phi_k(\boldsymbol{q}, \Omega) + \sum_{k',\sigma'} \frac{[f(\boldsymbol{k}, \boldsymbol{k}', \sigma, \sigma') + V(q)]\boldsymbol{v}'.\boldsymbol{q}\,\delta(\epsilon')}{\Omega - \boldsymbol{v}'.\boldsymbol{q}} \phi_{k'e}(\boldsymbol{q}, \Omega). \tag{6.110}$$

If the field is static

$$\phi_{ke}(\boldsymbol{q}) = \phi(\boldsymbol{q}) \left[1 + \sum_{k',\sigma'} f(\boldsymbol{k}, \boldsymbol{k}', \sigma, \sigma')\delta(\epsilon') + 2N(0)e^2/\epsilon_0 q^2\right]^{-1} \tag{6.111}$$

We therefore have the same form of screening of the field at large distances but with the screening length now defined in terms of Landau's function which gives the short-range effects of the interaction.

The frequencies of the collective modes are the eigenvalues of

$$\phi_{ke}(\boldsymbol{q}, \Omega) = \sum_{k',\sigma'} \frac{[f(\boldsymbol{k}, \boldsymbol{k}', \sigma, \sigma') + V(q)]\,\boldsymbol{v}'.\boldsymbol{q}\,\delta(\epsilon')}{\Omega - \boldsymbol{v}'.\boldsymbol{q}} \phi_{k'e}(\boldsymbol{q}, \Omega). \tag{6.112}$$

At long wavelengths (small q) the kernel is dominated by $V(q)$. Hence the frequency is the plasma frequency Ω_p given by

$$\Omega^2 = 2N(0)v_F^2 e^2/3\epsilon_0 \tag{6.113}$$

and is unchanged in lowest order. Strictly, equations (6.110) and (6.112) are valid only for small frequency and so cannot be used to determine the plasma frequency. However, the result does show that because of the Coulomb interaction the charged Fermi liquid cannot sustain zero sound.

These results can be derived from microscopic theory in two equivalent ways. In the first we follow Silin's suggestion explicitly. In an external field, the Hamiltonian is

$$H = \sum_{k,\sigma} \epsilon_{k0} c_{k\sigma}^+ c_{k\sigma} + \tfrac{1}{2} \sum_q{}' V(q)\rho_q \rho_{-q} + \sum_{k,q,\sigma} \phi_{kq\sigma} n_{kq\sigma} \tag{6.114}$$

where $V(q)$ may contain a short-range potential as well as the Coulomb interaction. We now rewrite it as

$$H = \sum_{k\sigma} \epsilon_{k0} c_{k\sigma}^{+} c_{k\sigma} + \sum_{k,q,\sigma} [\psi(q) + \phi_{kq\sigma}] n_{kq\sigma}$$

$$+ \tfrac{1}{2} \sum_q{}' V(q)[\rho_q - \langle\rho_q\rangle][\langle\rho_{-q}\rangle - \langle\rho_{-q}\rangle] - \tfrac{1}{2} \sum_q{}' V(q)\langle\rho_q\rangle\langle\rho_{-q}\rangle, \tag{6.115}$$

where $\psi(q)$ is the potential due to the average charge and is given by

$$\psi(q) = V(q)\langle\rho_{-q}\rangle. \tag{6.116}$$

The last term in the Hamiltonian (6.115) is a c-number and does not affect the response. If the third term is ignored we are calculating the one-particle Green's function using the Hartree approximation (see §3.7) in the presence of the field. The third term introduces all corrections not previously included. This means that provided ψ is calculated self-consistently at the end the contributions of Fig. 6.4 to Γ_0 do not appear. Thus the response to the effective potential $\psi + \phi$ can be calculated as in the last section. Hence if we write

$$\Gamma_0(\boldsymbol{k}, \omega, \sigma; \boldsymbol{k}', \omega'\sigma'; \boldsymbol{q}, \Omega) = \Gamma_0'(\boldsymbol{k}, \omega, \boldsymbol{\sigma}; \boldsymbol{k}'\omega', \sigma'; \boldsymbol{q}, \Omega) + V(\boldsymbol{q}), \tag{6.117}$$

the result in matrix notation is

$$P = p'[(Q+g) + (Q+g)\Gamma'(Q+g)](\psi + \phi), \tag{6.118}$$

where

$$\Gamma' = \Gamma_0' + \Gamma_0'(Q+g)\Gamma', \tag{6.119}$$

$$\psi = V[(Q+g) + (Q+g)\Gamma'(Q+g)](\psi + \phi) \tag{6.120}$$

and V is the matrix all the component elements of which are $V(q)$. Equations (6.118) and (6.119) are of the same form as those discussed in the last section and lead to the usual Fermi liquid response to the field $(\psi + \phi)$. Equation (6.120) then leads to (6.110) for the total effective field. The singular potential appears only in equation (6.120).

The alternative method is to use the formally correct equations (6.82) and (6.84) with q and Ω finite, of the last section. If we then split Γ_0 according to equation (6.117) and used the matrix formalism to eliminate Γ_0 and Γ_1 in favour of Γ_0' and Γ' defined by equation (6.117) and (6.119) we again find equations (6.118) and (6.119). Thus this straightforward approach is indeed equivalent to the previous one.

The electrons in normal metals are charged Fermi liquids to which the above theory should apply. Assuming the free-electron model, Silin (1958) has analysed some properties of metals on the basis of the theory. However, because of the crystal lattice, the system is not translationally invariant. This means that there is no simple relationship between the effective mass and Landau's function $f(\boldsymbol{k}, \boldsymbol{k}')$, and also that this function depends on $\boldsymbol{k}$ and $\boldsymbol{k}'$ separately and not simply on $|\boldsymbol{k} - \boldsymbol{k}'|$. Thus many more phenomenological parameters are required to describe electrons in metals and little use has been made of the theory in computations. Nevertheless, it provides a very useful framework for the understanding of the electronic properties of metals and in particular shows why the independent particle model of electrons in metals works so well, despite the strong interaction between the electrons.

References

Abel, W. R., Anderson, A. C. and Wheatley, J. C. (1966). *Phys. Rev. Lett.* **17**, 74.
Doniach, S. and Englesberg, S. (1966). *Phys. Rev. Lett.* **17**, 750.
Emery, V. J. (1968). *Phys. Rev.* **170**, 205.
Landau, L., (1957a). *Sov. Phys. JETP* **3**, 920.
Landau, L., (1957b). *Sov. Phys. JETP* **5**, 101.
Landau, L. (1959). *Sov. Phys. JETP* **8**, 70.
Leggett, A. J. (1965). *Phys. Rev.* **140A**, 1869.
Nozières, P. (1964). "Theory of Interacting Fermi Systems". Benjamin, New York.
Pethick, C. J. (1969). *Phys. Rev.* **177**, 391.
Rice, M. J. (1967). *Phys. Rev.* **159**, 153.
Silin, V. P. (1957). *Sov. Phys. JETP* **6**, 387, 985.
Silin, V. P. (1958). *Sov. Phys. JETP* **6**, 955 and **7**, 486.
Wheatley, J. C. (1975). In "The Helium Liquids" (ed. J. G. M. Armitage and I. E. Farquhar). Academic Press, London and New York.

Problems

1. Draw the first order diagrams which contribute to particle–hole scattering and deduce that to first order

$$\Gamma_1(\boldsymbol{k}, 0, \sigma; \boldsymbol{k}', 0, \sigma') = f(\boldsymbol{k}, \boldsymbol{k}', \sigma, \sigma') = V(\boldsymbol{k} - \boldsymbol{k}')\delta_{\sigma\sigma'} - V(0).$$

   To the same order, show that equation (6.96) is satisfied and illustrate the diagrams which have been summed to produce this result.

2. Draw the five diagrams which contribute to $\Gamma_0(\boldsymbol{k}, 0, \sigma; \boldsymbol{k}', 0, \sigma')$ to second order in the interaction $V(\boldsymbol{q})$.

Chapter 7

Electrons and Phonons

§7.1 Phonons

In this chapter we consider a number of problems concerning phonons and their interaction with electrons where Green's functions have been used with success. Indeed, because the inertia of the ions is important, the interaction between the electrons which is mediated by the phonons is not instantaneous but retarded. This makes Green's functions a particularly useful vehicle for describing them.

In §§7.1–7.3 we consider the phonons alone and look at their scattering of thermal neutrons, a prime method for investigating the spectrum of low lying excitations of a condensed system. The observations are expressed in terms of the structure factor $S(\boldsymbol{k}, \omega)$, which in turn is related to density–density Green's functions. We derive some general relations which the structure factor must satisfy.

In succeeding sections we investigate the interaction between electrons and phonons in a normal metal. We show that the interactions can be studied by a perturbation theory which is a simple generalization of that already given for a two-body interaction in Chapter 3. We show also that, because of the smallness of the ratio of electron mass to ion mass, phonon corrections to the electron–phonon vertex can be ignored. This simplifies applications enormously. Finally we calculate the renormalization of the effective mass of the electrons due to the phonons. In the next chapter we study the effect of the electron–phonon interaction on superconductivity.

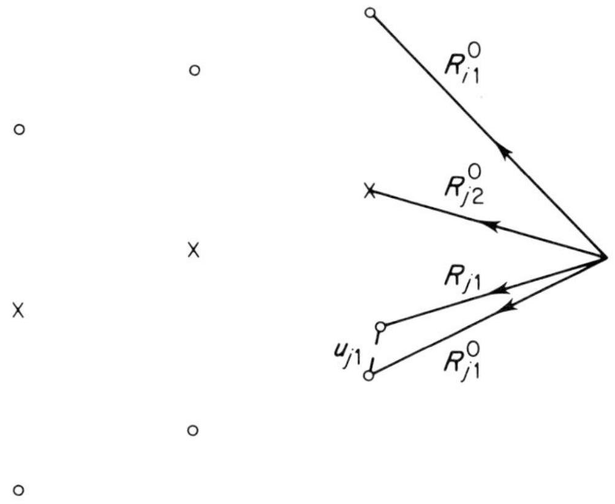

FIG. 7.1. Position vectors used in the description of the motion of the lattice. $R_{j\alpha}^0$ is the equilibrium position vector of an ion of type α, $R_{j\alpha}$ is the corresponding vector for the ion in motion and $u_{j\alpha}$ is its displacement from its equilibrium position.

We assume that the reader has a working knowledge of lattice vibrations and phonons in a crystalline lattice and refer him to standard texts such as Ashcroft and Mermin (1976) or Ziman (1960) for further information. We shall denote by $\boldsymbol{R}_{j\alpha}$ the position of the atom of type α in the jth cell at any time and by $\boldsymbol{R}_{j\alpha}^0$ the equilibrium position of the same atom. The equilibrium position of one of the atoms in each cell is taken as the reference atom and is denoted by $\boldsymbol{R}_j^0$. Thus

$$\boldsymbol{R}_j^0 = \boldsymbol{R}_{j\alpha}^0 \tag{7.1}$$

for one value of α. The displacement of an atom from its equilibrium position is denoted by $\boldsymbol{u}_{j\alpha}$, and is defined by

$$\boldsymbol{u}_{j\alpha} = \boldsymbol{R}_{j\alpha} - \boldsymbol{R}_{j\alpha}^0. \tag{7.2}$$

These vectors are illustrated in Fig. 7.1. The vectors of the reciprocal lattice are denoted by $\boldsymbol{K}$ or $\boldsymbol{K}_m$ where

$$\boldsymbol{K}_m \cdot (\boldsymbol{R}_i^0 - \boldsymbol{R}_j^0) = 2\pi m, \tag{7.3}$$

where m is an integer for all values of i and j.

In classical physics a lattice of N cells containing m atoms in each cell possesses $3mN$ normal modes of vibration. The general motion of the atoms is given by

$$\boldsymbol{u}_{j\alpha}(t) = \frac{1}{\sqrt{(NM_\alpha)}} \sum_{\boldsymbol{k},\lambda} Q_\lambda(\boldsymbol{k}, t)\boldsymbol{e}_\alpha^\lambda(\boldsymbol{k}) \exp(i\boldsymbol{k} \cdot \boldsymbol{R}_j^0), \tag{7.4}$$

where M_α is the mass of the atom at $\boldsymbol{R}^0_{j\alpha}$, $Q_\lambda(\boldsymbol{k}, t)$ gives the amplitude of a mode of wave vector $\boldsymbol{k}$ and has the sinusoidal form,

$$Q_\lambda(\boldsymbol{k}, t) = Q_\lambda(\boldsymbol{k}, 0) \exp[-i\omega_\lambda(\boldsymbol{k})t], \tag{7.5}$$

where $\omega_\lambda(\boldsymbol{k})$ is the frequency of the mode. For each value of $\boldsymbol{k}$ there exist $3m$ distinct modes denoted by different values of λ. The spectrum of $\omega_\lambda(\boldsymbol{k})$ as a function of $\boldsymbol{k}$ has, therefore, $3m$ branches. Of these three are acoustic and have

$$\operatorname*{Lim}_{k\to 0} \omega_\lambda(\boldsymbol{k})/k = \text{constant}. \tag{7.6}$$

The vectors $\boldsymbol{e}^\lambda_\alpha(\boldsymbol{k})$ are the (complex) polarization vectors of the different modes and can be chosen to satisfy the orthogonality relations

$$\sum_\alpha \boldsymbol{e}^\lambda_\alpha(\boldsymbol{k}) \boldsymbol{e}^{\lambda'}_\alpha(\boldsymbol{k}) = \delta_{\lambda,\lambda'}, \tag{7.7}$$

as well as

$$\boldsymbol{e}^\lambda_\alpha(\boldsymbol{k})^* = \boldsymbol{e}^\lambda_\alpha(-\boldsymbol{k}). \tag{7.8}$$

Then we must also have (to ensure that $\boldsymbol{u}$ is real),

$$Q^*_\lambda(\boldsymbol{k}, 0) = Q_\lambda(-\boldsymbol{k}, 0). \tag{7.9}$$

In the quantized form of the theory we write

$$Q_\lambda(\boldsymbol{k}) = [2\omega_\lambda(\boldsymbol{k})]^{-1/2}\{a_\lambda(\boldsymbol{k}) + a_\lambda(-\boldsymbol{k})^+\}, \tag{7.10}$$

where

$$[a_\lambda(\boldsymbol{k}), a_{\lambda'}(\boldsymbol{k}')^+] = \delta_{\lambda\lambda'}\delta_{\boldsymbol{k}\boldsymbol{k}'}, \qquad [a_\lambda(\boldsymbol{k}), a_{\lambda'}(\boldsymbol{k}')] = 0. \tag{7.11}$$

The Hamiltonian for the vibrations is then

$$H = \sum_{\boldsymbol{k},\lambda} \omega_\lambda(\boldsymbol{k})[a^+_\lambda(\boldsymbol{k})a_\lambda(\boldsymbol{k}) + \tfrac{1}{2}]. \tag{7.12}$$

The operators $a^+_\lambda(\boldsymbol{k})$ therefore create Bose excitations with energy $\omega_\lambda(\boldsymbol{k})$ (and momentum $\boldsymbol{k}$). These excitations are the phonons.

Because ions of the lattice, and therefore the phonons, usually interact with other excitations through the lattice displacements, it is often convenient to use special Green's functions, called phonon propagators, which depend on the normal coordinates Q. The temperature phonon propagator is

$$D_{\lambda\lambda'}(\boldsymbol{k}, \boldsymbol{k}', \tau - \tau') = 2[\omega_\lambda(\boldsymbol{k})\omega_{\lambda'}(\boldsymbol{k}')]^{1/2} \langle T\tilde{Q}_\lambda(\boldsymbol{k}, \tau)\tilde{Q}_{\lambda'}(\boldsymbol{k}', \tau')\rangle. \tag{7.13}$$

In the unperturbed system this is non-zero only if $\lambda = \lambda'$, $\boldsymbol{k} = -\boldsymbol{k}'$ and then

$$D_\lambda^0(k, \tau - \tau') = 2\omega(k)\langle T\tilde{Q}_\lambda(k, \tau)\tilde{Q}_\lambda(-k, \tau')\rangle \tag{7.14}$$

$$\begin{aligned} &= (1 - \exp(-\beta\omega_k))^{-1}\{\theta(\tau - \tau') \\ &\quad \times [\exp(-\beta\omega_k + \omega_k(\tau - \tau')) + \exp(\omega_k(\tau' - \tau))] \\ &\quad + \theta(\tau' - \tau)[\exp(\omega_k(\tau - \tau')) \\ &\quad + \exp(\omega_k(\tau' - \tau - \beta))]\}. \end{aligned} \tag{7.15}$$

In the last expression, we have omitted the branch label λ. The Fourier transform is

$$\bar{D}(k, \zeta_l) = 2\omega_k/(\omega_k^2 + \zeta_l^2), \qquad \zeta_l = 2n\pi/\beta. \tag{7.16}$$

This can be continued analytically to the function

$$D(k, \omega) = 2\omega_k/(\omega_k^2 - \omega^2) \tag{7.17}$$

which is analytic in the lower and upper half-planes and has poles at the phonon frequencies $\pm\, \omega_k$.

§7.2 Neutron scattering by the lattice

One important experimental method of examining the properties of the lattice is through neutron scattering. The cross section for elastic scattering is related to the structure of the lattice and that for inelastic scattering is related to the phonon spectrum. In neutron scattering a beam of neutrons with well defined momentum $\boldsymbol{k}$ is fired at the target and the spectrum of neutrons detected in various scattering directions is analysed. In this way the differential scattering cross section is obtained as a function of energy and scattering angle. In the case of scattering by a crystal, the interaction of the neutrons with the ions is weak and only single scattering by ions need be included.

According to standard quantum theory (Messiah 1961) the differential cross section for the scattering of neutrons from a state of momentum $\boldsymbol{k}$ to one of momentum $\boldsymbol{k}'$ per unit solid angle and per unit energy range is given by

$$\frac{d^2\sigma}{d\epsilon d\Omega} = \frac{k'}{k}\left(\frac{M}{2\pi}\right)^2 \sum_{i,f} |\langle \boldsymbol{k}'|\langle f|H|i\rangle|\boldsymbol{k}\rangle|^2\, \delta(\omega + \epsilon_i - \epsilon_f)\exp(-\beta\epsilon_i)Z^{-1}\ . \tag{7.18}$$

Here, H is the term of the Hamiltonian which describes the interaction between a neutron and the target, M is the mass of the neutron, $|i\rangle$ and $|f\rangle$ are the initial and final states of the target and

ϵ_i and ϵ_f are the corresponding energies. The loss in energy of the neutron is

$$(k^2/2M) - (k'^2/2M) = \epsilon_f - \epsilon_i. \tag{7.19}$$

In equation (7.18) we have ignored the spins of neutrons and ions. However, the result should be averaged over these spins, a step we shall take later on.

The interaction between neutrons and ions is of short range compared with the wavelength of the neutrons. This can therefore be treated as a delta-function and we can write

$$H = \sum_{j,\alpha} (2\pi/M) b_{j\alpha} \delta(\boldsymbol{r} - \boldsymbol{R}_{j\alpha}),$$

where $b_{j\alpha}$ is the scattering length for scattering by the ion at $\boldsymbol{R}_{j\alpha}$. Then

$$\langle \boldsymbol{k}'|H|\boldsymbol{k}\rangle = (2\pi/M) \sum_{j,\alpha} b_{j\alpha} \int d^3r \exp(-i\boldsymbol{k}'.\boldsymbol{r})\delta(\boldsymbol{r} - \boldsymbol{R}_{j\alpha}) \exp(i\boldsymbol{k}.\boldsymbol{r})$$

$$= (2\pi/M) \sum_{j,\alpha} b_{j\alpha} \exp(i\boldsymbol{K}.\boldsymbol{R}_{j\alpha}), \tag{7.20}$$

where

$$\boldsymbol{K} = \boldsymbol{k} - \boldsymbol{k}'$$

is the momentum transferred from neutron to ion.

We have to average the square of the modulus of (7.20) over spins. Now the spin dependence is in the scattering lengths $b_{j\alpha}$. Hence we have to perform the averages of $b^*_{j'\alpha'}b_{j\alpha}$. For non-magnetic materials spins on different sites are uncorrelated and

$$\overline{b^*_{j'\alpha'}b_{j\alpha}} = \overline{b^*_{j'\alpha'}}\,\overline{b_{j\alpha}} = b^*_{\alpha'}b_{\alpha}, \qquad j', \alpha' \neq j, \alpha, \tag{7.21}$$

where we allow for the fact that different kinds of atoms will have different scattering lengths. On the other hand, for $j', \alpha' = j, \alpha$ we have

$$\overline{b^*_{j\alpha}b_{j\alpha}} = |b^2_{\alpha}.| \tag{7.22}$$

We can therefore write

$$\overline{b^*_{j'\alpha'}b_{j\alpha}} = \delta_{jj'}\delta_{\alpha\alpha'}(|b^2_{\alpha}| - |b_{\alpha}|^2) + b^*_{\alpha'}b_{\alpha}. \tag{7.23}$$

Correspondingly the scattering cross section can be written as the sum of two parts referred to as the *coherent* and *incoherent* cross sections. Thus

$$\frac{d^2\sigma}{d\epsilon d\Omega} = \left(\frac{d^2\sigma}{d\epsilon d\Omega}\right)_{\text{coh}} + \left(\frac{d^2\sigma}{d\epsilon d\Omega}\right)_{\text{inc}}, \tag{7.24}$$

where

$$\left(\frac{d^2\sigma}{d\epsilon d\Omega}\right)_{\text{coh}} = \frac{k'}{k}\sum_{j\alpha}(|\overline{b_\alpha^2}| - |b_\alpha|^2)\sum_{i,f}|\langle f|\exp(i\boldsymbol{K}\cdot\boldsymbol{R}_\alpha)|i\rangle|^2 \times \delta(\omega+\epsilon_i-\epsilon_f)(\exp(-\beta\epsilon_i)/Z) \tag{7.25}$$

and

$$\left(\frac{d^2\sigma}{d\epsilon d\Omega}\right)_{\text{inc}} = \frac{k'}{k}\sum_{i,f}\left|\sum_{j,\alpha} b_\alpha\langle f|\exp(i\boldsymbol{K}\cdot\boldsymbol{R}_{j\alpha})|i\rangle\right|^2 \times \delta(\omega+\epsilon_i-\epsilon_f)(\exp(-\beta\epsilon_i)/Z). \tag{7.26}$$

If we write these results in terms of the displacements of the ions from their equilibrium positions $\boldsymbol{R}_{j\alpha}^0$ we find

$$\left(\frac{d^2\sigma}{d\epsilon d\Omega}\right)_{\text{coh}} = \frac{k'}{k}\sum_{j\alpha}(|\overline{b_\alpha^2}| - |b_\alpha|^2)\sum_{i,f}|\langle f|\exp(i\boldsymbol{K}\cdot\boldsymbol{u}_{j\alpha})|i\rangle|^2 \times \delta(\omega+\epsilon_i-\epsilon_f)(\exp(-\beta\epsilon_i)/Z) \tag{7.27}$$

and

$$\left(\frac{d^2\sigma}{d\epsilon d\Omega}\right)_{\text{inc}} = \frac{k'}{k}\sum_{i,f}\left|\sum_{j,\alpha} b_\alpha\exp(i\boldsymbol{K}\cdot\boldsymbol{R}_{j\alpha}^0)\langle f|\exp(i\boldsymbol{K}\cdot\boldsymbol{u}_{j\alpha})|i\rangle\right|^2 \times \delta(\omega+\epsilon_i-\epsilon_f)(\exp(-\beta\epsilon_i)/Z). \tag{7.28}$$

We see, therefore, that the coherent cross section does not depend explicitly on the equilibrium positions of the ions.

We now specialize these results to the case of the Bravais lattice with one kind of atom and refer the interested reader to Marshall and Lovesey (1971) for the generalization to other cases. There is then only one scattering length b; the cross sections become,

$$\left(\frac{d^2\sigma}{d\epsilon d\Omega}\right)_{\text{coh}} = \frac{k'}{k}(|\overline{b^2}| - |b|^2)\sum_{j}\sum_{i,f}|\langle f|\exp(i\boldsymbol{K}\cdot\boldsymbol{u}_j)|i\rangle|^2 \times (\exp(-\beta\epsilon_i)/Z)\delta(\omega+\epsilon_i-\epsilon_f)$$

and

$$\left(\frac{d^2\sigma}{d\epsilon d\Omega}\right)_{\text{inc}} = \frac{k'}{k}|b|^2\sum_{i,f}\left|\sum_{j}\exp(i\boldsymbol{K}\cdot\boldsymbol{R}_j^0)\langle f|\exp(i\boldsymbol{K}\cdot\boldsymbol{u}_j)|i\rangle\right|^2 \times \delta(\omega+\epsilon_i-\epsilon_f)(\exp(-\beta\epsilon_i)/Z). \tag{7.29}$$

These results are commonly written in the form

$$\left(\frac{d^2\sigma}{d\epsilon d\Omega}\right)_{\text{coh}} = N\frac{k'}{k}(|\overline{b^2}| - |b|^2)\,S(\boldsymbol{K},\omega) \tag{7.30}$$

and

$$\left(\frac{d^2\sigma}{d\epsilon d\Omega}\right)_{\text{inc}} = N\frac{k'}{k}|b|^2 S_i(\boldsymbol{K}, \omega), \tag{7.31}$$

where the structure factors $S(\boldsymbol{K}, \omega)$ and $S_i(\boldsymbol{K}, \omega)$ depend on the properties of the phonons only.

These structure factors are, in fact, correlation functions as we can reveal as follows. We consider the time Fourier transform of $S_i(\boldsymbol{K}, \omega)$,

$$\begin{aligned}
S_i(\boldsymbol{K}, t) &= (2\pi N)^{-1} \int_{-\infty}^{\infty} d\omega \exp(-i\omega t) S_i(\boldsymbol{K}, \omega) \\
&= (2\pi NZ)^{-1} \sum_{i,f} \sum_{jl} \langle i|\exp(-i\boldsymbol{K}\cdot\boldsymbol{R}_j)|f\rangle\langle f|\exp(i\boldsymbol{K}\cdot\boldsymbol{R}_l)|i\rangle \\
&\quad \times \exp[i\epsilon_i t - i\epsilon_f t - \beta\epsilon_i] \\
&= (2\pi NZ)^{-1} \sum \langle i|\exp(-\beta H)\exp(iHt) \\
&\quad \times \sum_j \exp(-i\boldsymbol{K}\cdot\boldsymbol{R}_j)\exp(-iHt)|f\rangle\langle f|\sum_l \exp(i\boldsymbol{K}\cdot\boldsymbol{R}_l)|i\rangle \\
&= (2\pi N)^{-1}\left\langle \sum_j \exp(-i\boldsymbol{K}\cdot\boldsymbol{R}_j(t)) \sum_l \exp(i\boldsymbol{K}\cdot\boldsymbol{R}_l)\right\rangle .
\end{aligned} \tag{7.32}$$

This is a correlation function (cf. §2.2) which is related to the density–density correlation function. In fact the particle density is

$$\rho(\boldsymbol{r}) = \sum_j \delta(\boldsymbol{r} - \boldsymbol{R}_j), \tag{7.33}$$

with space Fourier transform

$$\begin{aligned}
\rho_{\boldsymbol{k}} &= \int d^3r \exp(-i\boldsymbol{k}\cdot\boldsymbol{r}) \sum_j \delta(\boldsymbol{r} - \boldsymbol{R}_j) \\
&= \sum_j \exp(-i\boldsymbol{k}\cdot\boldsymbol{R}_j)
\end{aligned} \tag{7.34}$$

Hence

$$S_i(\boldsymbol{K}, t) = (2\pi N)^{-1}\langle\rho_{\boldsymbol{K}}(t)\rho_{-\boldsymbol{K}}(0)\rangle. \tag{7.35}$$

Similarly

$$S(\boldsymbol{K}, t) = (2\pi N)^{-1}\sum_j \langle\exp(-i\boldsymbol{K}\cdot\boldsymbol{u}_j(t))\exp(i\boldsymbol{K}\cdot\boldsymbol{u}_j)\rangle. \tag{7.36}$$

A full discussion of these results is very technical and for this the reader is referred to a specialist text (Marshall and Lovesey 1971). In this book we shall show how some of the limiting approximations arise. In the extreme limit of heavy ions, their recoil can be ignored and, therefore can be set equal to zero. This is the rigid lattice limit for the scattering and yields

$$S_i(K, t)|_{\text{rigid}} = (2\pi N)^{-1} \left| \sum_j \exp(iK \cdot R_j^0) \right|^2, \tag{7.37}$$

$$S(K, t)|_{\text{rigid}} = 1/2\pi. \tag{7.38}$$

The first of these is the standard Bragg scattering law. In both cases the structure factor S is independent of time and the scattering is elastic.

One can obtain an improved elastic scattering law, which takes account of the thermal and zero-point motion of the ions by replacing the structure factors by their limits as $t \to \infty$. In this limit the functions $\rho_K(t)$ and $\rho_{-K}(0)$ are independent and

$$S_i(K, t) \approx S_i(K, \infty) = (2\pi N)^{-1} \langle \rho_K(t) \rangle \langle \rho_{-K}(0) \rangle = (2\pi)^{-1} |\langle \rho_K(0) \rangle|^2 \tag{7.39}$$

Similarly

$$S(K, t) \approx S(K, \infty) = (2\pi N)^{-1} \sum_j |\langle \exp(iK \cdot u_j) \rangle|^2, \tag{7.40}$$

and the averages are taken over the unperturbed phonon Hamiltonian. The displacement vectors are all linear in the coordinates of harmonic oscillator variables. In this case it can be shown (Messiah 1961) that

$$\langle \exp(iK \cdot u_j) \rangle = \exp[-W(K)], \tag{7.41}$$

where

$$W(K) = \tfrac{1}{2} \langle (K \cdot u_j)^2 \rangle \tag{7.42}$$

and is independent of j. Hence.

$$S_i(K, \infty) = (2\pi N)^{-1} \left| \sum_i \exp(iK \cdot R_j^0) \right|^2 \exp[-2W(K)], \tag{7.43}$$

$$S(K, \infty) = (2\pi)^{-1} \exp[-2W(K)]. \tag{7.44}$$

Thus the effect of the thermal and zero-point motion is to reduce the structure factors by the factor $\exp[-2W(K)]$, which is known as the Debye–Waller factor. From the properties of harmonic oscillators it is not difficult to show that

$$W(K) = \frac{1}{4NM} \sum_{k,\lambda} \frac{|K \cdot e^\lambda(k)|^2}{\omega_\lambda(k)} [2n_\lambda(k) + 1]. \tag{7.45}$$

Although $W(K)$ becomes small as the momentum transfer tends to zero, it significantly improves the agreement between the theoretical and experimental cross sections.

The leading corrections to the results (7.43), (7.44) at low momentum transfers introduce single phonon emission and absorption processes. These terms arise as follows. Again exploiting the properties of harmonic oscillator wave functions (Messiah 1961) we have

$$\begin{aligned}
&\langle \exp(-i\boldsymbol{K}\cdot\boldsymbol{u}_j(t)\exp(i\boldsymbol{K}\cdot\boldsymbol{u}_l)\rangle \\
&= \langle \exp\{i\boldsymbol{K}\cdot[\boldsymbol{u}_l-\boldsymbol{u}_j(t)]\}\exp\{\tfrac{1}{2}[\boldsymbol{K}\cdot\boldsymbol{u}_j(t),\boldsymbol{K}\cdot\boldsymbol{u}_l]\}\rangle \\
&= \langle \exp\{i\boldsymbol{K}\cdot[\boldsymbol{u}_l-\boldsymbol{u}_j(t)]\}\rangle\exp\{\tfrac{1}{2}[\boldsymbol{K}\cdot\boldsymbol{u}_j(t),\boldsymbol{K}\cdot\boldsymbol{u}_l]\} \qquad (7.46)
\end{aligned}$$

since the commutator is a c-number. We can once more use the identity (7.41) to write

$$\begin{aligned}
&\langle \exp(-i\boldsymbol{K}\cdot\boldsymbol{u}_j(t))\exp(i\boldsymbol{K}\cdot\boldsymbol{u}_l)\rangle \\
&= \exp\{-\tfrac{1}{2}\langle\{\boldsymbol{K}\cdot[\boldsymbol{u}_l-\boldsymbol{u}_j(t)]\}^2\rangle\}\exp\{\tfrac{1}{2}[\boldsymbol{K}\cdot\boldsymbol{u}_j(t),\boldsymbol{K}\cdot\boldsymbol{u}_l]\} \\
&= \exp\{-\tfrac{1}{2}\langle[(\boldsymbol{K}\cdot\boldsymbol{u}_l)^2+(\boldsymbol{K}\cdot\boldsymbol{u}_j(t))^2-2\boldsymbol{K}\cdot\boldsymbol{u}_j(t)(\boldsymbol{K}\cdot\boldsymbol{u}_l)]\rangle \\
&= \exp[-2W(\boldsymbol{K})]\exp\langle[\boldsymbol{K}\cdot\boldsymbol{u}_j(t)(\boldsymbol{K}\cdot\boldsymbol{u}_l)]\rangle. \qquad (7.47)
\end{aligned}$$

The result depends on the correlation function $\langle\boldsymbol{K}\cdot\boldsymbol{u}_j(t)(\boldsymbol{K}\cdot\boldsymbol{u}_l)\rangle$ which can again be evaluated using the properties of harmonic oscillators. The result is

$$\begin{aligned}
\langle\boldsymbol{K}\cdot\boldsymbol{u}_j(t)(\boldsymbol{K}\cdot\boldsymbol{u}_l)\rangle &= \frac{1}{2NM}\sum_{\boldsymbol{k},\lambda}\frac{|\boldsymbol{K}\cdot\boldsymbol{e}^\lambda(\boldsymbol{k})|^2}{\omega_\lambda(\boldsymbol{k})} \\
&\quad\times\{\exp[i\boldsymbol{k}\cdot(\boldsymbol{R}_j^0-\boldsymbol{R}_l^0)]\exp[-i\omega_\lambda(\boldsymbol{k})t][1+n^\lambda(\boldsymbol{k})] \\
&\quad+\exp[-i\boldsymbol{k}\cdot(\boldsymbol{R}_j^0-\boldsymbol{R}_l^0)]\exp[i\omega_\lambda(\boldsymbol{k})t]\,n^\lambda(\boldsymbol{k})\}. \\
&\qquad (7.48)
\end{aligned}$$

The cross section is proportional to the frequency-dependent structure factor given by

$$\begin{aligned}
S_i(\boldsymbol{K},\omega) &= \int_{-\infty}^{\infty} dt\,\exp(i\omega t)\,S_i(\boldsymbol{K},t) \\
&= (2\pi N)^{-1}\int_{-\infty}^{\infty} dt\,\exp(i\omega t)\sum_{j,l}\exp[i\boldsymbol{K}\cdot(\boldsymbol{R}_j^0-\boldsymbol{R}_l^0)] \\
&\quad\times\exp[-2W(\boldsymbol{K})]\exp[\langle\boldsymbol{K}\cdot\boldsymbol{u}_j(t)(\boldsymbol{K}\cdot\boldsymbol{u}_l)\rangle]. \qquad (7.49)
\end{aligned}$$

At low momentum transfer this can be evaluated if the last

exponential is expanded in powers of the correlation function. The zero order term yields $S_i(K, \infty)$ and the elastic cross section. The first-order term is

$$S_i^1(K, \omega) = (2\pi N)^{-1} \int_{-\infty}^{\infty} dt \exp(i\omega t) \sum_{j,l} (2NM)^{-1} \exp(-2W(K))$$
$$\times \exp[iK(R_l^0 - R_j^0)] \sum_{k,\lambda} \frac{|K \cdot e^\lambda(k)|^2}{\omega_\lambda(k)}$$
$$\times \{\exp[ik \cdot (R_j^0 - R_l^0)] \exp[-i\omega_\lambda(k)t] [n^\lambda(k) + 1]$$
$$+ \exp[-ik \cdot (R_j^0 - R_l^0)] \exp[i\omega_\lambda(k)t] n^\lambda(k)\}.$$

The sums over j and l can be performed and lead to conservation of momentum up to a vector g of the reciprocal lattice. The integration over time leads to conservation of energy. Then

$$S_i^1(K, \omega) = (2MN)^{-1} \exp[-2W(K)] \sum_g \sum_{k,\lambda} \frac{|K \cdot e^\lambda(k)|^2}{\omega_\lambda(k)}$$
$$\times \{[n^\lambda(k) + 1] \delta(K - k - g) \delta[\omega - \omega_\lambda(k)]$$
$$+ n^\lambda(k) \delta(K + k - g) \delta[\omega + \omega_\lambda(k)]\}. \tag{7.50}$$

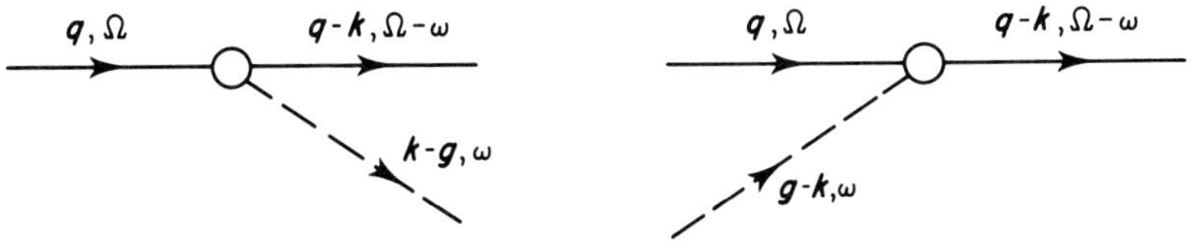

FIG. 7.2. Diagrams illustrating the inelastic scattering of neutrons by a crystal with the emission or absorbtion of a phonon. Momentum is conserved only to within a vector g of the reciprocal lattice.

The delta-functions in the first term in the final parenthesis show that this term refers to the emission of a single phonon. The second term refers to the absorption a single phonon. These processes are illustrated in Fig. 7.2.

Since it is possible to use monochromatic beams of neutrons and to measure the energies of detected neutrons, one can use the result (7.50) to determine the phonon spectrum. The first term in equation (7.50) corresponds to processes in which the neutron loses energy. It is, therefore distinguished from the elastic term. The delta functions ensure that the term is zero unless the momentum loss K and energy loss ω of the neutron obey the laws

$$K - k - g = 0, \tag{7.51}$$

$$\omega - \omega_\lambda(k) = 0. \tag{7.52}$$

Although equation (7.51) can be satisfied with any reciprocal lattice vector g, the values of k must lie in the first Brillouin zone. For a given value of k this fixes g. Thus four conditions have to be satisfied for the term to be non-zero. In practice the incident momentum of the neutron is fixed and the observation is made at a definite angle. This fixes the direction of the scattered neutron. Equations (7.51) and (7.52) then determine k and the energy of the scattered neutron. Thus in any given direction the scattered neutrons have discrete energies one for each value of λ. For each of these energies the equations (7.51), (7.52) determine k and $\omega_\lambda(k)$. In this way the phonon spectrum can be determined.

In practice, the later terms in the exponential series will produce a background on top of the discrete spectrum in each direction. The later terms arise from multi-phonon processes and do not themselves possess a discrete spectrum. Thus the first-order terms can be picked out from the background. This is a powerful technique for determining phonon spectra. An example of the results is shown in Fig. 7.3.

Phonons are not the only excitations of solids whose spectra can be determined by means of neutrons. Any boson excitation which can interact with neutrons can have its spectrum determined in this way. Spin waves (see Chapter 10) and phonons and rotons in liquid helium (see Chapter 9) are other examples.

§7.3 Structure factor, Green's functions and sum rule

The structure factor is the Fourier transform of a correlation function and so, according to the discussion in §2.5, is related to the spectral function. In fact the structure factor is not a symmetrical correlation function and so the relationship is not quite the same as that given in §2.5. When the operators are densities, the spectral function is

$$A(k, \omega) = Z^{-1}(1 - \exp(-\beta\omega)) \times \sum_{m,n} \exp(-\beta\epsilon_m)\, \rho_{kmn}\rho_{-knm}\, \delta(\omega - \epsilon_n + \epsilon_m). \tag{7.53}$$

Comparing this with equations (7.32) and (7.34) we see that

$$S_i(k, \omega) = N^{-1}(1 - \exp(-\beta\omega))^{-1} A(k, \omega). \tag{7.54}$$

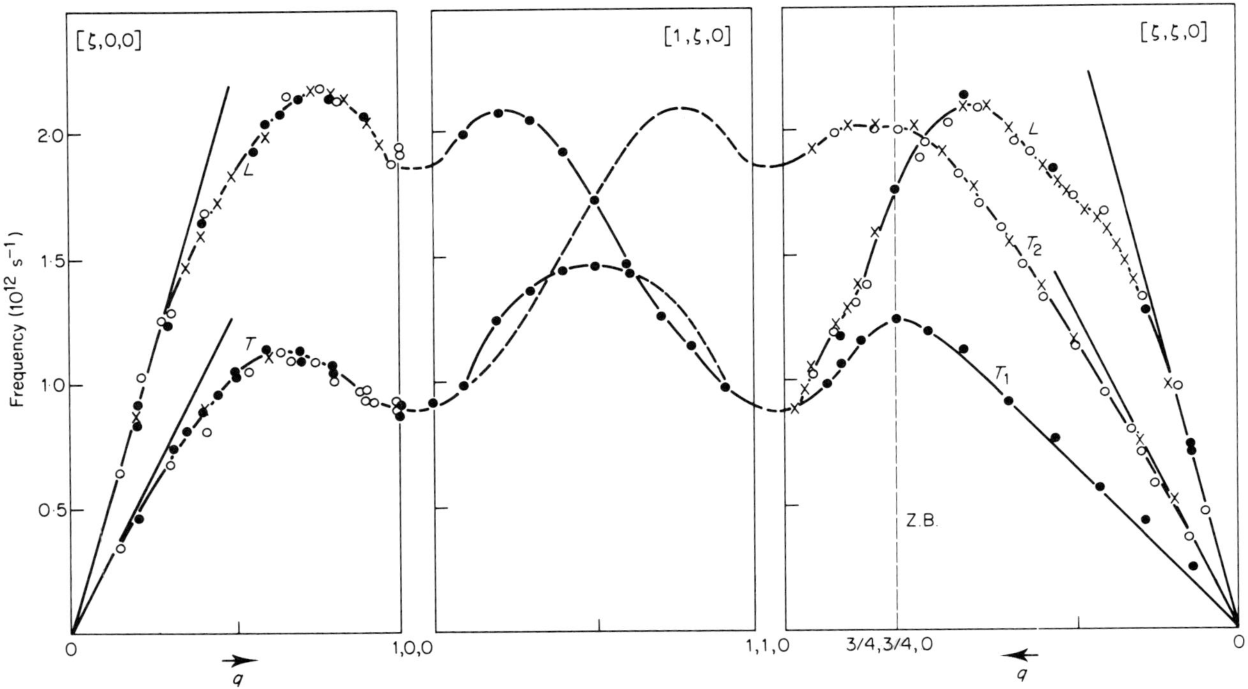

FIG. 7.3. The dispersion curves in three crystallographic directions for lead at 100 K as obtained by neutron scattering. The straight lines through the origin give the initial slopes as calculated from the elastic constants (after Brockhouse *et al.* 1962).

However,

$$A(\boldsymbol{k}, -\omega) = -A(\boldsymbol{k}, \omega), \tag{7.55}$$

whence

$$S_i(\boldsymbol{k}, -\omega) = \exp(-\beta\omega) S_i(\boldsymbol{k}, \omega). \tag{7.56}$$

For $\omega > 0$, $S(\boldsymbol{k}, \omega)$ gives the probability for scattering of neutrons with absorption of energy ω by the phonons. $S(\boldsymbol{k}, -\omega)$ similarly yields the probability for the inverse process. In thermodynamic equilibrium, the rates of these processes must balance. Since the ratio of the probability for occupation of the higher energy phonon states to that for occupation of the lower states is $\exp(-\beta\omega)$, equation (7.56) expresses this detailed balance.

We can use the connection between the structure factor and the spectral function to obtain a sum-rule for the structure factor. We start from the structure factor, $B(\boldsymbol{k}, \omega)$ which is related to a density–current-density correlation function and is defined by

$$B(\boldsymbol{k}, \omega) = Z^{-1}(1 - \exp(-\beta\omega)) \times \sum_{m,n} \exp(-\beta\epsilon_n) \boldsymbol{k} \cdot \boldsymbol{j}_{kmn} \rho_{-knm} \, \delta(\omega - \epsilon_n + \epsilon_m), \tag{7.57}$$

where $\rho(\boldsymbol{r})$ and $\boldsymbol{j}(\boldsymbol{r})$ are related by the conservation of charge,

$$\partial\rho/\partial t + \operatorname{div} \boldsymbol{j} = 0. \tag{7.58}$$

Thus, in terms of spatial Fourier transforms,

$$\boldsymbol{k} \cdot \boldsymbol{j}_k = i\partial\rho_k/\partial t = -[H, \rho_k] \tag{7.59}$$

and

$$\boldsymbol{k} \cdot \boldsymbol{j}_{kmn} = -\langle m | [H, \rho_k] | n \rangle = -(\epsilon_m - \epsilon_n)\rho_{kmn}. \tag{7.60}$$

In view of the delta-function in (7.57) this can be replaced by $\omega\rho_{kmn}$. Hence

$$B(\boldsymbol{k}, \omega) = \omega A(\boldsymbol{k}, \omega). \tag{7.61}$$

However according to equation (2.15), $B(\boldsymbol{k}, \omega)$ obeys the sum rule

$$\int_{-\infty}^{\infty} d\omega \, B(\boldsymbol{k}, \omega) = \langle [\boldsymbol{k} \cdot \boldsymbol{j}_k, \rho_{-k}] \rangle. \tag{7.62}$$

Now the potential energy depends only on the coordinates and not the momenta and consequently commutes with ρ_k. Thus

$$\boldsymbol{k} \cdot \boldsymbol{j}_k = -\left[\sum_j (p_j^2/2M), \sum_i \exp(i\boldsymbol{k} \cdot \boldsymbol{R}_j^0 + i\boldsymbol{k} \cdot \boldsymbol{u}_j)\right]$$

$$= -(2M)^{-1} \sum_i [\exp(i\boldsymbol{k} \cdot \boldsymbol{R}_j^0 + i\boldsymbol{k} \cdot \boldsymbol{u}_j)\boldsymbol{k} \cdot \boldsymbol{p}_i - \boldsymbol{k} \cdot \boldsymbol{p}_i \times \exp(i\boldsymbol{k} \cdot \boldsymbol{R}_j^0 + i\boldsymbol{k} \cdot \boldsymbol{u}_i)]$$

and

$$[\boldsymbol{k} \cdot \boldsymbol{j}_{\boldsymbol{k}}, \rho_{-\boldsymbol{k}}] = (k^2/M) \sum_i 1 = Nk^2/M, \tag{7.63}$$

where N is the number of ions. In deriving (7.63) we have used the result that

$$\left[p_{i\alpha}, \sum_j \exp(i\boldsymbol{k} \cdot \boldsymbol{R}_j^0 + i\boldsymbol{k} \cdot \boldsymbol{u}_j)\right]$$

$$= -i(\partial/\partial u_{i\alpha}) \sum_j \exp(i\boldsymbol{k} \cdot \boldsymbol{R}_j^0 + i\boldsymbol{k} \cdot \boldsymbol{u}_j)$$

$$= k_\alpha \exp(i\boldsymbol{k} \cdot \boldsymbol{R}_i^0 + i\boldsymbol{k} \cdot \boldsymbol{u}_j). \tag{7.64}$$

From equations (7.61), (7.62) and (7.63) we obtain

$$\int_{-\infty}^{\infty} \omega A(\boldsymbol{k}, \omega) d\omega = Nk^2/M. \tag{7.65}$$

If $A(\boldsymbol{k}, \omega)$ is replaced by $S(\boldsymbol{k}, \omega)$ through the use of (7.54) and detailed balance is also used, equation (7.65) leads to

$$\int_{-\infty}^{\infty} \omega S_i(\boldsymbol{k}, \omega) d\omega = k^2/2M. \tag{7.66}$$

This is a useful exact rule for checking both experimental results and theoretical models. It is known as the f-sum rule.

§7.4 The interaction between electrons and phonons in a metal

Green's functions have proved very useful in illuminating the subtleties of this problem. It is necessary to include the effect of the long range of the Coulomb interaction between the electrons (see Chapter 5) and between the electrons and the ions and also the Fermi liquid effects discussed in Chapter 6. Moreover, because the electrons have low excitation energies, the adiabatic approximation (Ziman 1960) which enables the electronic and lattice motions in insulators and semi-conductors to be separated is not valid. Nevertheless, the ratio of electron to ion mass m/M is still small in metals and this does allow some simplification.

If the interaction between the electrons and the ions is ignored, the electrons can be described by the charged Fermi liquid theory as in §6.3. This means that the single-particle Green's function for electrons is

$$G(\boldsymbol{k}, \omega) = [Z_c(\omega - \epsilon_{\boldsymbol{k}})]^{-1}, \tag{7.67}$$

where Z_c is the wave function renormalization due to the Coulomb interaction between the electrons and ϵ_k is the renormalized quasi-particle energy. In the same approximation the ions will vibrate about their equilibrium positions, interacting through the long-range Coulomb interaction and short-range forces. At long wavelengths the behaviour is dominated by the Coulomb interaction and the ions oscillate as a plasma, with the frequency

$$\Omega = (Z^2 e^2 n / M \epsilon_0)^{1/2}, \tag{7.68}$$

where n is the number density of ions. For a lattice of identical ions M is the mass of one ion and Ze its charge. In other cases these are appropriate averages.

The effect of the interaction of the ions with the electrons is two-fold. If the ions were held in their equilibrium positions, they would form a crystal lattice in which the electrons would move. The electron wave functions would change to Bloch wave functions and the Green's functions would alter correspondingly. This is an effect which can be allowed for by the use of pseudopotential theory (see, for example, Ashcroft and Mermin 1976). Two consequences follow. The energies ϵ_k change and reveal the usual band structure with energy gaps at the edge of the Brillouin zone. The wave functions become Bloch functions and the single-particle Green's functions

$$\langle T c_k c_{k'}^{+}(\tau) \rangle$$

will be non-zero wherever $(\boldsymbol{k} - \boldsymbol{k}')$ is a vector of the reciprocal lattice. For the most part we shall need only the diagonal Green's function $(\boldsymbol{k} = \boldsymbol{k}')$ and this can still be represented phenomenologically by equation (7.67) where both Fermi liquid and band structure effects are included. For a fuller treatment of this aspect the reader is referred to the review article of Scalapino (1969).

The second effect of the electron–ion interaction arises from the motion of the ions. This is dynamic and cannot be accounted for simply by a redefinition of electron parameters. Since the electrons and ions are charged, the interaction is long range. We know however, from the discussion of Chapter 5 that the interaction between one electron and the ions will be effectively screened by all the electrons. In this section and succeeding ones we wish to discuss this screening and the effects which remain after the screening has been included.

Formally, if we assume that the interaction between the electrons and the lattice is due to a potential, there will be an interaction term H_I in the Hamiltonian, where

$$H_I = \sum_{i,\alpha,\sigma} \int d^3r\, \psi_\sigma^+(r)\psi_\sigma(r)V(r-R_{i\alpha}). \tag{7.69}$$

Here, $V(r-R_{i\alpha})$ is the potential due to the ion, assumed rigid, which is at $R_{i\alpha}$. For small amplitudes of vibration H_I can be expanded in powers of $u_{i\alpha}$. Then

$$\begin{aligned} H_I = & \sum_{i,\alpha,\sigma} \int d^3r\, \psi_\sigma^+(r)\psi_\sigma(r)V(r-R_{i\alpha}^0) \\ & + \sum_{i,\alpha,\sigma} \int d^3r\, \psi_\sigma^+(r)\psi_\sigma(r)u_{i\alpha} . \nabla_{R_{i\alpha}} V(r-R_{i\alpha}) + \ldots \end{aligned} \tag{7.70}$$

The first term is the interaction with the rigid lattice which we are here accounting for by using the band energies and wave function and the second term is the interaction with the moving lattice. We assume that higher-order terms which are of the relative order of the ratio of ion displacement to ion separation are sufficiently small to be negligible. The electron–ion interaction is then given by the second term of equation (7.70).

We expand the second term of H_I by using the Bloch wave functions $\phi_{k\sigma}(r)$. Then

$$\psi_\sigma(r) = \sum_k c_{k\sigma}\phi_{k\sigma}(r), \tag{7.71}$$

where

$$\phi_{k\sigma}(r+R_l^0-R_{l'}^0) = \exp(ik.(R_l^0-R_{l'}^0))\phi_{k\sigma}(r). \tag{7.72}$$

It is convenient to denote the radius vectors for the atoms in one cell by R_α^0 and to measure the others from them. Then

$$R_{l\alpha}^0 = l + R_\alpha^0 \tag{7.73}$$

where l is a vector of the lattice (independent of α). Then

$$\phi_{k\sigma}(r+l) = \exp(ik.l)\phi_{k\sigma}(r) \tag{7.74}$$

and

$$\begin{aligned} & \int d^3r\, \phi_{k'\sigma}^*(r)\phi_{k\sigma}(r)\nabla_{R_{i\alpha}^0} V(r-R_{i\alpha}^0) \\ & = \int d^3r\, \phi_{k'\sigma}(r+l)\phi_{k\sigma}(r+l)\nabla_{R_\alpha^0} V(r-R_\alpha^0) \\ & = \exp(i(k-k').l)\, W_\alpha(k,k'), \end{aligned} \tag{7.75}$$

where

$$W_\alpha(k,k') = \int d^3r\, \phi_{k'\sigma}^*(r)\phi_{k\sigma}(r)\nabla_{K_\alpha^0} V(r-R_\alpha^0). \tag{7.76}$$

If $\boldsymbol{u}_{i\alpha}$ is expanded in terms of phonon operators according to equation (7.4) the electron–phonon interaction can be written

$$H_I' = \sum_{\boldsymbol{k},\boldsymbol{k}',\sigma} c_{\boldsymbol{k}'\sigma}^{+} c_{\boldsymbol{k}\sigma} \sum_{\boldsymbol{l},\alpha} W_\alpha(\boldsymbol{k},\boldsymbol{k}') \exp(i(\boldsymbol{k}-\boldsymbol{k}') \cdot \boldsymbol{l})(NM_\alpha)^{-1/2}$$

$$\times \sum_{\boldsymbol{q},\lambda} Q_\lambda(\boldsymbol{q}) e_\alpha^\lambda(\boldsymbol{q}) \exp(i\boldsymbol{q} \cdot \boldsymbol{l})$$

$$= \sum_{\boldsymbol{k},\boldsymbol{k}',\sigma} \sum_{\alpha,\lambda} c_{\boldsymbol{k}'\sigma}^{+} c_{\boldsymbol{k}\sigma} \boldsymbol{W}_\alpha(\boldsymbol{k},\boldsymbol{k}') \cdot \boldsymbol{e}_\alpha^\lambda(\boldsymbol{q}) Q_\lambda(\boldsymbol{q})(N/M_\alpha)^{1/2} \quad (7.77)$$

where, in the last sum, $\boldsymbol{q}$ lies in the first Brillouin zone and

$$\boldsymbol{q} = \boldsymbol{k}' - \boldsymbol{k} + \boldsymbol{K}, \quad (7.78)$$

where $\boldsymbol{K}$ is a vector of the reciprocal lattice. Since

$$Q_\lambda(\boldsymbol{q}) = [2\omega_\lambda(\boldsymbol{q})]^{-1/2} [a_\lambda(\boldsymbol{q}) + a_\lambda^{+}(-\boldsymbol{q})], \quad (7.79)$$

the interaction can also be written,

$$H_I' = \sum_{\boldsymbol{k},\boldsymbol{k}',\sigma,\lambda} g_{\boldsymbol{k},\boldsymbol{k}',\lambda} c_{\boldsymbol{k}'\sigma}^{+} c_{\boldsymbol{k}\sigma} [a_\lambda(\boldsymbol{q}) + a_\lambda^{+}(-\boldsymbol{q})], \quad (7.80)$$

with

$$g_{\boldsymbol{k},\boldsymbol{k}',\lambda} = \sum_\alpha \boldsymbol{W}_\alpha(\boldsymbol{k},\boldsymbol{k}') \cdot \boldsymbol{e}_\alpha^\lambda(\boldsymbol{q})[N/2M_\alpha \omega_\lambda(\boldsymbol{q})]^{1/2}. \quad (7.81)$$

We have obtained this result assuming that the ions are rigid. However, the form (7.93) is correct even if the ions deform as they move as long as only linear terms in the displacements of the ions are retained. The expression for $g_{\boldsymbol{k},\boldsymbol{k}',\lambda}$ will change with the model. It is useful to have an explicit form for $g_{\boldsymbol{k},\boldsymbol{k}',\lambda}$ in mind for purposes of orientation and we use the jellium model (as in Chapter 5). The lattice is modelled by a continuously distributed jelly of charge with the same average mass density as the original lattice; all its unperturbed frequencies are then given by (7.68). The electron states are then plane waves. A displacement $\boldsymbol{u}(\boldsymbol{R})$ of the jelly at $\boldsymbol{R}$ leads to a charge distribution $\rho_0 Ze \operatorname{div} \boldsymbol{u}$. This interacts with the electrons through the Coulomb interaction and leads to the interaction term in the Hamiltonian

$$\frac{\rho_0 Ze^2}{4\pi\epsilon_0} \sum_\sigma \int d^3r\, \psi_\sigma^{+}(\boldsymbol{r}) \psi_\sigma(\boldsymbol{r}) \frac{1}{|\boldsymbol{r}-\boldsymbol{r}'|} \operatorname{div} \boldsymbol{u}(\boldsymbol{r}') d^3r' \quad (7.82)$$

$$= \sum_{\boldsymbol{k},\boldsymbol{k}',\sigma} c_{\boldsymbol{k}'\sigma}^{+} c_{\boldsymbol{k}\sigma} [a_\lambda(\boldsymbol{q}) + a_\lambda^{+}(-\boldsymbol{q})] \frac{i\boldsymbol{q} \cdot \boldsymbol{e}_\lambda(\boldsymbol{q})}{(2\Omega NM)^{1/2}} \frac{Ze^2 \rho_0}{\epsilon_0 q^2}, \quad (7.83)$$

where

$$k' - k = q. \tag{7.84}$$

In this model, there is scattering only from the longitudinal lattice vibrations. Indeed in the simple form of the model we have outlined, only longitudinal vibrations exist. It follows from (7.83) that

$$g_{k,k'} = \frac{i(k'-k) \, . \, e_\lambda(k'-k) Ze^2 \rho_0}{(2\Omega NM)^{1/2} \epsilon_0 \, |k'-k|^2}. \tag{7.85}$$

In general, the total Hamiltonian is

$$\begin{aligned} H = & \sum_{k,\sigma} \epsilon_k c^+_{k\sigma} c_{k\sigma} + \sum_{q,\lambda} \omega_\lambda(q) [a^+_\lambda(q) a_\lambda(q) + \tfrac{1}{2}] \\ & + \tfrac{1}{2} \sum_{\substack{k,k',q \\ \sigma,\sigma'}} V(q) c^+_{k\sigma} c^+_{k'+q\sigma'} c_{k'\sigma'} c_{k+q\sigma} \\ & + \sum_{k,k',\sigma,\lambda} g_{k,k',\lambda} c^+_{k'\sigma} c_{k\sigma} [a_\lambda(q) + a^+_\lambda(-q)]. \end{aligned} \tag{7.86}$$

The first two terms are respectively the energies of the band electrons and free phonons. The third term is the Coulomb interaction between the electrons. Strictly it is the matrix element between Bloch states which should appear but it is usual to use the matrix element between plane wave states. This is correct at long wavelengths where the interaction is singular.

The electron–phonon term describes the interaction between fermions and bosons and is, therefore, different in form from the interaction from which we constructed perturbation theory in Chapter 3. However, as we show in the next section, the change that is required to the rules is quite minor. Here we simply state the rules for calculating Green's functions which involve no explicit phonon operators. Examples are the single-particle Green's function and the electric current-density, Green's function. Each diagram contains external lines corresponding to the operators in the Green's function. The Coulomb interaction is represented by dashed internal lines which meet an electron line at each end. The contribution from each dashed line and electron line is given as in §3.2, momentum and energy being conserved at each vertex. The effect of the phonon interaction is to introduce another effective electron–electron interaction which we shall denote by a wavy line as in Fig. 7.4.

Each wavy line represents the propagation of a phonon between interactions with electrons at the vertices. Energy is again conserved

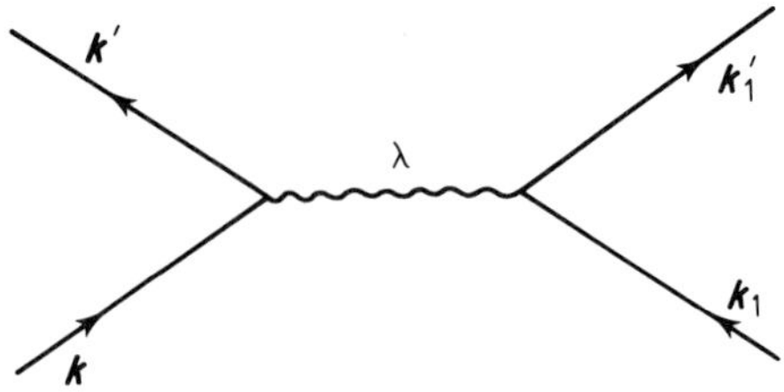

FIG. 7.4. A diagram representing the scattering of two electrons through the exchange of a phonon of polarization λ. Momentum is conserved to within a vector of the reciprocal lattice. The wavy line represents the phonon Green's function or propagator.

at each vertex but momentum is conserved only to within a vector of the reciprocal lattice, the phonon momentum always lying in the first Brillouin zone. For each vertex at which an electron is scattered by a phonon of polarization λ from $\boldsymbol{k}$ to $\boldsymbol{k}'$ there is a contribution of a factor $g_{\boldsymbol{k},\boldsymbol{k}',\lambda}$ to the term in the perturbation series. For each phonon propagator, $\boldsymbol{q}$, λ, there is a contribution $\mathscr{V}^{-1} D^0_\lambda(\boldsymbol{q}, \tau - \tau')$ as defined by equation (7.15) or $D(\boldsymbol{q}, \omega)$ as defined in equation (7.16) if energy variables are used throughout. Thus the Coulomb interaction $V(q)$ is simply replaced by the energy-dependent interaction $g_{\boldsymbol{k},\boldsymbol{k}'\lambda} D(\boldsymbol{q}, \omega) g_{\boldsymbol{k}_1,\boldsymbol{k}_2\lambda}$. The energy dependence arises because unlike the Coulomb interaction the interaction between the electrons mediated by the phonons takes time to travel from one position to another. The origin of the interaction is easy to understand. An electron moving through the lattice will polarize it. The polarization spreads through the lattice and eventually interacts with another electron and scatters it. Before applying perturbation theory we pause in the next section to justify the statements made in this paragraph. The reader more interested in the use than the proof can easily omit the next section.

§7.5 Perturbation theory for phonons

We turn now to the justification of the perturbation expansion and relate it to that already obtained for a two-particle interaction. The proof is so close to the previous one that we provide only an outline. To reduce the length of the following expressions we omit the Coulomb term which has already been dealt with and which can easily be added at the end. We work entirely with temperature Green's functions and, because the electron–phonon interaction is in general velocity-dependent, we employ the spatial Fourier transforms. We use the abbreviated notation

$$c_{k_1\sigma_1}(\tau_1) \equiv c(1) \tag{7.87}$$

with corresponding abbreviations for other quantities.

We can now follow the argument given in §3.2 to obtain the electron Green's function in the form,

$$G(1, 2) = -\frac{\langle TU(\beta)\hat{c}(1)\hat{c}(2)^{+}\rangle}{\langle TU(\beta)\rangle} \tag{7.88}$$

where the numerator is expanded in the form

$$-\langle TU(\beta)\,\hat{\psi}(1)\,\hat{\psi}(2)^{+}\rangle$$

$$= \sum_{n=0}^{\infty} \frac{(-1)^n}{n!} \int_0^\beta d\tau_n \ldots d\tau_1 \langle T[\hat{V}(\tau_n) \ldots V(\tau_1)\hat{c}(1)\hat{c}(2)^{+}]\rangle \tag{7.89}$$

and V is given by equation (7.86) as

$$V = \sum_{k,k',\sigma,\lambda} g_{k,k',\lambda}\, c_{k'\sigma}^{+} c_{k\sigma} \sqrt{2\omega_\lambda(\boldsymbol{q})}\; Q_\lambda(\boldsymbol{q}); \tag{7.90}$$

here $\boldsymbol{q}$ lies in the first Brillouin zone and is fixed by equation (7.78).

The averages in equation (7.89) are taken with respect to the system of independent electrons and phonons.

Hence the averages over the electron and phonon variables are independent and

$$(-1)^{n+1} \int d\tau_n' \ldots d\tau_1' \langle T\hat{V}(\tau_n') \ldots V(\tau_1')\hat{c}(1)\hat{c}(2)^{+}\rangle$$

$$= \int_0^\beta \prod_{i=1}^{n} \left\{ di'di''di''' \sum_{\lambda_i} g_{k_i' k_i'' \lambda_i} \delta(\tau_i' - \tau_i'')\,\delta(\tau_i' - \tau_i'') \right.$$

$$\times\, \delta(\boldsymbol{k}_i' - \boldsymbol{k}_i'' + \boldsymbol{q}_i''' - \boldsymbol{K}) \quad G_{0,n+1}(1, 1', 2' \ldots, n'; 2, 1_+'', 2_+'', \ldots, n_+'')$$

$$\times\, D(1''', 2''', \ldots, n'''), \tag{7.91}$$

where G_{0n} is the n-particle Green's function defined by equation (3.30) and

$$D(1''', 2''', \ldots, n''') = \left(\prod_i 2\omega_{\lambda_i}(q_i''') \right)^{1/2}$$

$$\times\, \langle \hat{Q}_{\lambda_1}(1''')\hat{Q}_{\lambda_2}(2''') \ldots \hat{Q}_{\lambda_n}(n''')\rangle. \tag{7.92}$$

We have already seen that the Green's function G_{0n} is related to single-particle Green's functions through Wick's theorem. Similarly we can show either directly or by splitting the operator Q into its components parts of creation and annihilation operators that D also satisfies a Wick's theorem of the form

$$D(1''', 2''', \ldots, n''')$$

$$= \sum_{\text{all pairings}} D(1''', 2''')D(3''', 4''') \ldots D([n-1]''', n''') \quad (7.93)$$

if n is even, and D is zero if n is odd. The result is that every term in the expansion of (7.91) can be represented by a Feynman diagram with n vertices (n even) of the kind described in §3.2. Instead of a dashed line to represent an interaction $V(\boldsymbol{q})$ we now have a wavy line to represent a phonon propagator. Note that we are already using the spatial Fourier transforms so

$$\begin{aligned} D(1,2) &= 2[\omega_{\lambda_1}(\boldsymbol{q}_1)\omega_{\lambda_2}(\boldsymbol{q}_2)]^{1/2} \langle \hat{Q}_{\lambda_1}(\boldsymbol{q}_1, \tau_1)\hat{Q}_{\lambda_2}(\boldsymbol{q}_2, \tau_2)\rangle \\ &= 2\omega_{\lambda_1}(\boldsymbol{q}_1) \langle \hat{Q}_{\lambda_1}(\boldsymbol{q}, \tau_2)\hat{Q}_{\lambda_1}(-\boldsymbol{q}_1, \tau_2)\rangle \, \delta_{\lambda_1\lambda_2} \, \delta_{\boldsymbol{q}_2, -\boldsymbol{q}_1} . \end{aligned} \quad (7.94)$$

The momentum $\boldsymbol{q}$ is ascribed to the wavy line. Similarly momenta are ascribed to the electron lines and, because of equation (7.78), momentum is conserved at each vertex to within a vector of the reciprocal lattice.

If we compare the term involving G_{2m+1} equation (7.91) with the corresponding equation (3.29) we see that they differ in sign by a factor $(-1)^m$. We can take care of this automatically by introducing a factor $D(\boldsymbol{q}, \tau_1 - \tau_2)$ for each wavy line carrying momentum $\boldsymbol{q}$, instead of the factor $[-V(\boldsymbol{q})]$ for the corresponding dashed line.

It follows in exactly the same way as in §3.2 that we can cancel the denominator of equation (7.88) by summing over connected diagrams only. Thus

$$G(1,2) = -\langle TU(\beta)\hat{c}(1)\hat{c}(2)^{+}\rangle_c . \quad (7.95)$$

Finally the combinatorial algebra is such that if only topologically distinct diagrams are included, the factors $(n!)$ in equation (7.89) are cancelled. The result is the rules given at the end of the previous section.

As with the electrons, it is possible to represent the corrections to the phonon propagator by means of a proper phonon self-energy. All insertions into a phonon propagator can be divided into two classes:

(i) insertions which *cannot* be divided into two parts by cutting a phonon line only;

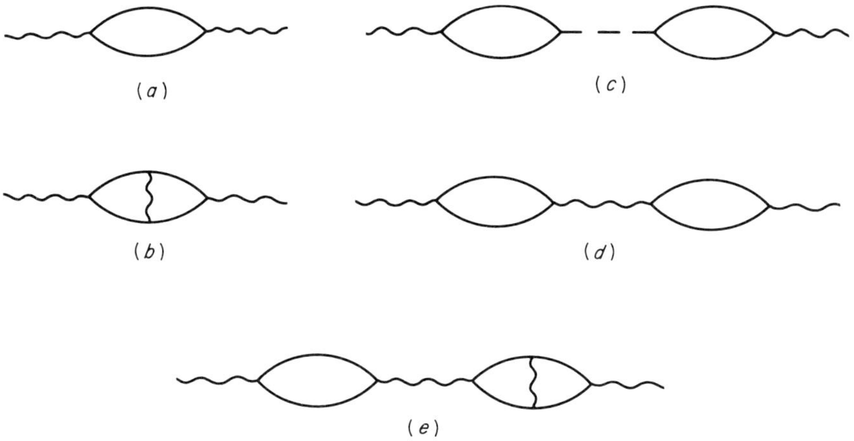

FIG. 7.5. Self-energy contributions to a phonon propagator or Green's function. Figs. (*a*), (*b*) and (*c*) show proper self-energy parts, (*d*) and (*e*) improper or reducible ones.

(ii) insertions which *can* be divided into two parts by cutting a phonon line only.

The former are proper self-energy parts and the latter improper self-energy parts. Figs. 7.5 (*a*), (*b*) and (*c*) show examples of proper self-energy parts, while (*d*) and (*e*) are examples of improper ones. Then, if $\Pi(\boldsymbol{q}, \omega)$ is the sum of the contribution of all proper self-energy parts, we have

$$D(\boldsymbol{q}, \omega) = D_0(\boldsymbol{q}, \omega) + D_0(\boldsymbol{q}, \omega)\Pi(\boldsymbol{q}, \omega)D(\boldsymbol{q}, \omega), \quad (7.96)$$

whence

$$D(\boldsymbol{q}, \omega)^{-1} = D_0(\boldsymbol{q}, \omega)^{-1} - \Pi(\boldsymbol{q}, \omega). \quad (7.97)$$

In the following sections we calculate $\Pi(\boldsymbol{q}, \omega)$.

§7.6 Screening of the electron–phonon interaction

We have seen in Chapters 5 and 6 that an important effect of the long range of the Coulomb interaction of the electrons is the screening of this Coulomb interaction itself and of other interactions. The effect on the electron–electron interaction was studied in Chapter 5 and the results obtained there can be used in this chapter. We turn now to the effect of this Coulomb interaction on the ions. We expect that it will screen both the electron–ion interaction and ion–ion

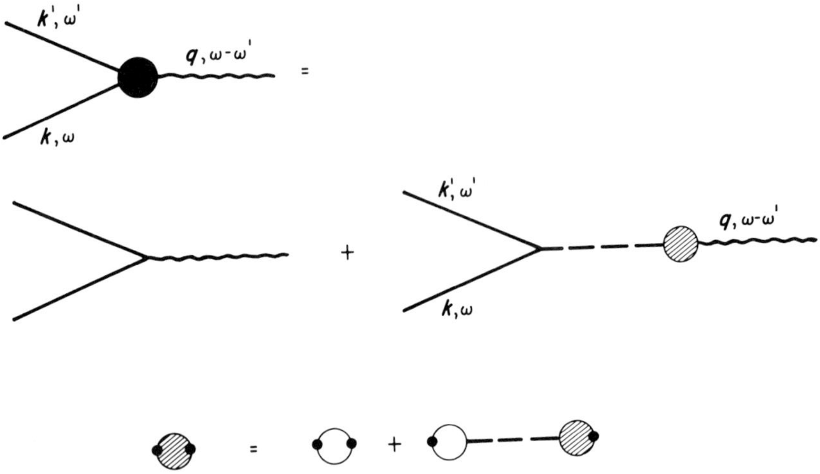

FIG. 7.6. The renormalization of the electron–phonon vertex by the Coulomb interaction. The unfilled circles represent proper polarization parts while the shaded circle represents the total screening.

interactions. Since the ions are themselves charged these interactions are both long-ranged. This is shown explicitly in expression (7.83) for the electron–phonon interaction in the case of the simplified model. A consequence of the long range on the ion–ion interaction is that, in the zeroth approximation (the non-interacting Hamiltonian), the lattice modes are plasma modes with frequencies which do not tend to zero in the long-wavelength limit (cf. equation (7.68)). We expect that the screening, which makes the effective ion–ion interaction short-range, will change the plasma modes into the observed acoustic modes. We see now how this arises in the theory.

We have noted that in perturbation theory the phonons appear through phonon propagators as shown in Fig. 7.4. The vertices at the ends of the propagators represent the electron–phonon interaction while the poles of the propagator yield the phonon frequencies. We take into account the screening of the electron–phonon interaction by inserting all possible polarization diagrams to renormalize the vertex as illustrated in Fig. 7.6. The effect is to replace a vertex $g_{kk'\lambda}$ by a renormalized vertex part $\bar{g}_{kk'\lambda}(\omega - \omega')$. Because $g_{kk'\lambda}$ depends on both $\boldsymbol{k}$ and $\boldsymbol{k}'$ the general result relates g to $\bar{g}$ through an integral equation. To display the main features here we confine ourselves to the jellium model (7.83) where there is only one longitudinal mode and g is a function of $\boldsymbol{k} - \boldsymbol{k}'$ only. Then

FIG. 7.7. The renormalization of the phonon Green's function by the Coulomb interaction. The thick wavy line represents the renormalized Green's function and the circle is the total polarization as in Fig. 7.6.

$$\bar{g}(\boldsymbol{k}-\boldsymbol{k}',\omega-\omega') = g(\boldsymbol{k}-\boldsymbol{k}') + V(\boldsymbol{k}-\boldsymbol{k}')\bar{\chi}(\boldsymbol{k}-\boldsymbol{k}',\omega-\omega')g(\boldsymbol{k}-\boldsymbol{k}') \tag{7.98}$$

and

$$\bar{\chi}(\boldsymbol{k}-\boldsymbol{k}',\omega-\omega') = \chi(\boldsymbol{k}-\boldsymbol{k}',\omega-\omega') + \chi(\boldsymbol{k}-\boldsymbol{k}',\omega-\omega')V(\boldsymbol{k}-\boldsymbol{k}')\bar{\chi}(\boldsymbol{k}-\boldsymbol{k}',\omega-\omega')\,. \tag{7.99}$$

Here, χ is the contribution from all proper polarization parts together and $\bar{\chi}$ the total polarization. From (7.98) (7.99) one finds that

$$\bar{g}(\boldsymbol{k}-\boldsymbol{k}',\omega-\omega') = g(\boldsymbol{k}-\boldsymbol{k}')/\epsilon(\boldsymbol{k}-\boldsymbol{k}',\omega-\omega') \tag{7.100}$$

where

$$\epsilon(\boldsymbol{q},\Omega) = 1 - V(\boldsymbol{q})\chi(\boldsymbol{q},\Omega). \tag{7.101}$$

As in Chapter 5 $\epsilon(\boldsymbol{q},\Omega)$ is the dielectric function of the electron gas and it can, if it is appropriate, be replaced by an empirical function in equation (7.100). In general, the energy Ω will be of the order of the Debye energy or less and this is much less than the energy (~ 10 eV) at which the frequency dependence of the dielectric constant becomes important. Thus with negligible error we can usually use the relation

$$\bar{g}(\boldsymbol{q}) = g(\boldsymbol{q})/\epsilon(\boldsymbol{q},0). \tag{7.102}$$

The renormalized phonon propagator is obtained by making all possible polarization insertions into the phonon line as in Fig. 7.7. We again write down the result for the renormalized propagator $D(\boldsymbol{q},\zeta)$ in terms of the bare one $D_0(\boldsymbol{q},\zeta)$ only for the same simple case. (In the general case, the propagators for phonons with different polarizations become connected. This means that one effect of screening is to mix the original modes.) Then

$$\bar{D}(\boldsymbol{q},\zeta) = D_0(\boldsymbol{q},\zeta) + \bar{D}_0(\boldsymbol{q},\zeta)\,|g(q)|^2\,\bar{\chi}(\boldsymbol{q},\zeta)\bar{D}(\boldsymbol{q},\zeta) \tag{7.103}$$

and this can be rewritten,

$$\begin{aligned}\bar{D}(\boldsymbol{q},\zeta) &= \bar{D}_0(q,\zeta)\,[1-|g(q)|^2\,\chi(q,\zeta)\,\bar{D}_0(q,\zeta)/\epsilon(q,\zeta)]^{-1}\\ &= 2\Omega[\Omega^2 - 2|g(q)|^2\,\Omega\chi(q,\zeta)/\epsilon(q,\zeta) + \zeta^2]^{-1}.\end{aligned} \tag{7.104}$$

From equations (7.68) and (7.84)

$$2|g(q)|^2/\Omega = e^2/\mathscr{V}\epsilon_0 q^2 = V(q), \tag{7.105}$$

the Coulomb interaction. Hence

$$\begin{aligned}\Omega^2 - 2|g(q)|^2\Omega\chi(q,\zeta)/\epsilon(q,\zeta) &= \Omega^2[1 - V(q)\chi(q,\zeta)/\epsilon(q,\zeta)] \\ &= \Omega^2/\epsilon(q,\zeta).\end{aligned} \tag{7.106}$$

At the poles of the phonon propagator ζ is approximately a phonon frequency and can again be ignored in the dielectric function. The renormalized phonon frequencies are therefore

$$\omega(q) = \Omega/\sqrt{\epsilon(q,0)}. \tag{7.107}$$

Since (cf. Chapter 5), as $q \to 0$,

$$\epsilon(q,0) \sim q^{-2},$$

$\omega(q)$ will have the proper dispersion relation for a sound wave. With the approximate form (5.43) for the dielectric function, we find that

$$\omega(q) = (m/3M)^{1/2}\, k_{\mathrm{F}} q/m, \tag{7.108}$$

and the velocity of sound is

$$v_s = (m/3M)^{1/2}\, v_{\mathrm{F}}. \tag{7.109}$$

For the simple metals this simple formula provides reasonable estimates for v_s. The result can also be derived macroscopically from a consideration of the oscillation of electron and ion fluids (Rickayzen 1965, Section 3.12). The phonon propagator can now be written

$$\bar{D}(q,\zeta) = \frac{\Omega}{\omega(q)} \frac{2\omega(q)}{\zeta^2 + \omega(q)^2}. \tag{7.110}$$

Apart from the first factor this expression has the same dependence on the renormalized frequency as the unperturbed propagator has on the bare frequency. In fact we can join the first factor to the renormalized vertex factors so that together they have the same dependence on the renormalized frequencies as the originals has on the bare frequencies. To see this, note that the phonon propagator is attached to two vertices as in Fig. 7.4. The unrenormalized contribution is

$$|g(q)|^2\bar{D}_0(q,\zeta). \tag{7.111}$$

The renormalized one is

$$|\bar{g}(q)|^2\bar{D}(q,\zeta) = \frac{|g(q)|^2}{\epsilon(q,0)^2} \frac{\Omega}{\omega(q)} D_R(q,\zeta)$$

with

$$D_R(q,\zeta) = \omega(q)/[\omega(q)^2 + \zeta^2], \tag{7.112}$$

the propagator for a renormalized phonon. We also write

$$\begin{aligned} g_R(q) &= \left[\frac{\Omega}{\omega(q)}\right]^{1/2} \frac{g(q)}{\epsilon(q,0)} \\ &= \frac{i\boldsymbol{q}\,.\,\boldsymbol{e}_\lambda(\boldsymbol{q})Ze^2n}{[2\omega(q)NM]^{1/2}\,q^2\epsilon_0\,\epsilon(\boldsymbol{q},0)}. \end{aligned} \tag{7.113}$$

This new renormalized vertex part depends on the screened ion–electron interaction,

$$Ze^2/[q^2\epsilon_0\,\epsilon(\boldsymbol{q},0)],$$

and is a function of the renormalized phonon frequencies. Thus we can use the renormalized phonon frequencies and renormalized vertex part and ignore the Coulomb corrections to these. We have shown this explicitly for the simple model but it is true in the general case (Scalapino 1969). We shall therefore use the renormalized functions for phonons henceforth without denoting them by a subscript R. The screened electron–phonon interaction is now quite short-ranged extending, in good metals, only over a few lattice spacings. The renormalized vertex and renormalized propagator as given by equations (7.113) and (7.112) agree with those obtained using the adiabatic approximation.

§7.7 Migdal's theorem

Approximate treatments of the electron–phonon interaction are much simplified by the use of a result due to Migdal (1958) and known as Migdal's theorem. In non-metals the electron and lattice motions are separated by using the Born–Oppenheimer approximation. Because of the small ratio of electron mass m to ion mass M the electrons move much faster than the ions. The electrons, therefore, move in a potential depending on the instantaneous positions of the ions while the ions are affected by an average potential due to the electrons. Mathematically, the approximation results in an expansion in $(m/M)^{1/2}$ and is valid because of the smallness of $\omega/\Delta E$ where ω is a phonon energy and ΔE the energy difference between electron states the phonon can connect. In metals this ratio of energies is no longer small because electrons can make transitions near the Fermi surface with ΔE tending to zero. Nevertheless, m/M is still very small in metals and Migdal's theorem provides a modified expansion in this parameter.

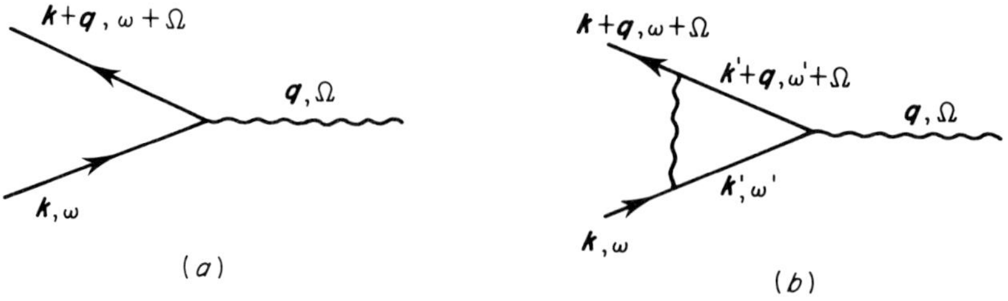

FIG. 7.8. The first order phonon correction to the electron–phonon vertex.

We begin by considering the simplest correction to the electron–phonon vertex illustrated in Fig. 7.8. The bare vertex is shown in Fig. 7.8(a) and the first correction in (b). The contribution of (a) is

$$g_{k,k+q,\lambda} \tag{7.114}$$

while the corresponding contribution of (b) to the vertex is

$$\beta^{-1} \sum_{k',\lambda',\omega'} g_{k',k'+q,\lambda} g_{k,k',\lambda'} g^*_{k'+q,k+q,\lambda'} D_{\lambda'}(k-k', \omega-\omega') \times G(k', \omega') G(k'+q, \omega'+\Omega). \tag{7.115}$$

We shall estimate the ratio in a rather crude way. First the dependence of g on k and k' is not important. In fact after the Coulomb screening is taken into account the electron–phonon interaction is short-range. We shall therefore replace $g_{k,k',\lambda}$ by an average g_q. Then the ratio of (7.115) to (7.114) is

$$\beta^{-1} |g_q|^2 \sum_{k',\lambda,\omega'} \frac{2\omega_{q,\lambda}}{\omega^2_{q,\lambda} + (\omega-\omega')^2} \frac{1}{i\omega' - \epsilon_{k'}} \frac{1}{i(\omega'+\Omega) - \epsilon_{k'+q}}. \tag{7.116}$$

This ratio depends on the parameters q, Ω and ω. Now ω is the energy of an electron measured relative to the Fermi surface and in the applications to follow is of the order of ω_D or less. In the case of the phonon self-energy discussed in the next section q and Ω are the wavelength and energy of the same phonon. For long-wavelength phonons $\Omega/q \sim v_s$ and for short-wavelength phonons the ratio is $\sim \omega_D/k_F$. In both cases this is much less than v_F. We therefore here consider the case, $\Omega/v_F q \ll 1$.

Now, if we could ignore the dependence of $\epsilon_{k'+q}$ on q in (7.116) we could evaluate the integral quite easily. We should have $\omega-\omega' \sim \omega_q$ and $\epsilon_{k'} \sim \Omega$, with the result for (7.116) of

$$\sim |g_q|^2 N(0)/\Omega .$$

This is of the order of magnitude of the mass enhancement parameter λ discussed in the next section where values are given. From there it is seen that $\lambda \leqslant 1$.

If we now include the dependence of $\epsilon_{k'+q}$ on q we shall have $\epsilon_{k'+q} \sim \Omega$ as well as $\epsilon_{k'} \sim \Omega$. This means that, if μ is defined by

$$\mu = \boldsymbol{k}' \cdot \boldsymbol{q}/k'q,$$

the integral over μ is significant over a range $\Delta\mu$ given by

$$\Delta\mu \sim \Omega/v_{\mathrm{F}}q .$$

Hence, the volume of phase space available is reduced by a factor $\Omega/v_{\mathrm{F}}q$ and the ratio (7.116) is

$$\frac{g^2 N(0)}{\Omega}\frac{\Omega}{v_{\mathrm{F}}q} \sim \lambda \frac{\Omega}{v_{\mathrm{F}}q} \ll 1 .$$

In fact, in the cases considered $\Omega/v_{\mathrm{F}}q \sim \sqrt{(m/M)}$ and we have the field theoretic analogue of the adiabatic approximation. The renormalization of the electron–phonon vertex can therefore be neglected. There are, however, some values of q and Ω ($\Omega/qv_{\mathrm{F}} \gg 1$) for which the adiabatic approximation breaks down. Full details of the calculation and of the exceptional cases are given in the paper of Migdal.

§7.8 Phonon and electron self-energies

As a result of Migdal's theorem the leading contribution to the phonon and electron proper self-energies are as shown in Fig. 7.9. We calculate these in turn.

It follows immediately that, if we use temperature Green's functions,

$$\bar{\Pi}(q, \Omega) = \beta^{-1} \sum_{\boldsymbol{k},\omega} |g_{\boldsymbol{k}\boldsymbol{k}'}|^2 \bar{G}(\boldsymbol{k}, \omega)\bar{G}(\boldsymbol{k}', \omega + \Omega), \qquad (7.117)$$

where

$$\boldsymbol{k}' - \boldsymbol{k} - \boldsymbol{q} = \boldsymbol{g}, \qquad (7.118)$$

a vector of the reciprocal lattice. We estimate this self-energy using the simple model (7.113). Then

$$\bar{\Pi}(\boldsymbol{q}, \Omega) = \beta^{-1} |g_q|^2 \sum_{\boldsymbol{k},\omega} \bar{G}(\boldsymbol{k}, \omega)\bar{G}(\boldsymbol{k} + \boldsymbol{q}, \omega + \Omega). \quad (7.119)$$

The sum over ω has already been evaluated in connection with Fermi

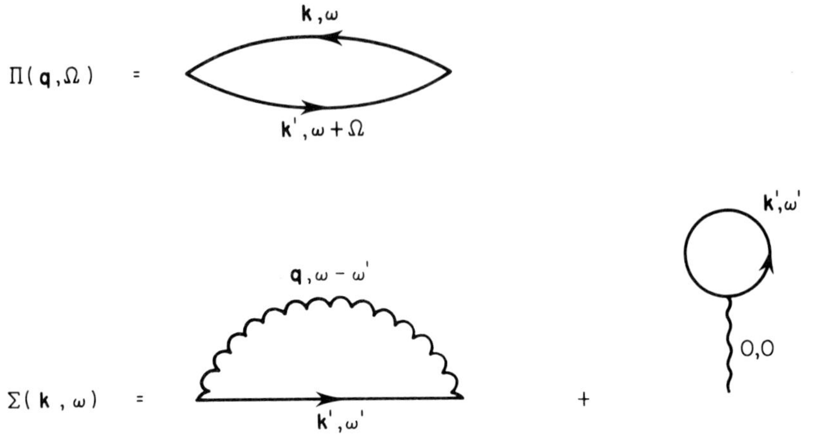

FIG. 7.9. The lowest-order contributions to the phonon self-energy Π and electron self-energy Σ.

liquid theory (equation (6.76)). With this result one obtains, for $q \ll k_F$,

$$\bar{\Pi}(q, \Omega) = N(0)\,|g_q|^2 \int_{-1}^{1} d\mu \frac{\boldsymbol{v} \cdot \boldsymbol{q}}{\boldsymbol{v} \cdot \boldsymbol{q} - i\Omega}.$$

If one considers how this will affect the lattice frequencies, one notes that for these, $i\Omega \sim v_s q$, and

$$\Pi/v_s q \sim \left(\frac{N(0)\,|g_q|^2}{v_s q}\right)\frac{v_s}{v_F}.$$

Hence this correction is of the order of the ratio of the velocity of sound to the Fermi velocity which is of the order of $\sqrt{(m/M)}$ and, therefore, is negligible. This confirms that the phonon frequencies can be obtained using the adiabatic approximation. For larger values of $q (q \sim k_F)$ renormalization does alter the phonon frequencies (Migdal 1958). In the case of optical phonons, there are no singularities at long wavelengths which could cause large effects. The self-energy correction is therefore small and does not change the character of the spectrum.

The self-energy of the electrons is given by the lower diagrams of Fig. 7.10. The second of these involves the propagator for a phonon with zero momentum and the electron-phonon matrix element g_{kk} with zero momentum transfer. Both of these factors are zero for acoustic phonons. Even for other phonons (Sham and Ziman 1963), the matrix element is zero or nearly so. Hence the contribution from this diagram is zero. The contribution from the first diagram yields

$$\bar{\Sigma}(\boldsymbol{k}, \omega) = \beta^{-1} \sum_{\boldsymbol{k}', \omega', \lambda} |g_{\boldsymbol{k},\boldsymbol{k}',\lambda}|^2 \bar{D}(\boldsymbol{q}, \omega - \omega') \bar{G}(\boldsymbol{k}', \omega') \tag{7.120}$$

where $\boldsymbol{q}$ is again given by equation (7.118).

Usually, we are interested in the self-energy when $\boldsymbol{k}$ lies near the Fermi surface and ω is small. Because the phonon energies are small compared with the Fermi energy, $\boldsymbol{k}'$ will also lie near the Fermi surface and ω' will be small. Then the dependence of $\bar{\Sigma}$ on ϵ_k is small and can be ignored. For a non-isotropic metal $\Sigma(\boldsymbol{k}, \omega)$ will still depend on the direction of $\boldsymbol{k}$. For metals with a sufficient density of impurities (dirty metals), the wave functions are spread over the Fermi surface and $\Sigma(\boldsymbol{k}, \omega)$ is independent of $\boldsymbol{k}$ and involves only the average of (7.120) over the Fermi surface (Anderson 1959). We restrict the discussion to this case which also includes the case of a pure metal with a spherical Fermi surface. Then

$$\bar{\Sigma}(\omega) = \beta^{-1} \int \frac{d\Omega_k}{4\pi} \sum_{\boldsymbol{k}', \omega', \lambda} |g_{\boldsymbol{k}\boldsymbol{k}'\lambda}|^2 \bar{D}(\boldsymbol{q}, \omega - \omega') \bar{G}(\boldsymbol{k}', \omega'). \tag{7.121}$$

At low temperatures ($T \ll \Theta_{\mathrm{D}}$) the values of ω' are so close together that the sum over ω' can be replaced by an integral according to

$$\beta^{-1} \sum_{\omega'} \rightarrow (2\pi)^{-1} \int_{-\infty}^{\infty} d\omega'.$$

Hence,

$$\bar{\Sigma}(\omega) = \int \frac{d\Omega_k}{4\pi} \sum_{\boldsymbol{k}', \lambda} |g_{\boldsymbol{k},\boldsymbol{k}',\lambda}|^2 \int_{-\infty}^{\infty} \frac{d\omega'}{2\pi} \frac{2\omega_{q,\lambda}}{\omega_{q,\lambda}^2 + (\omega - \omega')^2} \times \frac{1}{(i\omega' - \epsilon_{k'} - \bar{\Sigma}(\omega'))}. \tag{7.122}$$

We are usually interested in electrons with $\omega \lesssim \omega_{\mathrm{D}}$, the Debye frequency. For these, $\bar{\Sigma}(\omega)$ can be expanded in powers of ω in the denominator to yield

$$\bar{\Sigma}(\omega') = \bar{\Sigma}(0) + i\omega' \frac{\partial \Sigma(0)}{i \partial \omega'} = \bar{\Sigma}(0) + i\omega'(1 - Z), \tag{7.123}$$

where we know from the general analytic properties of the Green's function that the real part of Z is positive. The first term of (7.123) leads to a slight shift in the chemical potential which can be absorbed in a redefinition of ϵ_k. If we put

$$\tilde{\epsilon}_k = \epsilon_k / Z, \tag{7.124}$$

the renormalized energy, and integrate over ω' we obtain from (7.122),

$$\bar{\Sigma}(\omega) = \int \frac{d\Omega_k}{4\pi} \sum_{k',\lambda} \frac{|g_{k,k',\lambda}|^2}{Z} \times \left\{ \frac{\theta(-\tilde{\epsilon}_{k'})}{i\omega + \omega_{q,\lambda} - \tilde{\epsilon}_{k'}} + \frac{\theta(\tilde{\epsilon}_{k'})}{i\omega - \omega_{q,\lambda} - \tilde{\epsilon}_{k'}} \right\}.$$

This is straightforwardly analytically continued to

$$\Sigma(\omega) = \int \frac{d\Omega_k}{4\pi} \sum_{k',\lambda} \frac{|g_{k,k',\lambda}|^2}{Z} \times \left\{ \frac{\theta(-\tilde{\epsilon}_{k'})}{\omega + \omega_{q,\lambda} - \tilde{\epsilon}_{k'} + i\delta} + \frac{\theta(\tilde{\epsilon}_{k'})}{\omega - \omega_{q,\lambda} - \tilde{\epsilon}_{k'} + i\delta} \right\}. \tag{7.125}$$

Since $g_{kk'\lambda}$ is approximately independent of $\tilde{\epsilon}_k$ and $\tilde{\epsilon}_{k'}$ this result can be written as

$$\Sigma(\omega) = \int_0^\infty d\tilde{\epsilon} \int_0^\infty d\omega_0 \alpha^2(\omega_0) F(\omega_0) \times \{(\omega + \omega_0 + \tilde{\epsilon} + i\delta)^{-1} + (\omega - \omega_0 - \tilde{\epsilon} + i\delta)^{-1}\}, \tag{7.126}$$

where

$$\alpha^2(\omega_0) F(\omega_0) = N(0) \sum_\lambda \int \frac{d\Omega_k}{4\pi} \int \frac{d\Omega_{k'}}{4\pi} |g_{k,k',\lambda}|^2 \delta(\omega_0 - \omega_{q,\lambda}). \tag{7.127}$$

In deriving this result we have changed the sign of $\tilde{\epsilon}$ in the first term of equation (7.125). Because of the factor Z in equation (7.125), the density of states, $N(0)$, in (7.127) is that for the unperturbed band. The relevant properties of the phonons are contained in the single function $\alpha^2(\omega_0)F(\omega_0)$ which is conveniently thought of as comprising the two factors

$$F(\omega_0) = \sum_{\lambda,q} \delta(\omega_0 - \omega_{q,\lambda}), \tag{7.128}$$

the density of states of phonons, and $\alpha^2(\omega_0)$ which is defined by equations (7.127) and (7.128). The latter represents an appropriate average of the electron–phonon coupling constant. The function $\alpha^2 F(\omega_0)$ plays a central role in the theory of strong-coupling superconductors (cf. Scalapino 1969).

TABLE 7.1. Empirical values of the mass-enhancement factor λ found from Θ_D, electronic specific heat and the superconducting transition temperature (taken from McMillan 1968).

Element	Be	Al	Zn	Ga	Cd	In
λ	0.23	0.38	0.38	0.40	0.38	0.69
Element	Sn	Hg	Tl	Pb	Ti	V
λ	0.60	1.00	0.71	1.12	0.38	0.60
Element	Zr	Nb	Mo	Ru	Hf	Ta
λ	0.41	0.82	0.41	0.38	0.34	0.65
Element	W	Re	Os	Ir		
λ	0.28	0.46	0.39	0.34		

From equations (7.123), (7.124) and (7.126) we can obtain an explicit form for Z and so the effective mass. We have

$$1 - Z = \frac{\partial \Sigma(0)}{\partial \omega} = \int_0^\infty d\tilde{\epsilon} \int_0^\infty d\omega_0 \alpha^2(\omega_0) F(\omega_0)$$

$$\times \left\{ \frac{\partial}{\partial \tilde{\epsilon}} [(\omega_0 + \tilde{\epsilon} + i\delta)^{-1}] + \frac{\partial}{\partial \tilde{\epsilon}} [(\omega_0 + \tilde{\epsilon} - i\delta)^{-1}] \right\}$$

$$= -2 \int_0^\infty d\omega_0 \alpha^2(\omega_0) F(\omega_0)/\omega_0 .$$

Hence,

$$m^*/m = 1 + \lambda = 1 + 2 \int_0^\infty d\omega_0 \alpha^2(\omega_0) F(\omega_0)/\omega_0 . \tag{7.129}$$

The dimensionless parameter λ is positive and increases the effective mass. Its value provides an estimate of the strength of electron–phonon coupling. Calculated and empirical values of λ are shown in Table 7.1. As is pointed out in the next chapter transition temperatures in superconductors depend strongly on λ, and they are used in its determination (see Scalapino 1969).

The self-energy $\Sigma(\omega)$, given by equation (7.126), contains an imaginary part which as we have shown (see §3.5) is related to the lifetime τ of the electrons. In fact

$$\tau^{-1} = -2 \operatorname{Im} \Sigma$$

$$= 2\pi \int_0^\infty d\tilde{\epsilon} \int_0^\omega d\omega_0 \alpha^2(\omega_0) F(\omega_0) \delta(\omega - \omega_0 - \tilde{\epsilon})$$

$$= 2\pi \int_0^\omega d\omega_0 \alpha^2(\omega_0) F(\omega_0). \tag{7.130}$$

Equation (7.130) is essentially Fermi's golden rule with $\alpha^2(\omega_0)$ the average matrix element for the emission of a phonon of energy ω_0. As is to be expected only low-energy phonons contribute to the lifetime of low-energy electrons. In fact, an electron has to be able to emit a phonon of lower energy in order to decay. At low frequencies, the Debye part of the spectrum dominates and

$$\alpha^2(\omega_0)F(\omega_0) \propto \omega_0^2 .$$

Hence at low temperatures and for low energies

$$\tau^{-1}(\omega) \propto \omega^3 . \tag{7.131}$$

This means that omitting τ^{-1} on the right-hand side of equation (7.122) is consistent with the final result.

The lifetime of the electrons is important for the transport properties of metals such as the electrical and thermal conductivities. These can be studied by the methods of Chapter 4. For details the reader is referred to the original work of Tewordt (1963) and Ambegaokar and Tewordt (1964).

In this section we have ignored the effect of the Coulomb interaction on the self-energy. This can however be included with only a small formal change. We begin by calculating the effect of the Coulomb interaction on the one-particle Green's function. This calculation was carried out in §§5.2 and 6.2 and led to the result (equation (6.70))

$$\bar{G}^{-1}(k, \omega) = Z_c(i\omega - \epsilon_k) \tag{7.132}$$

where Z_c is the wave function renormalization and ϵ_k the renormalized energy. We now calculate the phonon contribution to the self-energy, using the Green's functions given by equation (7.132) as the "bare" ones. Equation (7.122) is then replaced by

$$\bar{\Sigma}_{\text{ph}}(\omega) = \int \frac{d\Omega_k}{4\pi} \sum_{k'\lambda} |g_{kk'\lambda}|^2 \int \frac{d\omega'}{2\pi} \frac{2\omega_{q\lambda}}{\omega_{q\lambda}^2 + (\omega - \omega')^2} \times \frac{1}{Z_c(i\omega' - \epsilon_{k'}) - \bar{\Sigma}_{\text{ph}}(\omega')} \tag{7.133}$$

If we now put

$$\tilde{\Sigma}(\omega) = \bar{\Sigma}_{\text{ph}}(\omega')/Z_c ,$$

then $\tilde{\Sigma}$ satisfies the same equation as Σ except that g is replaced by g/Z_c. With this same replacement we obtain the same formula for the effective mass.

Other contributions to the study of the effect of phonons on the properties of electrons in normal metals are contained in Prange and Kadanoff (1964) and Sham and Ziman (1963).

References

Ambegaokar, V. and Tewordt, L. (1964). *Phys. Rev.* **134**, A805.
Anderson, P. W. (1959). *J. Phys. Chem. Solids* **11**, 26.
Ashcroft, N. W. and Mermin, N. D. (1976). "Solid State Physics", Holt, Rhinhart & Winston, Inc., New York.
Brockhouse, B. N., Arase, T., Caglioti, C., Rao, K. R. and Woods, A. D. B. (1962). *Phys. Rev.* **128**, 1099.
McMillan, W. L. (1968). *Phys. Rev.* **167**, 331.
Marshall, W. and Lovesey, S. W. (1971) "Theory of Thermal Neutron Scattering", Oxford University Press, Oxford.
Messiah, A. (1961). "Quantum Mechanics", Vol. 1. North-Holland, Amsterdam.
Migdal, A. (1958). *Sov. Phys. JETP* **7**, 916.
Prange, R. E. and Kadanoff, L. P. (1964). *Phys. Rev.* **134**, A566.
Rickayzen, G. (1965). "Theory of Superconductivity". Wiley, New York.
Scalapino, D. J. (1969). "Superconductivity" (Ed. R. D. Parkes), Vol. 1. Dekker, New York.
Sham, L. J. and Ziman, J. M. (1963). *Solid State Phys.* **15**, 221.
Tewordt, L. (1963) *Phys. Rev.* **129**, 657.
Ziman, J. M. (1960). "Electrons and Phonons". Oxford University Press, Oxford.

Problems

1. Prove equations (7.88) and (7.93).
2. Write down the equation of motion for the single-particle electron temperature Green's function for the electron–phonon Hamiltonian (7.86). (Ignore the Coulomb term.) By following the argument in §2.8, show that the thermodynamic potential can be expressed in terms of this Green's function through the equation

$$\Omega(1) - \Omega(0) = \int_0^1 \frac{d\Lambda}{\Lambda} \left[\sum_k \left(\frac{d}{d\tau} + \epsilon_k \right) G(k, \tau) + \delta(\tau) \right]_{\tau \to 0-}$$

   where the electron–phonon interaction is multiplied by a factor Λ.
3. Show that the result (7.130) for the lifetime of an electron agrees with that obtained using Fermi's Golden Rules to lowest order in the electron–phonon interaction.

Chapter 8

Superconductivity

§8.1 Introduction

In this book we do not have the space to provide the background to the theory of superconductivity or a general description of the observed properties of superconductors. We list at the end of this chapter a number of books which provide such an introduction. We shall restrict ourselves to the microscopic description of the superconductive state, especially through the use of Green's functions. However, for the reader with no background knowledge we list here some of the most important properties of superconductors; some are the result of direct observation while others are derived more indirectly from the interpretation of many experiments.

(1) Superconductors are usually normal metals at room temperature, nearly always not very good conductors, which become superconductors at very low temperatures. The highest known transition temperature T_c at the present time is 22 K for Nb_3Ge.

(2) In the superconductive state, superconductors carry an electric current with no measurable resistance, and multiply-connected superconductors can, without any EMF, carry an electric current which persists with no noticeable decay. The magnetic flux inside a hollow cylinder carrying a persistent current is quantized in units of π/e (or h/2e in conventional units).

(3) A finite amount of energy is required to excite quasi-particles at low temperatures. The minimum energy required is of the order of $k_B T_c$.

(4) A bulk long thin specimen of superconductor placed in a weak magnetic field in a direction parallel to the lines of force excludes the magnetic field. This effect is usually known as the Meissner effect named after one of the discoverers. The magnetic field does penetrate the superconductor to a small depth, the penetration depth, which is of the order of 10 nm.

(5) With the geometry of (4), their behaviour in large magnetic fields divides superconductors into two classes. Type I superconductors allow no magnetic penetration of flux until a critical field H_c is reached. Above the field H_c superconductivity is destroyed and the magnetic field penetrates completely.

Type II superconductors allow partial penetration of the magnetic field at a field H_{c1} and complete penetration only at a larger field H_{c2}. The different behaviour of Type I and Type II superconductors is illustrated in Fig. 8.1 where the magnetic moment of the sample is plotted against the external field.

(6) The transition temperature depends on the isotopic mass M of the lattice and obeys a relation

$$T_c \propto M^{\beta}.$$

For many superconductors β is close to -0.5. This effect is known as the isotope effect.

§8.2 Instability of the normal state

One important idea in the development of the theory of superconductivity was the recognition that the electron–phonon interaction could be important. This was proposed theoretically by Fröhlich (1950) at the same time as the isotope effect was being observed (Maxwell 1950, Reynolds *et al.* 1950) to confirm it. A second important idea was due to Cooper (1956) who showed that an attractive interaction between electrons would make the Fermi sea of electrons unstable. He showed specifically that the ground state of a pair of electrons added to the Fermi sea would be a resonant bound state with energy less than $2E_F$. A pair of electrons of this kind is now called a Cooper pair.

Cooper's instability shows up in the repeated scattering of a pair of electrons. We can display it by looking at the response of the superconductor to a static external field which adds a pair of electrons to a superconductor. Such a hypothetical field can be destribed by adding to the Hamiltonian a term

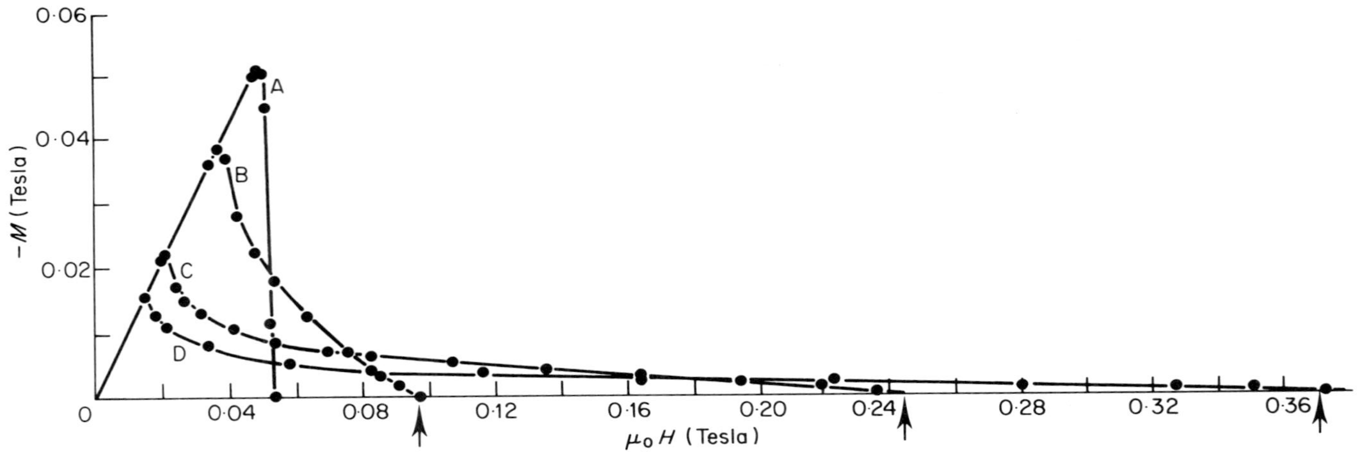

FIG. 8.1. Magnetization curves of annealed polycrystalline lead and lead–indium alloys taken in ascending field at 4.2 K. A: lead; B: lead–2.08 wt % indium; C: lead–8.23 wt % indium; D: lead–20.4 wt % indium. Only curve A shows a pure Meissner effect (after Livingstone 1963).

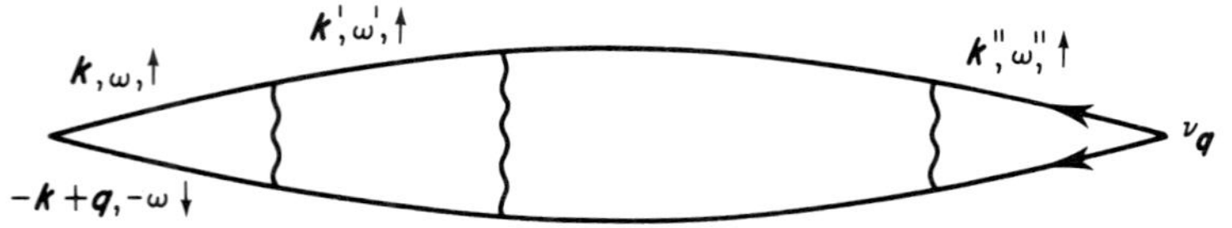

FIG. 8.2. Ladder diagrams representing the response of a metal to a pairing field ν.

$$\sum_{k,q} (\nu_q c_{k\uparrow}^+ c_{-k+q\downarrow}^+ \ \nu_q^* c_{-k+q\downarrow} c_{k\uparrow}), \tag{8.1}$$

where we have assumed that the added electrons have spin zero. It is possible to have a similar instability in a state of spin 1, but for all superconductors the instability in spin zero appears to be dominant. In liquid ^{3}He, however, another Fermi liquid, the instability is in the state of spin 1 (Leggett 1975). We look at the response of the system by examining the expectation value of the operator

$$\sum_k c_{k+q\downarrow} c_{k\uparrow}$$

after a long time. This estimates the amplitude of the wave function of the additional electrons.

According to the rules of perturbation theory the simplest diagrams which contribute to this expectation value and contain the repeated scattering of pairs of electrons are the ladder diagrams, illustrated in Fig. 8.2. From these diagrams we deduce that

$$\left\langle \sum_k c_{-k+q\downarrow} c_{k\uparrow} \right\rangle = \beta^{-1} \sum_{k,\omega} G(k, i\omega) G(-k+q, i\omega) \Lambda(k, q, i\omega) \nu_q , \tag{8.2}$$

where

$$\Lambda(k, q, i\omega) = 1 + \beta^{-1} \sum_{k',\omega'} |g_{kk'}|^2 D(k-k', i\omega - i\omega') \times G(k', i\omega) G(-k'+q, -i\omega') \Lambda(k', q', i\omega'). \tag{8.3}$$

In order to get at the essentials quickly let us simplify equation (8.3) drastically. First, the electron–phonon interaction is short-range and we can model it by taking $|g^2|$ to be constant. Moreover, because of the short range of the interaction, the electrons tend to be scattered across the Fermi surface and the short-wavelength phonons are most important. We therefore replace the phonon frequencies by an average frequency ω_D which is of the order of the Debye frequency. The form of the phonon propagator is now

$$D(k-k', i\omega - i\omega') = 2\omega_D / [\omega_D^2 + (\omega - \omega')^2]. \tag{8.4}$$

As usual we are interested in the behaviour of electrons near the Fermi surface for which $\omega, \omega' \sim 0$. The phonon propagator provides a cut-off at energy ω_D. We therefore make the model replacement

$$|g_{kk'}^2|\,D(k-k', i\omega - i\omega') = \begin{cases} V, & \text{a constant, for } |\omega|, |\omega'| < \omega_D \\ 0, & \text{for } |\omega| \text{ or } |\omega'| > \omega_D . \end{cases} \tag{8.5}$$

This simplification is equivalent to that used by Cooper and was introduced by Bardeen (1951) at an earlier date. It reproduces with remarkable accuracy the properties of phonons required in the theory of superconductivity. The interaction V is then a parameter of the theory which measures the strength of the electron–phonon coupling.

With the replacement of (8.5), $\Lambda(\boldsymbol{k}, \boldsymbol{q}, i\omega)$ becomes independent of $\boldsymbol{k}$, ω for $|\omega| < \omega_D$ and is given by,

$$\Lambda(\boldsymbol{q}) = \left[1 - \beta^{-1} \sum_{\boldsymbol{k}, |\omega| < \omega_D} G(\boldsymbol{k}, i\omega) G(-\boldsymbol{k}+\boldsymbol{q}, -i\omega)\right]^{-1}. \tag{8.6}$$

Now we can show that $\Lambda(\boldsymbol{q})$, and hence the response given by (8.2), becomes infinite at sufficiently low temperatures. In fact $\Lambda(\boldsymbol{q})$ becomes infinite when the denominator vanishes. For $\boldsymbol{q} = 0$, this occurs when

$$\begin{aligned} 1 &= V\beta^{-1} \sum_{\omega} \int d^3k\, G(\boldsymbol{k}, i\omega) G(-\boldsymbol{k}, -i\omega) \\ &= V\beta^{-1} \sum_{|\omega| < \omega_D} N(0) \int_{-\infty}^{\infty} d\epsilon/(\omega^2 + \epsilon^2)^{-1} \\ &= N(0)V(2\pi/\beta) \sum_{\omega=0}^{\omega_D} (|\omega|)^{-1} \\ &= N(0)V \sum_{n=0}^{\beta\omega_D/2\pi} (n + \tfrac{1}{2})^{-1}. \end{aligned} \tag{8.7}$$

At very low temperatures $\beta\omega_D$ is large and the sum is

$$\gamma + \ln(2\beta\omega_D/\pi)$$

where γ is Euler's constant (Abramowitz and Stegun 1965). It follows that $\Lambda(0)$ diverges at a temperature T_c given by

$$\gamma + \ln(2\omega_D/\pi k_B T_c) = [N(0)V]^{-1}.$$

Inverting this equation we find the explicit form for T_c,

$$T_c = 1.13(\omega_D/k_B) \exp[-1/N(0)V]. \tag{8.8}$$

No matter how weak the coupling is, provided that it is positive, the temperature T_c exists. At the temperature T_c the response to a uniform field ν becomes infinite. This means that any small fluctuation in the state of a pair of electrons is unstable and below T_c we should expect a new state. The situation is analogous to that of ferromagnetic materials. At high temperatures they show a finite response to weak magnetic fields. As the temperature falls, the response, given by the magnetic susceptibility, rises and becomes infinite at the Curie temperature. Below the Curie temperature the material possesses a spontaneous magnetic moment which is non-zero even in the absence of an external magnetic field. In the same way we can expect that the quantities $\langle c^+_{k\uparrow} c^+_{-k\downarrow}\rangle$ which become infinite in a superconductor in a small external field at the transition temperature will spontaneously be non-zero below the transition temperature in the absence of an external field. The superconducting state is characterized by this feature.

An alternative way of stating this result is that as T_c is approached from above fluctuations in the quantity $c^+_{k\uparrow} c^+_{-k\downarrow}$ which are given by $\langle c_{-k\downarrow} c_{k\uparrow} c^+_{k\uparrow} c^+_{-k\downarrow}\rangle$ become large, tending to infinity at the limit. As a result, below T_c any fluctuations build up in such a way that $\langle c^+_{k\uparrow} c^+_{-k\downarrow}\rangle$ is finite.

There are a number of comments to make about this result.

(1) The Hamiltonian conserves number, yet it is being asserted that the expectation value of an operator which does not conserve number is non-zero. The expectation value

$$\begin{aligned}\langle c^+_{k\uparrow} c^+_{-k\downarrow}\rangle &= \mathrm{Tr}\,[\exp(-\beta H) c^+_{k\downarrow} c^+_{-k\downarrow}] \\ &= \sum_\alpha \langle\alpha|\exp(-\beta H) c^+_{k\downarrow} c^+_{-k\downarrow}|\alpha\rangle \qquad (8.9)\end{aligned}$$

can, in fact, be shown to be strictly zero if the sum is taken over all states $|\alpha\rangle$. For in this case the states can be chosen to be simultaneous eigenstates of H and N_{op} and each term in the sum of (8.9) is zero. The only way we can obtain non-zero is by restricting the states over which we sum. We shall see (§8.6) that by assuming that (8.9) is non-zero we restrict the states $|\alpha\rangle$ in such a way that they are described by a phase ϕ. States with different values of ϕ are macroscopically different and would take astronomical times to transform into each other. It is therefore consistent in discussing thermodynamic quantities to fix the phase. This is a common characteristic of phase transitions and is referred to as symmetry breaking (see also §12.3 for a general discussion of this point).

In the case of magnetism the symmetry is invariance under some or all rotations of spin. Nevertheless, below T_c a magnet has a magnetic moment and the net spin of an electron is in a definite direction. This is again because states with different directions of the magnetic moment take astronomical times to change into each other. It is therefore appropriate to restrict the sum-over-states to states which are accessible in microscopic times.

In the case of superconductivity, the invariance is invariance of the Hamiltonian under the unitary transformation $\exp(iN\chi)$. We shall see that the effect of this transformation on the states $|\alpha\rangle$ is to change their phase by 2χ. It follows that the ground state is infinitely degenerate, different values of the phase yielding different ground states.

Of course it must be, and is, possible to deal with an isolated superconductor by using basic states in which N is conserved. With such states $\langle c_{k\uparrow}^{+} c_{-k\downarrow}^{+}\rangle$ is zero. However, for an isolated superconductor this is not a measurable quantity. One finds, with a suitable choice of states, that all the measurable results are unaltered (to order N^{-1}) by such an approach. However, this approach is mathematically more complicated and deficient in cases where $\langle c_{k\uparrow}^{+} c_{-k\downarrow}^{+}\rangle$ is measurable. In these cases (tunnelling of electrons between metals), only small fluctuations in N are needed to fix the phase which is, therefore a better quantum number than N.

(2) We have looked at $\Lambda(\boldsymbol{q})$ only in the case $\boldsymbol{q} = 0$. It can be shown that for $\boldsymbol{q} \neq 0$, Λ diverges at a lower temperature. As temperature decreases, it is $\Lambda(0)$ which first diverges and we expect the homogeneous superconductor to be in a uniform state in the new phase.

(3) We began this discussion by considering the effect of a hypothetical pairing field ν_q in analogy with a magnetic field $\boldsymbol{H}$. However, the pairing field is not as universal and easily observable as a magnetic field; the latter is long-range and can be generated by varying electric currents as well as by bar magnets. Pairing fields can be produced by joining two superconductors so that electrons can tunnel between them. If the transition temperatures of the isolated superconductors are T_{c1} and T_{c2} with $T_{c1} < T_{c2}$ then for temperatures T such that $T_{c1} < T < T_{c2}$, the second superconductor provides a pairing field for the first.

§8.3 Thermodynamics of a superconductor

(a) *Simple model*

We begin with the simple model described by the interaction of equation (8.5) which leads to the repeated scattering of pairs of electrons. In this section we shall use the model Hamiltonian

$$H_M = \sum_{k,\sigma} \epsilon_k c_{k\sigma}^+ c_{k\sigma} - \tfrac{1}{2} V \sum_{\substack{k,k',q \\ \sigma\sigma'}} c_{k\sigma}^+ c_{-k+q\sigma'}^+ c_{-k'+q\sigma'} c_{k'\sigma} \quad (8.10)$$

which produces the diagram shown in Fig. 8.2 with the interaction (8.5), for studying superconductivity. In the next section we shall go back to the phonon interaction for which the simple interaction is a model. We have introduced the negative sign in equation (8.10) because a positive phonon interaction corresponds to an attractive two-particle interaction [cf. equation (7.5)]. In order to use the cut-off in equation (8.5) we assume that the interaction in (8.10) is summed only over electron states within the energy ω_D of the Fermi surface. The Hamiltonian (8.10) is the one originally studied by Bardeen, Cooper and Schrieffer (1957).

To find the properties of the Hamiltonian we adopt a self-consistent approach analogous to the Hartree approximation (see §§3.8 and 6.1) which has already been used to study the normal state. In the Hartree approximation we assume that the operators $n_{k\sigma} (= c_{k\sigma}^+ c_{k\sigma})$ are not very different from their expectation values. In the present case we assume that the operators $c_{k\sigma}^+ c_{-k+q\sigma}^+$ are not very different from their expectation values. Assuming a uniform state with pairing of opposite spins, we put

$$\langle c_{k\uparrow}^+ c_{q-k\downarrow}^+ \rangle = b_k^* \delta_{q,0}, \qquad \langle c_{k\sigma}^+ c_{q-k\sigma}^+ \rangle = 0. \quad (8.11)$$

In the Hamiltonian (8.10) we write

$$c_{k\sigma}^+ c_{q-k\sigma'}^+ = \langle c_{k\sigma}^+ c_{q-k\sigma'}^+ \rangle + (c_{k\sigma}^+ c_{q-k\sigma'}^+ - \langle c_{k\sigma}^+ c_{q-k\sigma'}^+ \rangle)$$

and neglect terms quadratic in the fluctuations represented by the last bracket. Then we find that

$$H = \sum_{k\sigma} \epsilon_k c_{k\sigma}^+ c_{k\sigma} - \sum_k (c_{k\uparrow}^+ c_{-k\downarrow}^+ \Delta + c_{-k\downarrow} c_{k\uparrow} \Delta^*) + \tfrac{1}{2} \sum_k b_k^* \Delta, \quad (8.12)$$

where

$$\Delta = V \sum_k b_k = V \langle \psi_\downarrow(r) \psi_\uparrow(r) \rangle. \quad (8.13)$$

The field Δ is an internal pairing field to be obtained self-consistently from equations (8.11) and (8.13). (For comparison with other authors, note that Δ is often defined with the opposite sign to that given in equation (8.13).)

The equations of motion of the electron operators are now

$$\begin{aligned} dc_{k\uparrow}(\tau)/d\tau &= -\epsilon_k c_{k\uparrow}(\tau) + \Delta c^+_{-k\downarrow}(\tau), \\ dc^+_{-k\downarrow}(\tau)/d\tau &= \epsilon_k c^+_{-k\downarrow}(\tau) + \Delta^* c_{k\uparrow}(\tau), \\ dc^+_{k\uparrow}(\tau)/d\tau &= \epsilon_k c^+_{k\uparrow}(\tau) - \Delta^* c_{-k\downarrow}(\tau), \\ dc_{-k\downarrow}(\tau)/d\tau &= -\epsilon_k c_{-k\downarrow}(\tau) - \Delta c^+_{k\uparrow}(\tau), \end{aligned} \tag{8.14}$$

where we have used $c(\tau)$ and $c^+(\tau)$ for the τ-dependent operators, as no confusion is likely to arise.

The equations for the Green's functions are

$$\left(\frac{d}{d\tau} + \epsilon_k\right) G_\uparrow(\boldsymbol{k}, \tau) = -\delta(\tau) + \Delta F(\boldsymbol{k}, \tau), \tag{8.15}$$

where

$$F(\boldsymbol{k}, \tau) = -\langle T c^+_{-k\downarrow}(\tau) c^+_{k\uparrow}(0)\rangle$$

is a new type of Green's function which results from the pairing. It is often called the anomalous Green's function. The equation of motion for this Green's function is

$$\left(\frac{d}{d\tau} - \epsilon_k\right) F(\boldsymbol{k}, \tau) = \Delta^* G_\uparrow(\boldsymbol{k}, \tau). \tag{8.16}$$

It is straightforward to solve equations (8.15) and (8.16). We find it convenient, especially for later developments, to introduce a notation due to Nambu (1960). The operators in the Hamiltonian are paired together in a way which is made evident if we write

$$\alpha_k = \begin{pmatrix} c_{k\uparrow} \\ c^+_{-k\downarrow} \end{pmatrix}, \qquad \alpha_k^+ = (c^+_{k\uparrow} \quad c_{-k\downarrow}), \tag{8.17}$$

where α_k is a column vector (which destroys momentum $\boldsymbol{k}$ and spin $\uparrow$). The equations of motion are then

$$d\alpha_k(\tau)/d\tau = -\epsilon_k \tau_3 \alpha_k(\tau) + (\Delta\tau_+ + \Delta^*\tau_-)\alpha_k(\tau), \tag{8.18}$$

where we use the Pauli spin matrices

$$\tau_1 = \begin{pmatrix} 0 & 1 \\ 1 & 0 \end{pmatrix}, \qquad \tau_2 = \begin{pmatrix} 0 & -i \\ i & 0 \end{pmatrix}, \qquad \tau_3 = \begin{pmatrix} 1 & 0 \\ 0 & -1 \end{pmatrix},$$

$$\tau_\pm = \tfrac{1}{2}(\tau_1 \pm i\tau_2).$$

The single-particle Green's function is now a 2×2 matrix

$$
\begin{aligned}
G(\boldsymbol{k}, \tau) &= -\langle T\alpha_k(\tau)\alpha_k^+\rangle \\
&= -\left\langle T\begin{pmatrix} c_{k\uparrow}(\tau) \\ c^+_{-k\downarrow}(\tau)\end{pmatrix} (c^+_{k\uparrow}\, c_{-k\downarrow})\right\rangle \\
&= -\begin{pmatrix} \langle Tc_{k\uparrow}(\tau)c^+_{k\uparrow}\rangle & \langle Tc_{k\uparrow}(\tau)c_{-k\downarrow}\rangle \\ \langle Tc^+_{-k\downarrow}(\tau)c^+_{k\uparrow}\rangle & \langle Tc^+_{-k\downarrow}(\tau)c_{-k\downarrow}\rangle \end{pmatrix}. \qquad (8.19)
\end{aligned}
$$

Since

$$\{\alpha_k, \alpha_k^+\} = 1, \qquad (8.20)$$

the (matrix) equation of motion for the new Green's function is

$$\left(\frac{d}{d\tau} + \epsilon_k \tau_3\right) G(\boldsymbol{k}, \tau) = -\delta(\tau) + (\Delta\tau_+ + \Delta^*\tau_-). \qquad (8.21)$$

The Fourier transform of $G(\boldsymbol{k}, \tau)$ is still defined by

$$G(\boldsymbol{k}, \tau) = \beta^{-1} \sum_{\zeta_l} \bar{G}(\boldsymbol{k}, \zeta_l) \exp(-i\zeta_l \tau),$$

and satisfies

$$[-i\zeta_l + \epsilon_k \tau_3 - \Delta\tau_+ - \Delta^*\tau_-]\, \bar{G}(\boldsymbol{k}, \zeta_l) = -1. \qquad (8.22)$$

Hence,

$$\bar{G}(\boldsymbol{k}, \zeta_l) = [i\zeta_l - \epsilon_k\tau_3 + \Delta\tau_+ + \Delta^*\tau_-]^{-1},$$

where the inverse is taken in the matrix sense. If we multiply (8.22) by $[i\zeta_l + \epsilon_k \tau_3 - \Delta\tau_+ - \Delta^*\tau_-]$ we obtain the Fourier transform explicitly in the form

$$\bar{G}(\boldsymbol{k}, \zeta_l) = \frac{i\zeta_l + \epsilon_k\tau_3 - \Delta\tau_+ - \Delta^*\tau_-}{-\zeta_l^2 - \epsilon_k^2 - |\Delta|^2} \qquad (8.23)$$

and

$$G(\boldsymbol{k}, \omega) = \frac{\omega + \epsilon_k\tau_3 - \Delta\tau_+ - \Delta^*\tau_-}{\omega^2 - \epsilon_k^2 - |\Delta|^2}. \qquad (8.24)$$

This has poles where

$$\omega = \pm E_k = \pm(\epsilon_k^2 + |\Delta|^2)^{1/2}. \qquad (8.25)$$

These are the new energies of quasi-particles in the superconductor. Provided that $\Delta \neq 0$, there is a gap $|\Delta|$ in the quasi-particle spectrum.

The parameter, or molecular field, Δ is still unknown and we must apply the self-consistency conditions (8.13) to find it. In terms of the Green's function,

$$b_k = \mathrm{Tr}\, \tau_- G(\boldsymbol{k}, 0_-)$$

$$= \beta^{-1} \sum_{\zeta_l} \frac{\Delta \exp(-i\zeta_l 0_-)}{\zeta_l^2 + \epsilon_k^2 + |\Delta|^2}$$

$$= \frac{\Delta}{2\pi i} \int_c \frac{d\omega \exp(-\omega 0_-)}{\exp(\beta\omega) + 1} \frac{1}{\omega^2 - \epsilon^2 - |\Delta|^2},$$

where we use the method of Appendix B to evaluate the sum over ζ_l. Hence

$$\begin{aligned} b_k &= \Delta\{-[2E_k(e^{\beta E_k} + 1)]^{-1} + [2E_k(e^{-\beta E_k} + 1)]^{-1}\} \\ &= \Delta[1 - 2f(E_k)]/2E_k, \end{aligned} \tag{8.26}$$

and

$$\Delta = V\Delta \sum_k [1 - 2f(E_k)]/2E_k. \tag{8.27}$$

Equation (8.27) is the equation for the gap and was derived by Bardeen, Cooper and Schrieffer (1957) in the paper (known as BCS) which is the cornerstone of the theory of superconductivity.

Equation (8.27) does not determine the phase of Δ which is an arbitrary constant that does not affect the thermodynamics of the superconductor. We shall, therefore, take Δ real when this is possible. Nevertheless, phase differences do give rise to observable physical effects and the phase of Δ is the phase referred to at the end of the previous section and in §8.7. With Δ real,

$$G(k, \omega) = \frac{\omega + \epsilon_k \tau_3 - \Delta\tau_1}{\omega^2 - E_k^2}. \tag{8.28}$$

At the absolute zero of temperature, the gap equation becomes

$$\begin{aligned} 1 &= \tfrac{1}{2}N(0)V \int_{-\omega_D}^{\omega_D} d\epsilon/[\epsilon^2 + \Delta^2(0)]^{1/2} \\ &= N(0)V \sinh^{-1}[\omega_D/\Delta(0)], \end{aligned}$$

whence

$$\Delta(0) = \omega_D [\sinh(1/N(0)V)]^{-1}. \tag{8.29}$$

In practice, $N(0)V$ is commonly much less than unity and $[N(0)V]^{-1}$ is large; in these cases

$$\Delta(0) = 2\omega_D \exp[-1/N(0)V]. \tag{8.30}$$

As the temperature rises, because of the factor $(1 - 2f)$ in the integral of equation (8.27), the gap decreases until the temperature is reached at which $\Delta \to 0$. This temperature T_c is given by

$$1 = N(0)V \int_0^{\omega_D} \frac{1-2f(\epsilon)}{\epsilon}\, d\epsilon$$

$$= N(0)V \left\{ \ln(\omega_D/k_B T_c) \tanh(\omega_D/2kT_c) - \int_0^{\omega_D} \ln(\epsilon/k_B T_c) \frac{\partial}{\partial \epsilon} [\tanh(\epsilon/2k_B T_c)]\, d\epsilon \right\}. \qquad (8.31)$$

For weak-coupling superconductors $k_B T_c \ll \omega_D$ and the equation becomes

$$1 = N(0)V \left\{ \ln(\omega_D/k_B T_c) - \int_0^{\infty} dx \ln x \frac{d}{dx} [\tanh(x/2)] \right\}.$$

The last integral can be expressed in terms of Euler's constant and one obtains, as is to be expected, a transition temperature given by equation (8.8). For weak coupling superconductors the result explains why T_c, which is of the order of 10 K or less, is usually found to be much less than the Debye temperature which is usually of the order of 100 K or more. Typical values of T_c and θ_D are shown in Table 8.1.

TABLE 8.1.† Measured values of T_c and θ_D

Element	T_c expt (K)	θ_D (K)
Zn	0.9	235
Cd	0.56	164
Hg	4.16	70
Al	1.2	375
In	3.4	109
Tl	2.4	100
Sn	3.75	195
Pb	7.22	96
Ti	0.4	430
Zr	0.55	265
V	4.9	338
Nb	8.8	320
Ta	4.4	230
Mo	0.92	360
U	1.1	200

†Taken from Meservey and Schwartz (1969), by courtesy of Marcel Dekker Inc.

Equations (8.8) and (8.30) together make one clear prediction from the theory, namely

$$2\Delta(0)/k_B T_c = 3.53. \qquad (8.32)$$

Since the numerator $2\Delta(0)$ is the energy gap observed in absorption and emission experiments because two quasi-particles are created together (see §8.3) both numerator and denominator of the ratio can be measured directly. Observed values of the ratio are displayed in Table 8.2. In view of the simplicity of the theory and the drastic assumptions made in using the simple form of interaction, the agreement is remarkable.

TABLE 8.2.† Measured values of $2\Delta(0)/kT_c$ (BCS theoretical value = 3.53)

Superconductor	Tunnelling measurements	Thermodynamic measurements
Al	4.2 ± 0.6	3.53
	2.5 ± 0.3	
	2.8 − 3.6	
	3.37 ± 0.1	
Cd	3.2 ± 0.1	3.44
Ga		3.52, 3.50, 3.48
Hg (α)	4.6 ± 0.1	3.95
In	3.63 ± 0.1	3.65
	3.45 ± 0.07	
	3.61	
La	1.65 − 3.0 (fcc)	3.72 (fcc) (d-hep)
	3.2	
Nb	3.84 ± 0.06	3.65
	3.6	
	3.6	
Pb	4.29 ± 0.04	3.95
	4.38 ± 0.01	
Sn	3.46 ± 0.1	3.61, 3.57
	3.10 ± 0.05	
	3.51 ± 0.18	
	2.8 − 4.06	
	3.1 − 4.3	
Ta	3.60 ± 0.1	3.63
	3.5	
	3.65 ± 0.1	
Tl	3.57 ± 0.05	3.63
	3.9	
V	3.4	3.50
Zn	3.2 ± 0.1	3.44

†Taken from Mersevey and Schwartz (1969), by courtesy of Marcel Dekker Inc.

At other temperatures, equation (8.27) must be solved numerically for the gap parameter. The coupling constant can, however, be eliminated using integration by parts. Then

$$[N(0)V]^{-1} = \sinh^{-1}[\omega_D/\Delta(T)] + 2\int_0^{\omega_D} d\epsilon \sinh^{-1}[\epsilon/\Delta(T)]\, \partial f(E)/\partial\epsilon$$

$$= \ln[2\omega_D/\Delta(T)] + 2\int_0^{\infty} d\epsilon \sinh^{-1}[\epsilon/\Delta(T)]\, \partial f(E)/\partial\epsilon$$

$$= \ln[2\omega_D/\Delta(T)] + F[\Delta(T)/T].$$

Using equation (8.8) to eliminate the coupling constant we obtain the universal equation

$$\Delta(T) = 1.77k_B T_c \exp\{F[\Delta(T)/T]\}. \qquad (8.33)$$

With equation (8.32) this shows that $\Delta(T)/\Delta(0)$ is a universal function of the reduced temperature T/T_c. A plot of this function is shown in Fig. 8.3. Shown in the same figure are values of the gap in Pb, In, and Sn taken from tunnelling measurements.

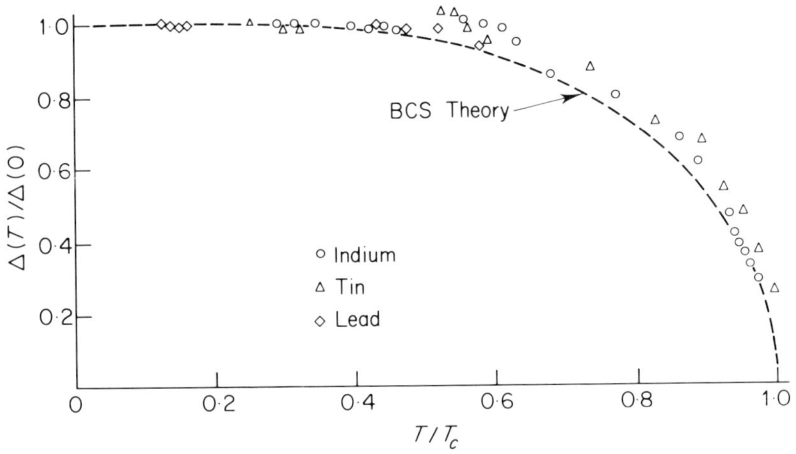

FIG. 8.3. The energy gaps of Pb, Sn and In films, obtained by tunnelling methods, as a function of reduced temperature. The dotted curve is the universal curve, due to BCS, obtained from equation (8.33) (after Giaever and Megerle 1961).

The density of states in one spin of the new quasi-particles is given by

$$N(\omega) = -\frac{1}{\mathscr{V}\pi}\sum_k \mathrm{Im}\,[G_{Rk\uparrow\uparrow}(k,\omega)]$$

$$= -\frac{1}{\mathscr{V}\pi}\sum_k \mathrm{Im}\,\frac{\omega+\epsilon_k}{[(\omega+i\delta)^2 - E_k^2]}$$

$$= N(0) \int_{-\infty}^{\infty} d\epsilon \{\tfrac{1}{2}[1+(\epsilon/E)]\,\delta(\omega-E)+\tfrac{1}{2}[1-(\epsilon/E)]\,\delta(\omega+E)\}.$$

For $\omega > 0$, this becomes

$$N(\omega)/N(0) = \int_0^{\infty} d\epsilon\,\delta(\omega - E),$$

where we have added together the contributions from negative and positive ϵ. Hence

$$\frac{N(\omega)}{N(0)} = \int_{\Delta}^{\infty} \frac{E\,dE\,\delta(\omega - E)}{\epsilon} = \frac{\omega}{(\omega^2 - \Delta^2)^{1/2}}, \qquad \omega \geqslant \Delta. \quad (8.34)$$

This is shown in Fig. 8.4. Immediately above the gap the density of states becomes infinite. As the energy increases further the density of states decreases asymptotically to that for the normal metal. This density of states is responsible for many of the properties of superconductors and is observed directly in the tunnelling of quasi-particles (see §8.7).

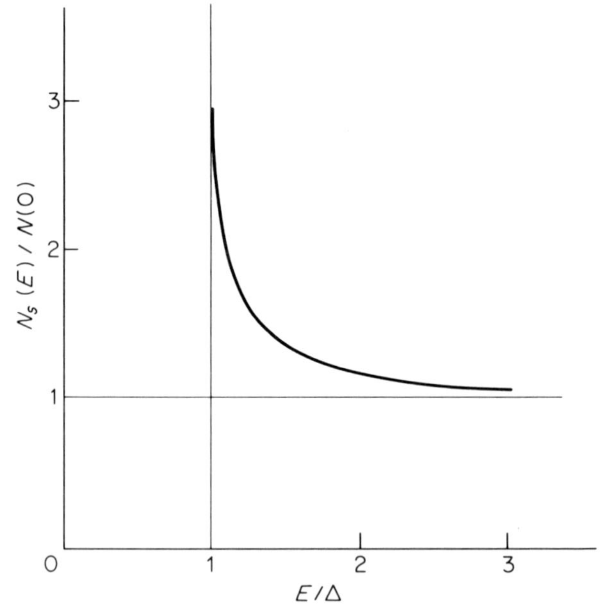

FIG. 8.4. Ratio of the density of states in the superconducting state to that in the normal state plotted against E/Δ.

We showed in Chapter 2 that the thermodynamic potential can be related to the one-particle Green's function through equation (2.97) where the Green's function is determined in terms of a variable interaction λV. The thermodynamic potential is a function of the parameter λ and is equal to that in the normal state when $\lambda = 0$ and to that in the superconducting state when $\lambda = 1$. The difference in these potentials is known from a thermodynamic argument (see the general references at the end of the chapter) to be related to the critical magnetic field, H_c by

$$\tfrac{1}{2}\mu_0 H_c^2 = \Omega(0) - \Omega(1).$$

From equation (2.97) we then obtain (with an extra factor of 2 for the two spins),

$$\tfrac{1}{2}\mu_0 H_c^2 = \beta^{-1} \sum_l \sum_k \int_0^1 \lambda^{-1} d\lambda \{\mathrm{Tr}\,[\tfrac{1}{2}(1+\tau_3)(-i\zeta_l + \epsilon_k) G(k, \zeta_l)]$$

$$+ 1\} \exp(i\zeta_l 0_+)$$

$$= \beta^{-1} \sum_l \sum_k \int_0^1 \frac{d\lambda}{\lambda} \frac{\Delta^2}{\zeta_l^2 + \epsilon_k^2 + \Delta^2} \exp(i\zeta_l 0_+). \tag{8.35}$$

The gap equation, however, is (cf. equation (8.27) and before)

$$1 = \lambda V \beta^{-1} \sum_l \sum_k (\zeta_l^2 + \epsilon_k^2 + \Delta^2)^{-1} \exp(i\zeta_l 0_+). \tag{8.36}$$

Hence,

$$\tfrac{1}{2}\mu_0 H_c^2 = \int_0^1 d\lambda \Delta^2(T, \lambda)/\lambda^2 V. \tag{8.37}$$

But from equation (8.36)

$$-d\lambda/\lambda^2 V = -d\Delta^2 \beta^{-1} \sum_l \sum_k [(\zeta_l^2 + \epsilon_k^2 + \Delta^2)^2]^{-1}.$$

Hence

$$\tfrac{1}{2}\mu_0 H_c^2(T) = \beta^{-1} \sum_l \sum_k \int_0^{\Delta^2(T)} \frac{y\,dy}{(\zeta_l^2 + \epsilon^2 + y)^2}$$

$$= \beta^{-1} \sum_l \sum_k \left\{ \ln\left[\frac{\zeta_l^2 + \epsilon_k^2 + \Delta^2}{\zeta_l^2 + \epsilon_k^2}\right] - \frac{\Delta^2}{\zeta_l^2 + \epsilon_k^2 + \Delta^2} \right\}. \tag{8.38}$$

Since Δ vanishes for $|\epsilon_k| > \omega_D$, the sum over ϵ cuts off at ω_D. When T is zero this can be evaluated explicitly and yields the BCS result

$$\tfrac{1}{2}\mu_0 H_c^2(0) = \tfrac{1}{2} N(0) \Delta^2(0). \tag{8.39}$$

For other temperatures it can be evaluated numerically.

Other thermodynamic functions such as the entropy and specific heat can be obtained by differentiation with respect to T from equation (8.38). All are found to agree satisfactorily with observation. A detailed comparison of theory and observation is given in the paper of Meservey and Schwartz (1969).

(b) *Fuller treatment of the interaction*

The treatment given in the last two pages suffers from two significant defects. It ignores the Coulomb interaction, even the screened Coulomb interaction, entirely and it simplifies the interaction due to the phonons. To deal with these we have to go back to the original Hamiltonian, equation (7.85), which we repeat here,

$$H = \sum_{k,\sigma} \epsilon_k c^+_{k\sigma} c_{k\sigma} + \sum_{q,\lambda} \omega_\lambda(q)[a^+_\lambda(q) a_\lambda(q) + \tfrac{1}{2}]$$

$$+ \tfrac{1}{2} \sum_{k,k',q} V(q) c^+_{k\sigma} c^+_{k'+q\sigma'} c_{k'\sigma'} c_{k+q\sigma}$$

$$+ \sum_{k,k',\sigma,\lambda} g_{k,k',\lambda} c^+_{k',\sigma} c_{k,\sigma} [a_\lambda(q) + a^+_\lambda(-q)]. \qquad (8.40)$$

We saw that one effect of the Coulomb interaction is to renormalize the phonon propagators and the electron–phonon interaction. In the following when we draw diagrams we assume that this renormalization has been taken into account; the phonon propagators will be renormalized ones, the electron–phonon vertex will be the renormalized one and the Coulomb interaction the screened one. The residual electron–phonon interaction is weak but nevertheless produces the pairing that was discussed in the last subsection. We therefore wish to treat this interaction by a Hartree approximation appropriate to the pairing.

We begin by introducing the operators α_k defined by equation (8.17), into the Hamiltonian. Since

$$c^+_{k'\uparrow} c_{k\uparrow} + c^+_{-k\downarrow} c_{-k'\downarrow} = \alpha^+_{k'} \tau_3 \alpha_k + \delta_{kk'}$$

the Hamiltonian can be written, apart from constants which have no dynamic effects and are the same in the normal and superconducting states, as

$$H = \sum_k \epsilon_k \alpha^+_k \tau_3 \alpha_k + \sum_{q,\lambda} \omega_\lambda(q)[a^+_\lambda(q) a_\lambda(q) + \tfrac{1}{2}]$$

$$+\tfrac{1}{2}\sum_{k,q,k'} V(q)(\alpha_k^+ \tau_3 \alpha_{k+q})(\alpha_{k'+q}^+ \tau_3 \alpha_{k'})$$

$$+\sum_{k,q,\lambda} g_{k,k',\lambda}(\alpha_{k'}^+ \tau_3 \alpha_k)[a_\lambda(q) + a_\lambda^+(-q)]. \qquad (8.41)$$

We again treat the interaction terms by perturbation theory. The rules are the same as given in § 7.4 except for the following changes:

(1) A bare electron propagator now represent the matrix

$$\bar{G}_0(k, \zeta_l) = -\int_0^\beta \langle T\alpha_k(\tau)\alpha_k^+\rangle_0 \exp(i\zeta_l\tau)d\tau = [i\zeta_l - \epsilon_k\tau_3]^{-1}$$
$$= (i\zeta_l + \epsilon_k\tau_3)/(-\zeta_l^2 - \epsilon_k^2) \qquad (8.42)$$

(as given by equation (8.23) with $\Delta = 0$).

(2) Bare electron–phonon and electron–Coulomb vertices carry an extra factor τ_3.

(3) The order of matrices along an electron line follows the order of the corresponding vertices and propagators along the line.

(4) For any closed electron line, we take the trace over the product of matrices represented by the line.

The lowest-order contributions to the electron self-energy (matrix), $\Sigma(k, \zeta)$ are given by the Hartree–Fock diagrams shown in Fig. 8.5. From them we obtain

$$\Sigma(k,\zeta) = \beta^{-1}\sum_{k',\zeta'}\Big[-V_s(k-k', \zeta-\zeta')$$
$$+\sum_\lambda |g_{Rkk'\lambda}|^2 D_\lambda(k-k', \zeta-\zeta')\Big]\tau_3 \bar{G}(k', \zeta')\tau_3, \qquad (8.43)$$

$\Sigma(k,\zeta) =$

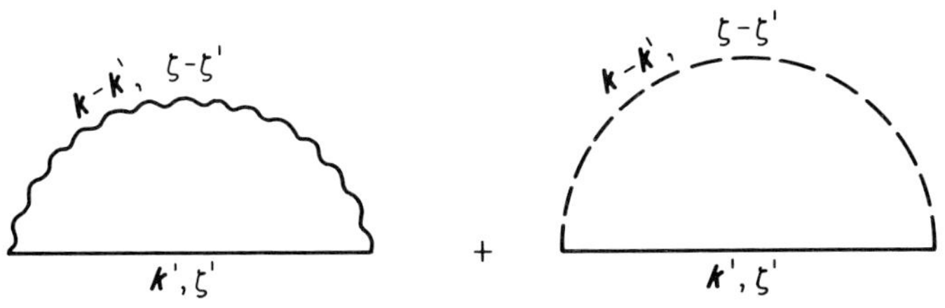

FIG. 8.5. Hartree–Fock contributions to the electron self-energy.

where, as explained above V_s is the screened Coulomb interaction, D the propagator for renormalized phonons and $g_{Rkk'\lambda}$ the screened electron–phonon interaction. $\bar{G}$ is the full electron Green's function given by

$$\bar{G}(k, \zeta)^{-1} = i\zeta_l - \epsilon_k \tau_3 - \Sigma(k, \zeta), \tag{8.44}$$

which depends on Σ. Equation (8.43) is, therefore, a self-consistent equation for Σ and is a generalization of the gap equation. It has solutions of the form

$$\Sigma(k, \zeta) = [1 - Z(k, \zeta)]\, i\zeta + \phi(k, \zeta)\tau_1 + X(k, \zeta)\tau_3, \tag{8.45}$$

which describe a superconducting state and generalize the BCS solution. We refer the reader to the review article of Scalapino (1969) for a detailed discussion of the solution and a comparison of the results with experiment. One characteristic of the solutions is that there still exists a gap parameter but that it is now a function of the energy ω. Further, for many isotropic metals the transition temperature depends on the phonons only through the mass-enhancement factor λ given in §7.8. The transition temperatures can therefore be used to determine λ.

§8.4 Effects of external fields

With Green's functions, the technique for finding the effects of external fields on superconductors is completely analogous to that for the normal state. As an example we consider the important case of the response to an external electromagnetic field. This encompasses the Meissner effect, superconductivity, infrared transmission and absorption and the anomalous skin effect. Since there is a gap in the density of states we expect striking differences from the properties of normal metals. For example, at low frequencies it will be impossible to excite across the gap and we should expect absorption to fall. It will actually be zero at the absolute zero of temperature. At higher temperatures some quasi-particles which can absorb small amounts of energy will be excited. There will, therefore, be some absorption at low energies.

For simplicity we shall ignore the effects of band structure and use

$$\epsilon_k = (k^2/2m) - \mu.$$

In the presence of a transverse electromagnetic field described by a vector potential $A(r, t)$, the kinetic energy term in the Hamiltonian becomes

$$K = \sum_\sigma \int d^3 r \psi_\sigma^+(r)(1/2m)(-i\nabla - eA)^2 \psi(r), \tag{8.46}$$

and the operator for current density is [see equation (4.58)]

$$\begin{aligned} j_{\rm op}(r) &= (e/2m) \sum_\sigma \{\psi_\sigma^+(-i\nabla - eA)\psi_\sigma + [(i\nabla - eA)\psi_\sigma^+]\psi_\sigma\} \\ &\equiv j_i(r) - (e^2 A/m) \sum_\sigma \psi_\sigma^+(r)\psi_\sigma(r). \end{aligned} \tag{8.47}$$

From equation (8.46) we see that the Hamiltonian contains a term linear in A,

$$-\int A(r, t) \cdot j_1(r)\, d^3 r. \tag{8.48}$$

According to the analysis in §4.3, the current density produced by a field of wave vector q, frequency ω, is given by

$$J_\alpha(q, \omega) = -\sum_\beta K_{\alpha\beta}(q, \omega) A_\beta(q, \omega), \tag{8.49}$$

where

$$K_{\alpha\beta}(q, \omega) = (ne^2/m)\delta_{\alpha\beta} + G^j_{R\,\alpha\beta}(q, \omega), \tag{8.50}$$

and $G^j_{R\,\alpha\beta}(q, \omega)$ is the space and time Fourier transform of the retarded Green's function,

$$G^j_{R\,\alpha\beta}(1, 1') = -i\langle[j_{1\alpha}(1), j_{1\beta}(1')]\rangle\, \theta(t - t'). \tag{8.51}$$

According to the general theory $G^j_{R\,\alpha\beta}(q, \omega)$ is obtained by analytic continuation from $G^j_{\alpha\beta}(q, i\zeta)$, the Fourier transform of the temperature Green's function

$$G^j_{\alpha\beta}(1, 1') = -\langle T\tilde{j}_{1\alpha}(1)\tilde{j}_{1\beta}(1')\rangle. \tag{8.52}$$

We continue in analogy with the discussion of §4.3, except that we now have to expand (8.52) in terms of the unperturbed Green's functions for the superconducting state which we found in the last section. This is most easily executed in terms of the matrix Green's functions and operators α_k which we introduced. In terms of them

$$\begin{aligned} j_1(r) &= \sum_q j_1(q) \exp(iq \cdot r), \\ j_1(q) &= (e/2m) \sum_{k,\sigma} (2k + q) c^+_{k\sigma} c_{k+q\sigma} \\ &= (e/2m) \sum_k (2k + q)[c^+_{k\uparrow} c_{k+q\uparrow} - c^+_{-k-q\downarrow} c_{-k\downarrow}] \end{aligned}$$

$$= (e/2m) \sum_{k} (2\boldsymbol{k} + \boldsymbol{q})[c_{k\uparrow}^{+} c_{k+q\uparrow} + c_{-k\downarrow} c_{-k-q\downarrow}^{+} - \delta_{q,0}]$$

$$= (e/2m) \sum_{k} (2\boldsymbol{k} + \boldsymbol{q}) \alpha_{k}^{+} \alpha_{k+q}. \tag{8.53}$$

The Kronecker delta yields zero when the sum over $\boldsymbol{k}$ is performed.

As in §4.3, the leading contribution to (8.52) is given by the diagram in Fig. 8.6, where the solid lines represent the matrix Green's functions. Arising from equation (8.53), there is a factor of $e(2\boldsymbol{k} + \boldsymbol{q})/2m$ at each vertex. Further because we have a closed electron loop we have to take the trace of the matrices involved. Hence

$$G_{\alpha\beta}^{j}(\boldsymbol{q}, i\zeta) = \frac{e^2}{4m^2\beta\mathscr{V}} \sum_{k,\lambda} (2k_\alpha + q_\beta)(2k_\alpha + q_\beta) \times \operatorname{Tr} \bar{G}(\boldsymbol{k}, \omega_\lambda)\bar{G}(\boldsymbol{k} + \boldsymbol{q}, \omega_\lambda + \zeta). \tag{8.54}$$

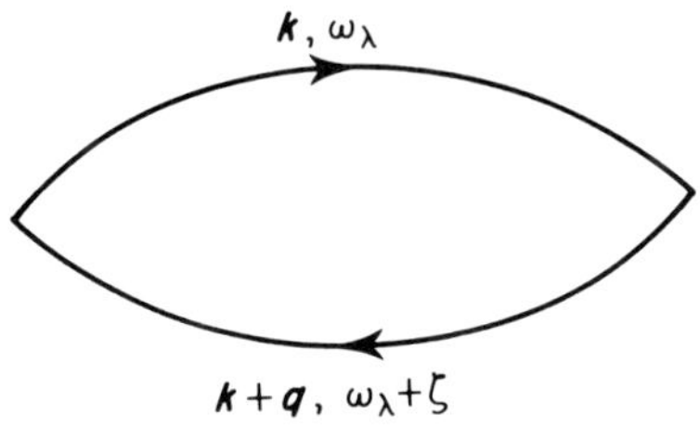

FIG. 8.6. The leading contribution to the current–current correlation function.

We can perform the sum over λ and the analytic continuation by introducing again the spectral form,

$$\bar{G}(\boldsymbol{k}, \zeta) = -\int_{-\infty}^{\infty} \frac{d\omega}{\pi} \frac{\operatorname{Im} G(\boldsymbol{k}, \omega + i\delta)}{i\zeta - \omega}. \tag{8.55}$$

Then

$$G_{\alpha\beta}^{j}(\boldsymbol{q}, i\zeta) = \frac{e^2}{4m^2\beta\mathscr{V}} \sum_{k,\lambda} (2k_\alpha + q_\alpha)(2k_\beta + q_\beta) \times \int_{-\infty}^{\infty} \frac{dx}{\pi} \int_{-\infty}^{\infty} \frac{dy}{\pi} \frac{\operatorname{Tr} \operatorname{Im} G(\boldsymbol{k}, x + i\delta) \operatorname{Im} G(\boldsymbol{k} + \boldsymbol{q}, y + i\delta)}{(i\omega_\lambda - x)(i\omega_\lambda + i\zeta - y)}$$

$$= \frac{e^2}{4m^2\mathscr{V}} \sum_{k} (2k_\alpha + q_\alpha)(2k_\beta + q_\beta) \int_{-\infty}^{\infty} \frac{dx}{\pi} \int_{-\infty}^{\infty} \frac{dy}{\pi}$$

$$\times \frac{\mathrm{Tr}\,\mathrm{Im}\,G(\boldsymbol{k},x+i\delta)\,\mathrm{Im}\,G(\boldsymbol{k}+\boldsymbol{q},y+i\delta)[f(x)-f(y)]}{x-y+i\zeta}$$

and

$$G^{j}_{\alpha\beta}(\boldsymbol{q},\omega) = \frac{e^2}{4m^2\mathscr{V}}\sum_{\boldsymbol{k}}(2k_\alpha+q_\alpha)(2k_\beta+q_\beta)\int_{-\infty}^{\infty}\frac{dx}{\pi}\int_{-\infty}^{\infty}\frac{dy}{\pi}\frac{f(x)-f(y)}{x-y+\omega}$$

$$\times\,\mathrm{Tr}\,\mathrm{Im}\,G(\boldsymbol{k},x+i\delta)\,\mathrm{Im}\,G(\boldsymbol{k}+\boldsymbol{q},y+i\delta). \qquad (8.56)$$

Equations (8.48), (8.49) and (8.56) contain linear response of a pure superconductor to a general electromagnetic field. Here we shall consider only the case of a static field. Further discussion can be found in Rickayzen (1965, 1969).

Let us specialize now to the simple model of BCS. Then

$$\mathrm{Im}\,G(\boldsymbol{k},x+i\delta) = \mathrm{Im}\frac{x+\epsilon_{\boldsymbol{k}}\tau_3+\Delta\tau_1}{(x+i\delta)^2-E_{\boldsymbol{k}}^2}$$

$$= -\pi\frac{(x+\epsilon_{\boldsymbol{k}}\tau_3+\Delta\tau_1)}{2E_{\boldsymbol{k}}}[\delta(x-E_{\boldsymbol{k}})-\delta(x+E_{\boldsymbol{k}})]. \qquad (8.57)$$

If this is used for the two spectral functions in (8.56) we obtain

$$G^{j}_{R\alpha\beta}(\boldsymbol{q},\omega) = \frac{e^2}{8m^2\mathscr{V}}\sum_{\boldsymbol{k}}(2k_\alpha+q_\alpha)(2k_\beta+q_\beta)\left\{\left(1+\frac{\epsilon\epsilon^1+\Delta^2}{EE^1}\right)\right.$$

$$\times\left[\frac{f(E)-f(E^1)}{E-E^1+\omega}-\frac{f(-E)-f(-E^1)}{E^1-E+\omega}\right]$$

$$\left.+\left(1-\frac{\epsilon\epsilon^1+\Delta^2}{EE^1}\right)\left[\frac{f(E)-f(-E^1)}{E+E^1+\omega}+\frac{f(-E)-f(E^1)}{-E-E^1+\omega}\right]\right\}, \qquad (8.58)$$

where we use the abbreviated notation

$$\epsilon = \epsilon_{\boldsymbol{k}}, \qquad \epsilon^1 = \epsilon_{\boldsymbol{k}+\boldsymbol{q}}, \qquad E = E_{\boldsymbol{k}}, \qquad E^1 = E_{\boldsymbol{k}+\boldsymbol{q}}.$$

The denominators in equation (8.58) show that the first terms arise from transitions in which quasi-particles are scattered by the field while the last terms arise from transitions in which pairs of quasi-particles are created or recombine. The factors

$$1+(\epsilon\epsilon^1+\Delta^2)/EE^1 \quad \text{and} \quad 1-(\epsilon\epsilon^1+\Delta^2)/EE^1$$

are squares of matrix elements for these processes. In the limit of $\Delta \to 0$ the analogous processes are the scattering of electrons and holes and the creation or recombination of electron–hole pairs.

We look first at the response to a static magnetic field. This is given by $\omega = 0$ and should explain the Meissner effect. In fact the existence of a Meissner effect depends on the behaviour of $K(q, 0)$ at long wavelength, $q \to 0$. In this limit

$$\lim_{q \to 0} G^{j}_{R\alpha\beta}(q, 0) = \frac{2e^2}{m^2 \mathscr{V}} \sum_{k} k_\alpha k_\beta \frac{\partial f(E)}{\partial E} = \frac{2e^2}{3m^2 \mathscr{V}} \delta_{\alpha\beta} \sum_{k} k^2 \frac{\partial f(E)}{\partial E}.$$

Hence

$$\lim_{q \to 0} K_{\alpha\beta}(q, 0) = (n_s e^2/m)\delta_{\alpha\beta}$$

where n_s is the density of superconducting electrons (see §8.5 below) and is given by

$$n_s = n + (2/3m\mathscr{V}) \sum_{k} k^2 \, \partial f(E)/\partial E. \tag{8.59}$$

If n_s is not zero a Meissner effect exists (see Rickayzen (1965), p. 48). At the absolute zero of temperature, the second term in (8.59) is zero and

$$n_s = n.$$

Hence, a Meissner effect exists at absolute zero. As the temperature rises, the magnitude of the second term increases, partly because the distribution increases and partly because the gap falls. However, the term is negative, so n_s decreases with increasing temperature until at the transition temperature n_s is zero.

If the important values of q are much less than Δ/v_F, $K_{\alpha\beta}(0, 0)$ can be used for finding the effect of static magnetic fields. The vector potential within the superconductor then satisfies

$$\nabla^2 \boldsymbol{A} = -\boldsymbol{J}/\mu = (n_s e^2/m\mu)\boldsymbol{A}. \tag{8.60}$$

The magnetic field then decays within a distance λ of the surface of the material, where λ is called the penetration depth and is given by

$$\lambda = (m\mu/n_s e^2)^{1/2}.$$

Since the important values of q in the solution of (8.60) are of the order of λ^{-1}, this solution requires

$$\lambda^{-1} \ll \Delta/v_F. \tag{8.61}$$

This limit is known as the London limit. It is actually rarely satisfied for pure superconductors although it is for superconductors with a sufficiently large concentration of impurities. The length $\pi v_F/\Delta$ is called the coherence length of the material and is denoted by

$$\zeta_0 = v_F/\pi\Delta.$$

Condition (8.61) can then be written

$$\lambda \gg \zeta_0.$$

When this inequality is not satisfied it is necessary to use the full form for $K(q, 0)$ together with equation (8.79) to determine the penetration of the field into the superconductor.

The long-wavelength result

$$\boldsymbol{J} = -(n_s e^2/m)\boldsymbol{A}, \tag{8.62}$$

is not gauge-invariant, a point that was a matter of controversy at one time. It is not gauge-invariant because, contrary to equation (8.62), a static longitudinal vector potential does not contribute to the magnetic field and produces no current. Formally, this can be seen as follows. A longitudinal vector potential can be written as the gradient of a scalar function

$$\boldsymbol{A} = \nabla\chi.$$

But $\boldsymbol{A}$ appears in the Hamiltonian and current density only in the forms,

$$(-i\nabla - e\boldsymbol{A})\psi \quad \text{and} \quad (i\nabla - e\boldsymbol{A})\psi^+,$$

which now become

$$(-i\nabla - e\nabla\chi)\psi \text{ and } (i\nabla - e\nabla\chi)\psi^+.$$

But χ can be eliminated from these expressions if we make the unitary transformation

$$\psi = \phi \exp(ie\chi), \qquad \psi^+ = \exp(-ie\chi)\phi^+.$$

The operators ϕ obey the same continuation relations as ψ, and, since ψ appears originally in the remainder of the Hamiltonian only in the form

$$\psi^+\psi = \phi^+\phi,$$

the Hamiltonian is the same function as the ϕ's as it was of the ψ's. Hence $\nabla\chi$ is eliminated from the Hamiltonian and the current density and so it will induce no electric current. Equation (8.80) cannot, therefore, be correct when $\boldsymbol{A} = \nabla\chi$.

In fact, we have not been entirely consistent. Although the longitudinal vector potential has no physical effect it does affect the gap parameter. If we consider the simple model,

$$\begin{aligned}\Delta &= V\sum_{\boldsymbol{k}} \langle c_{-\boldsymbol{k}\downarrow} c_{\boldsymbol{k}\uparrow}\rangle = V\langle\psi_\downarrow(\boldsymbol{r})\psi_\uparrow(\boldsymbol{r})\rangle \\ &= V\langle\phi_\downarrow(\boldsymbol{r})\phi_\uparrow(\boldsymbol{r})\rangle \exp(2ie\chi).\end{aligned}$$

Hence the introduction of the longitudinal vector potential leads to a change in the phase of Δ. The gap parameter Δ is, therefore, not a gauge-invariant quantity although

$$(-i\nabla - 2e\boldsymbol{A})\Delta$$

is. This suggests that if we were to calculate the one-particle Green's functions in the presence of the vector potential using the diagrams of Fig. 8.7 but with the bare electron lines standing for propagators of non-interacting electrons in the presence of the vector potential, gauve invariance would look after itself. Expanding the result to a first order in $\boldsymbol{A}$ should lead to a gauge-invariant result for $\boldsymbol{A}$. This method was discussed in §3.8 in connection with conservation laws. In fact, it can be shown that gauge invariance is responsible for conservation of electric charge so that it is not surprising that we should be led to the same method.

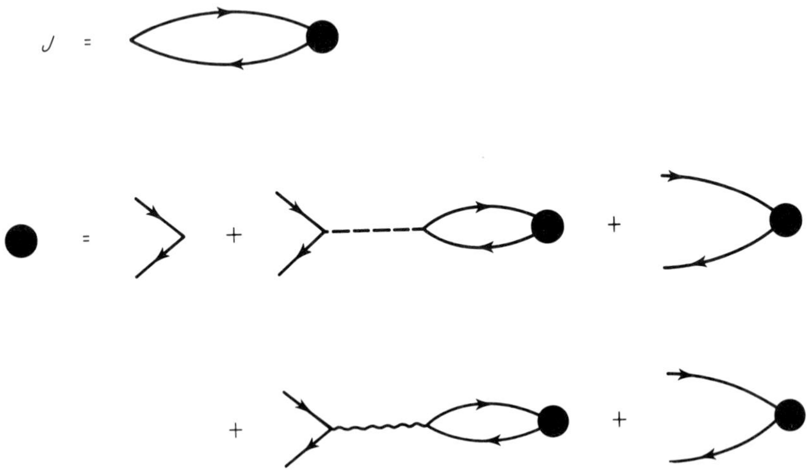

FIG. 8.7. Contributions to J which are consistent with the Ward identity (see §3.9) and gauge invariance.

From the discussion in §3.8 it is clear that the advocated method implies that the response to the vector potential should be calculated by replacing the contribution of Fig. 8.6 by that of Fig. 8.7 which does include that of Fig. 8.6. Contributions from diagrams of the form (*a*) vanish for symmetry reasons because the first electron loop that contributes to J then contains one vector operator at one vertex and scalar operators elsewhere. The effect of Fig. 8.7 on the current density has been calculated by a number of authors (Anderson 1958, Rickayzen 1959, Nambu 1960, Bogoliubov *et al.* 1958). The result has been shown to be gauge-invariant explicitly. It has also been shown that the correction to the previous result for the effect of a *transverse* electromagnetic field is negligible.

For the further analysis of the electromagnetic properties of superconductors as well as for discussions of other reversible processes, the reader is referred to the many books on the subjects listed at the end of this chapter.

§8.5 Effects of impurities

The effects of impurities are particularly important in the case of superconductivity; we cannot be sure that the conductivity is infinite without considering their effect. Furthermore, magnetic impurities which can flip the spins of electrons interfere with the pairing and reduce the gap parameter. A sufficient concentration of magnetic impurities leads to gapless superconductivity. Here, we wish to take into account these effects in the simplest possible way. We shall, therefore, include only s-wave scattering by the impurities. The interaction term due to the impurities is, therefore,

$$\begin{aligned} H' &= U_1 \sum_i \Psi^+(\boldsymbol{R}_i)\Psi(\boldsymbol{R}_i) + U_2 \sum_\beta \Psi^+(\boldsymbol{R}_\beta)\boldsymbol{S}_\beta \cdot \sigma\Psi(\boldsymbol{R}_\beta) \\ &= \int d^3 r V_1(\boldsymbol{r})\Psi^+(\boldsymbol{r})\Psi(\boldsymbol{r}) + \int d^3 r \boldsymbol{V}_2(\boldsymbol{r}) \cdot \Psi^+(\boldsymbol{r})\boldsymbol{\sigma}\Psi(\boldsymbol{r}); \end{aligned}$$

$$V_1(\boldsymbol{r}) = U_1 \sum_i \delta(\boldsymbol{r}-\boldsymbol{R}_i), \qquad \boldsymbol{V}_2(\boldsymbol{r}) = U_2 \sum_\beta \boldsymbol{S}_\beta \delta(\boldsymbol{r}-\boldsymbol{R}_\beta). \tag{8.63}$$

Here, we assume that we have non-magnetic impurities of one type, at positions $\boldsymbol{R}_i$ and magnetic impurities of one type with spins $\boldsymbol{S}_\beta$ at positions $\boldsymbol{R}_\beta$. We have also written the field operators in the matrix form

$$\Psi = \begin{pmatrix} \psi\uparrow \\ \psi\downarrow \end{pmatrix}. \tag{8.64}$$

(Some authors use $V_2' \boldsymbol{S} \cdot \boldsymbol{s}$ for the interaction with magnetic impurities. Comparison with (8.63) shows that $V_2' = 2V_2$).

We take the Hamiltonian for the BCS conductor in the simple BCS form

$$H = \sum \epsilon_k c_{k\sigma}^+ c_{k\sigma} - \sum (\Delta c_{k\uparrow}^+ c_{-k\downarrow}^+ + \Delta c_{-k\downarrow} c_{k\uparrow}), \quad (8.65)$$

where Δ has to be found self-consistently from

$$\Delta = V \sum_k \langle c_{-k\downarrow} c_{k\uparrow} \rangle. \quad (8.66)$$

The Hamiltonian (8.65) can also be written in the form

$$H = \int \Psi^+ \epsilon(-i\nabla)\Psi d^3r - \Delta \int d^3r [\psi_\uparrow^+(r) + \psi_\downarrow(r)\psi_\uparrow(r)]. \quad (8.67)$$

The equations of motion are therefore,

$$\frac{d\Psi}{d\tau}(r) = -\epsilon(-i\nabla)\Psi(r) + \Delta \begin{pmatrix} \psi_\downarrow^+ \\ -\psi_\uparrow^+ \end{pmatrix} - V_1(r)\Psi - V_2 \cdot \boldsymbol{\sigma} \Psi,$$

$$\frac{d\Psi^+}{d\tau}(r) = \epsilon(-i\nabla)\Psi^+(r) + \Delta(-\psi_\downarrow \psi_\uparrow) + V_1(r)\Psi^+ + \Psi^+ V_2 \cdot \boldsymbol{\sigma}. \quad (8.68)$$

In terms of Ψ^*, the transpose of Ψ^+, these equations can be rewritten

$$\frac{d\Psi}{d\tau}(r) = -\epsilon(-i\nabla)\Psi(r) + i\Delta\sigma_2\Psi^* - V_1(r)\Psi - V_2 \cdot \boldsymbol{\sigma}\Psi,$$

$$\frac{d\Psi^*}{d\tau} = \epsilon(-i\nabla)\Psi^* - i\Delta\sigma_2\Psi + V_1(r)\Psi^* + V_2 \cdot \tilde{\boldsymbol{\sigma}}\Psi^*. \quad (8.69)$$

This can be further reduced formally to one matrix equation if we introduce (Ambegaokar and Griffin 1965) the four-component vector

$$\Phi = \begin{pmatrix} \Psi \\ \Psi^* \end{pmatrix}, \qquad \Psi^* = \begin{pmatrix} \psi_\uparrow^+ \\ \psi_\downarrow^+ \end{pmatrix}. \quad (8.70)$$

If the matrices $\boldsymbol{\tau}$ act on Φ, treating Ψ and Ψ^* each as single components, equation (8.69) becomes

$$\frac{d\Phi}{d\tau} = -\epsilon(-i\nabla)\tau_3\Phi - \tau_2\sigma_2\Delta\Phi - V_1(r)\tau_3\Phi - V_2 \cdot \boldsymbol{\alpha}\Phi, \quad (8.71)$$

where

$$\boldsymbol{\alpha} = \tfrac{1}{2}[\boldsymbol{\sigma}(1+\tau_3) - \tilde{\boldsymbol{\sigma}}(1-\tau_3)] = \tfrac{1}{2}\boldsymbol{\sigma}(1+\tau_3) + \tfrac{1}{2}\sigma_2\boldsymbol{\sigma}\sigma_2(1-\tau_3)$$
$$= \sigma_2 \boldsymbol{j} + \tau_3(\sigma_1 \boldsymbol{i} + \sigma_3 \boldsymbol{k}). \quad (8.72)$$

The corresponding Green's function satisfies

$$[(d/d\tau) + \epsilon(1)\tau_3 + \tau_2\sigma_2\Delta + V_1(r_1)\tau_3 + V_2(r_1).\boldsymbol{\alpha}]\,G(1,2) = -\delta(1,2) \tag{8.73}$$

and can formally be solved in the same way as for the properties of impurities in a normal metal (see Chapter 4).

We treat the impurity terms in (8.73) by perturbation theory and obtain the correction to the self-energy in the Born approximation as illustrated in Fig. 8.8. [It is not difficult to go beyond the Born approximation for the non-magnetic impurities; beyond the Born approximation for the magnetic impurities requires considerations of the Kondo effect (Kondo 1969).] The unperturbed Green's function G_0 is the solution of (8.73) with V_1 and V_2 zero. Its Fourier transform is

$$\bar{G}_0(k,\zeta) = (i\zeta - \epsilon_k\tau_3 - \Delta\tau_2\sigma_2)^{-1}. \tag{8.74}$$

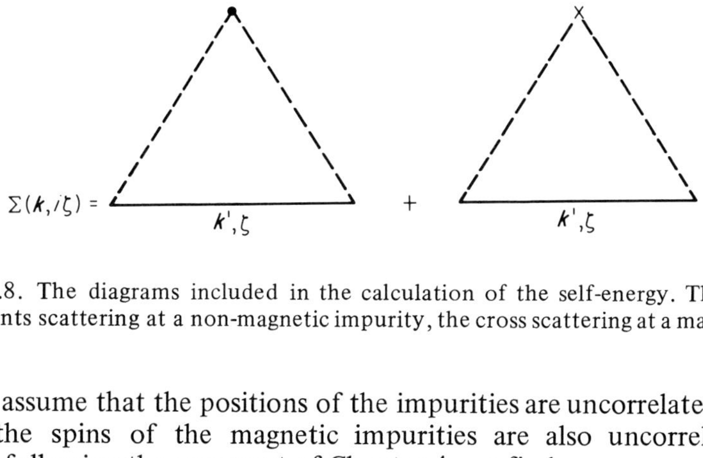

FIG. 8.8. The diagrams included in the calculation of the self-energy. The dot represents scattering at a non-magnetic impurity, the cross scattering at a magnetic one.

We assume that the positions of the impurities are uncorrelated and that the spins of the magnetic impurities are also uncorrelated. Then, following the argument of Chapter 4, we find

$$\Sigma(k,i\zeta) = N_i|U_1|^2 \sum_{k'} \tau_3\bar{G}(k',\zeta)\tau_3 + \tfrac{1}{3}S(S+1)N_\beta|U_2|^2 \sum_{k',i,j} \alpha_i\bar{G}(k',\zeta)\alpha_j \tag{8.75}$$

where N_i is the density of non-magnetic impurities, N_β the density of magnetic ones and we have used the result for the average over magnetic spins,

$$\langle S_{\beta i}S_{\beta j}\rangle = \tfrac{1}{3}S(S+1)\delta_{ij}. \tag{8.76}$$

As before, the renormalization of the τ_3 term in G is negligible. It is therefore possible to write

$$G^{-1} = G_0^{-1} - \Sigma = (i\tilde{\zeta} - \epsilon_k \tau_3 - \tilde{\Delta}\tau_2\sigma_2)^{-1}, \tag{8.77}$$

where

$$\begin{aligned}(i\zeta - i\tilde{\zeta}) - (\Delta - \tilde{\Delta})\tau_2\sigma_2 &= -\frac{1}{2\tau_a}\frac{\tau_3(i\tilde{\zeta} + \tilde{\Delta}\tau_2\sigma_2)\tau_3}{(\tilde{\zeta}^2 + \tilde{\Delta}^2)^{1/2}} \\ &\quad - \frac{1}{6\tau_b}\frac{\sum_i \alpha_i(i\tilde{\zeta} + \tilde{\Delta}\tau_2\sigma_2)\alpha_i}{(\tilde{\zeta}^2 + \tilde{\Delta}^2)^{1/2}} \\ &= -\frac{1}{2\tau_a}\frac{i\tilde{\zeta} - \tilde{\Delta}\tau_2\sigma_2}{(\tilde{\zeta}^2 + \tilde{\Delta}^2)^{1/2}} - \frac{1}{2\tau_b}\frac{(i\tilde{\zeta} + \tilde{\Delta}\tau_2\sigma_2)}{(\tilde{\zeta}^2 + \tilde{\Delta}^2)^{1/2}},\end{aligned} \tag{8.78}$$

with

$$\tau_a^{-1} = N_i \pi N(0)|U_1|^2, \qquad \tau_b^{-1} = N_\beta \pi S(S+1) N(0)|U_2|^2. \tag{8.79}$$

Notice that in the two numerators on the right-hand side of equation (8.78), $\tilde{\Delta}$ has opposite signs showing that the effects of the two kinds of impurity are different. The difference can be traced to the fact that the interaction with non-magnetic impurities is invariant under time reversal but that with magnetic impurities is not (De Gennes 1966). In the former case, it is still possible to pair time reversed states (even if momentum is not conserved) but in the latter it is not. Anderson (1959) was the first to point out that, because of this, non-magnetic impurities will not change the transition temperature nor the gap of an isotropic superconductor.

From equation (8.78) we see that

$$\begin{aligned}\tilde{\zeta} &= \zeta + \frac{1}{2}\left(\frac{1}{\tau_a} + \frac{1}{\tau_b}\right)\frac{\tilde{\zeta}}{(\tilde{\zeta}^2 + \tilde{\Delta}^2)^{1/2}} \\ \tilde{\Delta} &= \Delta + \frac{1}{2}\left(\frac{1}{\tau_a} + \frac{1}{\tau_b}\right)\frac{\tilde{\Delta}}{(\tilde{\zeta}^2 + \tilde{\Delta}^2)^{1/2}}.\end{aligned} \tag{8.80}$$

The equations are closed by the gap equation

$$\begin{aligned}\Delta &= V\sum_k \langle c_{-k\downarrow} c_{k\uparrow}\rangle \\ &= \tfrac{1}{4} V \sum_{k,\zeta} \operatorname{Tr} \sigma_2\tau_2 G(k, i\zeta)\end{aligned}$$

$$= \beta^{-1}\pi VN(0) \sum_{\tilde{\xi}} [\tilde{\Delta}/(\tilde{\Delta}^2 + \tilde{\xi}^2)^{1/2}]. \tag{8.81}$$

It is usual to introduce the ratio

$$u = \tilde{\xi}/\tilde{\Delta}. \tag{8.82}$$

Then, from (8.80), u is given by

$$\xi/\Delta = u\{1 - [\tau_b \Delta(1 + u^2)^{1/2}]^{-1}\} \tag{8.83}$$

and the gap equation becomes

$$\Delta = \pi VN(0)/\beta \sum_u (1 + u^2)^{-1/2}. \tag{8.84}$$

If no magnetic impurities are present τ_b, the spin-flip lifetime becomes infinite and

$$u = \xi/\Delta.$$

The gap equation is then independent of the impurities and, hence, the gap and transition temperature are independent of impurities. This was the result to which Anderson first drew attention.

When τ_b is finite both the transition temperature and gap are affected as are other properties. We mention here a few of the results and refer the reader to the review article of Maki (1969) for further details. The transition temperature T_c is obtained by linearizing the gap equation. One finds that

$$\ln (T_c/T_{c0}) = \psi(\tfrac{1}{2}) - \psi(\tfrac{1}{2} + \rho_c) \tag{8.85}$$

where T_{c0} is the transition temperature of the pure superconductor and $\psi(x)$ is the digamma function

$$\psi(x) = \frac{d}{dx} \ln \Gamma(x)$$

$$= -\gamma + \sum_{n=1}^{\infty} \frac{x}{n(n+x)} \qquad (x \neq -1, -2, -3 \ldots). \tag{8.86}$$

The parameter ρ_c is defined by

$$\rho_c = (2\pi\tau_b T_c)^{-1}, \tag{8.87}$$

and is a measure of the strength of depairing. The magnetic impurities reduce T_c until a concentration is reached for which

$$\tau_b^{-1} = \pi T_{c0}/2\gamma, \tag{8.88}$$

at which concentration T_c is zero and superconductivity is destroyed.

To discuss the energy gap we have to consider the analytic continuation of $\bar{G}(\boldsymbol{k}, \zeta)$ to $G(\boldsymbol{k}, \omega)$. This can be accomplished by putting

$$\omega = i\zeta, \qquad \tilde{\omega} = i\tilde{\zeta}, \qquad V = iu,$$

then

$$G(\boldsymbol{k}, \omega) = (\tilde{\omega} - \epsilon_k \tau_3 - \tilde{\Delta}\tau_2\sigma_3),$$

$$V = \tilde{\omega}/\tilde{\Delta},$$

$$\omega/\Delta = V\{1 - [\tau_b \Delta\sqrt{(1 - V^2)}]^{-1}\}, \tag{8.89}$$

and the density of states is

$$N(\omega) = N(0) \operatorname{Im} [V/\sqrt{(1 - V^2)}]$$
$$= N(0)\tau_b \Delta \operatorname{Im} V. \tag{8.90}$$

In determining V from equation (8.89) we have to remember that G is an analytic function of ω in the upper half-plane. For $\tau_b \Delta < 1$, this means that V has a finite positive imaginary part no matter how small ω is. Hence, there is no gap in the density of states even though the gap parameter is not zero. This is one example of gapless superconductivity, the existence of Cooper pairs (and so of superconductivity) without the existence of an energy gap in the spectrum of excitations. Thus an energy gap is not essential for the existence of Cooper pairs. The density of states, for various values of $\tau_b \Delta$ is shown in Fig. 8.9.

Finally, we consider the electromagnetic response as given (see §8.4) by

$$\boldsymbol{J} = -K(\boldsymbol{q}, \omega)\boldsymbol{A}(\boldsymbol{q}, \omega). \tag{8.91}$$

The Meissner effect and superconductivity are given by this result in the limit $q \to 0$, $\omega \to 0$. When impurities are present one obtains the same result whatever order the limits are taken in (Evans and Rickayzen 1965), and one finds

$$K(0, 0) = n_s e^2/m \tag{8.92}$$

with finite density of superconducting electrons n_s given by (Skalski *et al.* 1964)

$$n_s = \frac{n}{\Delta}\int_\Delta^\infty d\omega \tanh(\tfrac{1}{2}\beta\omega) \operatorname{Re}\left\{(1 - V^2)\left[(V^2 - 1)^{1/2} + \left(\frac{1}{\tau_a} + \frac{1}{\tau_b}\right)\frac{1}{\Delta}\right]\right\}^{-1}. \tag{8.93}$$

When $\tau_a, \tau_b \to \infty$, this reduces to the result of §8.3. Since n_s is finite as long as Δ is finite, it follows that superconductivity exists as long as there is a pairing and even when there is no energy gap. The result

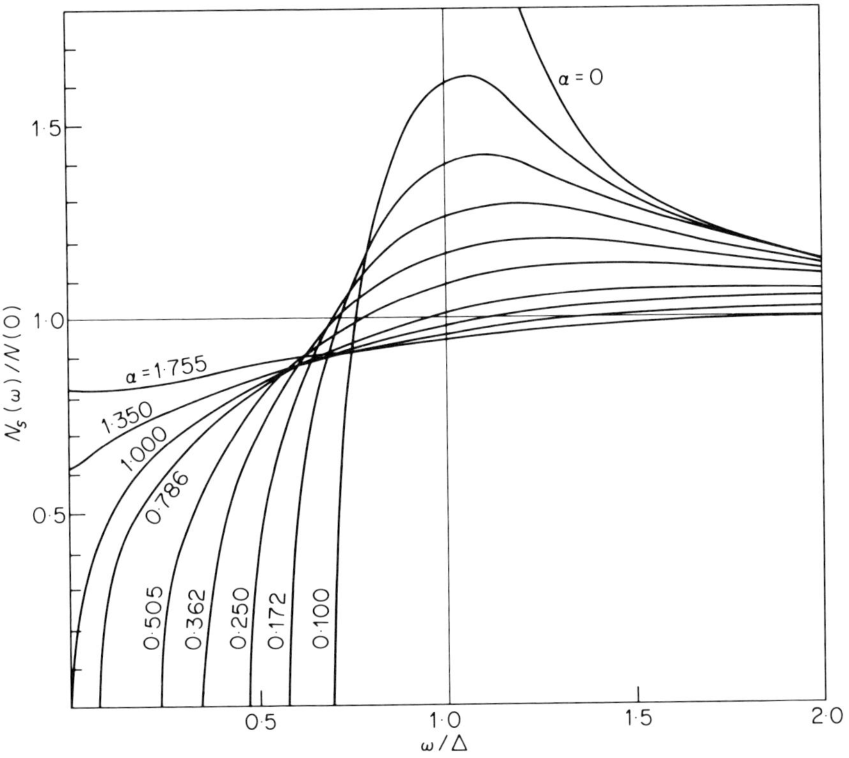

FIG. 8.9. The density of states $N_s(\omega)$ in a paramagnetic alloy as a function of ω/Δ for different values of $\alpha = (\tau_b \Delta)^{-1}$. The case $\alpha = 0$ corresponds to non-magnetic impurities only.

can be understood as follows. Because of the pairing, the BCS state is a rather complicated coherent state. When an external field accelerates all the electrons, the electrons will be paired in the accelerated frame. Random impurities will scatter electrons, possibly breaking up some of the pairs, but the electrons as a whole will come to a metastable state with the new pairing. Just because the impurities scatter randomly, they will not reduce all the pairs to rest at the same time. In the (nearly) translationally invariant superconductor electrons in states of total momentum zero are paired in thermodynamic equilibrium. Under the effect of a field, they are paired with total momentum

$$\boldsymbol{P} = -e\boldsymbol{A}. \tag{8.94}$$

Then

$$\partial \boldsymbol{P}/\partial t = e\boldsymbol{E},$$

i.e. the pairs are freely accelerated by the field. However, $n_s \neq n$ (except in a pure superconductor at the absolute zero of temperature) because some pairs break up.

§8.6 Phase of the gap parameter and flux quantization

We pointed out in the last section that there exist many metastable states of a superconductor corresponding to different momenta of the paired electron states. If a long-wavelength, low-frequency electromagnetic field is present, there is a metastable current given by

$$\boldsymbol{J} = -(n_s e/m)A. \tag{8.95}$$

We saw in §8.4, however, that this expression is not gauge-invariant. We also saw that Δ is not a gauge-invariant quantity but that

$$(-i\boldsymbol{\nabla} - 2e\boldsymbol{A})\Delta$$

is. If we write

$$\Delta = |\Delta| \exp(i\phi), \tag{8.96}$$

the gauge-invariant quantity can be written,

$$(\boldsymbol{\nabla}\phi - 2e\boldsymbol{A})|\Delta| - i\boldsymbol{\nabla}|\Delta|.$$

Only the first of these terms depends on the gauge. Indeed the argument in §8.4 shows that

$$(\boldsymbol{\Delta}\phi - 2e\boldsymbol{A}) \tag{8.97}$$

is gauge-invariant. Different values of this quantity, the generalized momentum of a pair, correspond to different metastable states. As the Hamiltonian depends only on this combination, the current carried in one of these states (for sufficiently slowly varying ϕ and A) is

$$\boldsymbol{J} = (n_s/2m)(\boldsymbol{\nabla}\phi - 2e\boldsymbol{A}). \tag{8.98}$$

In any physical situation the appropriate values of $\nabla\phi$ and $\boldsymbol{A}$ are determined by the boundary conditions. We apply the equation to the case of a superconducting ring much thicker than the penetration depth, carrying a persistent current, see Fig. 8.10.

The supercurrent flows only near the inner surface of the ring creating a magnetic field inside the hole. Well within the ring the current is zero. Hence

$$\boldsymbol{A} = \nabla\phi/2e. \tag{8.99}$$

If we integrate this along a closed path well within the ring but surrounding the hole, we find that the trapped magnetic flux Φ is given by

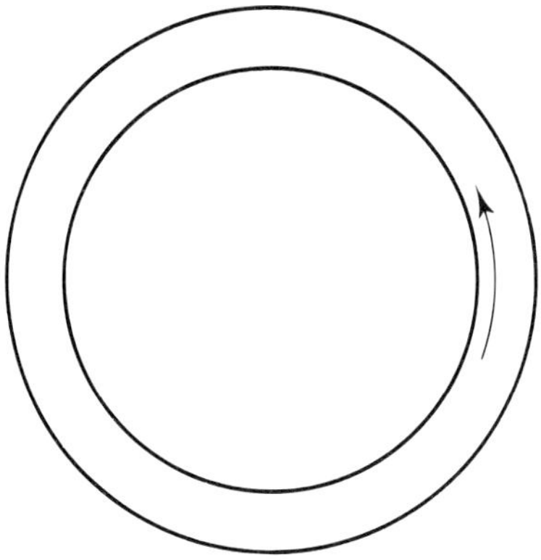

FIG. 8.10. The cross section of a hollow cylinder used for displaying the quantization of magnetic flux.

$$\Phi = \int A \, . \, ds = \frac{1}{2e} \int \nabla \phi \, . \, ds$$
$$= [\phi]/2e, \qquad (8.100)$$

where $[\phi]$ is the change in ϕ in going round the hole. However, Δ (which comes from the wave function) is single-valued. Hence

$$[\phi] = 2\pi n,$$

where n is an integer, and

$$\Phi = \pi n/e. \qquad (8.101)$$

Magnetic flux in the ring is quantized in units of π/e. This is an important example of macroscopic quantization and has now been well confirmed by experiment. In conventional units the quantum of flux is

$$\pi\hbar/e = h/2e = 2.07 \times 10^{-15} \text{ weber}. \qquad (8.102)$$

Flux quantization generally plays an important part in the properties of superconductors in magnetic fields, notably in the case of Type-II superconductors (see Fetter and Hohenberg (1969) for a review) and in Josephson tunnelling which we mention in the following section.

§8.7 Tunnelling of electrons

When two metals are separated by a thin potential barrier such as a thin oxide layer, it is possible for electrons to tunnel through the barrier and for electric currents to flow. If one of the metals is a superconductor the tunnelling current depends on the density of states in the superconductor and this density of states can be studied directly in tunnelling experiments. The experiments are so sensitive that, for the strong-coupling superconductors it is possible (McMillan and Rowell 1969) for the results to be used to help determine the phonon density of states of the superconductor. If both metals are superconducting it is possible for pairs of electrons to tunnel through the barrier. This tunnelling was predicted by Josephson (1962) and leads to a number of interesting effects and applications.

The tunnelling is always weak and can be treated by perturbation theory assuming an interaction term in the Hamiltonian,

$$H' = \sum_{k,q,\sigma} (T_{kq} c_{k\sigma}^{+} d_{q\sigma} + T_{kq}^{*} c_{k\sigma} d_{q\sigma}^{+}). \tag{8.103}$$

Here the first term transfers an electron from the right-hand metal to the left with probability amplitude T_{kq}. The second term represents the inverse process. We show the barrier potential schematically in Fig. 8.11. With this geometry, the amplitude for transfer conserves momentum in directions in the plane of the barrier and is a strongly peaked function of momentum, being large when the momentum is normal to the barrier (McMillan and Rowell 1969). For electrons near the Fermi surface, this means that in T_{pq} we have $k_x \approx k_{FL}$, $q_x \approx q_{FR}$, where k_{FL} and q_{FR} are the Fermi momenta in the two metals. Then T_{pq} is a strong function of $\boldsymbol{p}_\perp (= \boldsymbol{q}_\perp)$ the momentum in the plane of the barrier.

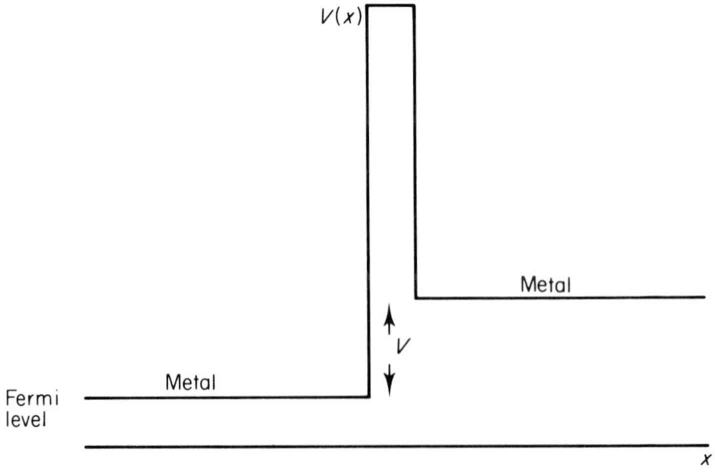

FIG. 8.11. The potential function for electrons in two metals separated by an oxide layer and at a potential difference V.

The tunnelling current I from right to left is the rate at which the number of electrons N_L in the left-hand metal increases times the electronic charge. Hence

$$\begin{aligned} I_{\text{op}} &= e\dot{N}_L \\ &= ie\left[H_0 + H', \sum_{k\sigma} c^+_{k\sigma} c_{k\sigma}\right] \\ &= ie\left[H', \sum_{k\sigma} c^+_{k\sigma} c_{k\sigma}\right], \end{aligned}$$

since the unperturbed Hamiltonian conserves the number of particles on each side of the barrier. Hence

$$I = -ie \sum_{k,\sigma,q} (T_{k,q} c^+_{k\sigma} d_{q\sigma} - T^*_{k,q} d^+_{q\sigma} c_{k\sigma}). \tag{8.104}$$

If the barrier is invariant under time reversal

$$T^*_{-k,-q} = T_{kq}$$

and

$$I = -ie \sum_{k,q} (T_{kq} \alpha^+_k \beta_q - T^*_{kq} \beta^+_q \alpha_k) \tag{8.105}$$

where

$$\alpha_k = \begin{pmatrix} c_{k\uparrow} \\ c_{-k\downarrow} \end{pmatrix}, \qquad \beta_q = \begin{pmatrix} d_{q\uparrow} \\ d^+_{-q\downarrow} \end{pmatrix}. \tag{8.106}$$

In terms of the same operators

$$H' = \sum_{k,q} (T_{kq} \alpha^+_k \tau_3 \beta_q + T^*_{kq} \beta^+_q \tau_3 \alpha_k). \tag{8.107}$$

To first order in H' the mean electric current from right to left is (see equation 1.27)

$$I(t) = \int_{-\infty}^{\infty} G(t', t)\, dt', \tag{8.108}$$

where

$$G(t', t) = -i\, \langle [\hat{H}(t'), \hat{I}_{\text{op}}(t)] \rangle\, \theta(t - t') \tag{8.109}$$

The Green's function is to be found in the usual way by analytic continuation from the corresponding temperature Green's function.

If the right-hand metal is at higher electric potential V than the left-hand metal, one finds by the methods used in this book that

$$G_1(t, t') = e \int_{-\infty}^{\infty} \frac{d\omega}{2\pi} \exp\,[i\omega(t' - t)] \int_{-\infty}^{\infty} \frac{dx}{\pi} \int_{-\infty}^{\infty} \frac{dy}{\pi}$$

$$\times \sum_{k,q} \operatorname{Tr} |T_{kq}|^2 \{\operatorname{Im} [G_L(\boldsymbol{q}, x + i\delta)] \tau_3 \exp(ieVt'\tau_3)$$
$$\times \operatorname{Im} [G_R(\boldsymbol{k}, y + i\delta)] \exp(-ieVt\tau_3)\} \frac{f(x) - f(y)}{x - y + \omega}. \quad (8.110)$$

There are then two kinds of contributions to the result; there are those that arise from the diagonal parts (in τ-space) of the Green's functions and give rise to normal single-particle tunnelling which also occurs in the normal state and those that arise from the off-diagonal parts of the Green's functions. The latter correspond to the tunnelling of pairs of electrons in the condensed phase. They have no analogue in the normal state and give rise to the Josephson effects.

If one of the metals, say the right-hand one is in the normal state, there are no Josephson effects and one finds, for the differential conductance

$$\frac{dI_s}{dV} = -4\pi e |T|^2 \int_{-\infty}^{\infty} dx\, N_L(x) \frac{\partial f(x + eV)}{\partial V} \quad (8.111)$$

where $|T|^2$ is a constant of the metal which depends on the matrix elements, and $N_L(x)$ is the density of states in the left-hand metal, defined by

$$N_L(x) = \int d^3q \operatorname{Tr} [\operatorname{Im} G_L(\boldsymbol{q}, x + i\delta)(1 + \tau_3)/2]. \quad (8.112)$$

As the temperature tends to zero

$$\frac{-\partial f}{\partial V} f(x + eV) \rightarrow e\delta(x + eV)$$

and

$$dI_s/dV = 4\pi e^2 |T|^2 N_L(eV). \quad (8.113)$$

The unknown constant of proportionality can be removed if one observes the ratio of the conductance in the superconducting state to that in the normal state

$$\frac{[dI_s/dV]_s}{[dI_s/dV]_n} = \frac{N_L(eV)}{N(0)}. \quad (8.114)$$

Hence the measurement of the differential conductance yields a direct measure of the density of states of the left-hand metal. This is how the curves illustrated in Fig. 8.3 were obtained. For further details the reader is referred to the review articles of McMillan and Rowell (1969). For a discussion of the beautiful Josephson effects the reader is referred to the review articles of Josephson (1969) and Mercerau (1969).

References

General

(a) *On superconductivity*

de Gennes, P. G. (1966). "Superconductivity of Metals and Alloys". Benjamin, New York.
Kuper, C. G. (1969). "Theory of Superconductivity". Clarendon, Oxford.
Lynton, E. A. (1969). "Superconductivity", 3rd Edn. Metheun, London.
Parks, R. D. (ed.) (1969). "Superconductivity". Dekker, New York.
Rickayzen, G. (1965). "Theory of Superconductivity". Wiley, New York.
Rose-Innes, A. C. and Rhoderick, E. H. (1969). "Introduction to Superconductivity". Pergamon Press, Oxford.
Saint-James, D., Sarma, G. and Thomas, E. J. (1969). "Type-II Superconductivity". Pergamon Press, Oxford.
Schrieffer, J. R. (1964). "Theory of Superconductivity". Benjamin, New York.
Solymar, L. (1972). "Superconductive Tunnelling and Applications". Chapman and Hall, London.
Williams, J. E. C. (1970). "Superconductivity and its Applications". Pion, London.

(b) *On superfluid ^3He*

Armitage, J. G. M. and Farquhar, I. E. (eds.) (1975). "The Helium Liquids". Academic Press, London and New York.
Leggett, A. J. (1975). *Rev. Mod. Phys.* **47**, 331.

Special

Abramowitz, M. and Stegun, I. A. (1965). "Handbook of Mathematical Functions". Dover, London.
Ambegaokar, V. and Griffin, A. (1965). *Phys. Rev.* **137**, A1151.
Anderson, P. W. (1958). *Phys. Rev.* **112**, 1900.
Anderson, P. W. (1959). *J. Phys. Chem. Solids* **11**, 26.
Bardeen, J. (1951). *Rev. Mod. Phys.* **23**, 261.
Bardeen, J., Cooper, L. N. and Schrieffer, J. R. (1957). *Phys. Rev.* **108**, 1175.
Bogoliubov, N. N., Tolmachev, V. V. and Shirkov, D. V. (1958). "A New Method in the Theory of Superconductivity". Acad. Aci., USSR, translation, Consultants Bureau, New York, 1959.
Cooper, L. N. (1956). *Phys. Rev.* **104**, 1189.
Evans, W. A. B. and Rickayzen, G. (1965). *Ann. Phys.* **33**, 275.
Fetter, A. L. and Hohenberg, P. C. (1969). In Parks (1969), Vol. 2, Chapter 14.
Frölich, H. (1950). *Phys. Rev.* **79**, 845.
Giaever, I. and Megerle, K. (1961). *Phys. Rev.* **122**, 1101.
Josephson, B. D. (1962). *Phys. Lett.* **1**, 251.
Josephson, B. D. (1969). In Parks (1969), Vol. 1, Chapter 9.
Kondo, J. (1969). "Solid State Physics, Advances and Applications" (Ed. H. Ehrenreich, F. Seitz and D. Turnbull), Vol. 23, p. 183. Academic Press, New York and London.

Livingstone, J. D. (1963). *Phys. Rev.* **129**, 1943.
McMillan, W. L. and Rowell, J. M. (1969). In Parks (1969), Vol. 1, Chapter 11.
Maki, K. (1969). In Parks (1969), Vol. 2, Chapter 18.
Maxwell, E. (1950). *Phys. Rev.* **78**, 477.
Mercereau, J. E. (1969). In Parks (1969), Vol. 1, Chapter 8.
Meservey, R. and Schwartz, B. B. (1969). In Parks (1969), Vol. 1, Chapter 3.
Nambu, Y. (1960). *Phys. Rev.* **117**, 648.
Reynolds, C. A., Serin, B., Wright, W. H. and Nesbitt, L. B. (1950). *Phys. Rev.* **78**, 487.
Rickayzen, G. (1959). *Phys. Rev.* **115**, 795.
Rickayzen, G. (1969). In Parks (1969), Vol. 1, Chapter 2.
Scalapino, D. J. (1969). In Parks (1969), Vol. 1, Chapter 10.
Skalski, S., Bebeder-Matibeti, O. and Weiss, P. R. (1964). *Phys. Rev.* **136**, A1500.

Problems

1. By following the argument given in either §8.1 or §8.2 show that degenerate fermions interacting via a two-particle interaction become unstable to the formation of Cooper pairs with parallel spins at the temperature T_c for which the equation in Δ_k,

$$\Delta_k = \sum_{k'} [V(k-k') - V(k+k')] \frac{\Delta_{k'}}{2\epsilon_k'} \tanh(\epsilon_{k'}/2kT_c),$$

first possesses a non-trivial solution.

2. Bogoliubov, Shirkov and Tolmachev introduced a simple model to take account of the different effects of the Coulomb interaction and electron–phonon interaction. In the model

$$V_{kk'} = V_{1k_1k'} + V_{2k_1k'}$$

$$V_{1kk'} = -V_1 < 0 \text{ for } |\epsilon_k|, |\epsilon_{k'}| < \omega_{\mathrm{D}},\ V_{1kk'} = 0, \text{ otherwise}$$

$$V_{2kk'} = V_2 > 0, \text{ for } |\epsilon_k|, |\epsilon_{k'}| < E_c,\ \omega_{\mathrm{D}} < E_c,\ V_2 < V_1$$

$$= 0, \text{ otherwise}$$

$$N(\epsilon) = N(0),\ |\epsilon_k| < E_c.$$

Show that in this model $\Delta(\epsilon_k)$ has different values in the energy ranges $|\epsilon_k| < \omega_{\mathrm{D}}$ and $\omega_{\mathrm{D}} < |\epsilon_k| < E_c$. Find an expression for T_c in the weak coupling limit.

3. An external sound field ϕ can be assumed to be classical and to interact with the electrons through a term in the Hamiltonian

$$H' = \sum_q \rho_q \phi_q \exp(i\omega_q t) + \rho_{-q}\phi_q^* \exp(-i\omega_q t),$$

$$\rho_q = \sum_{k,\sigma} c_{k\sigma}^+ c_{k+q\sigma}.$$

Deduce that the rate at which energy is absorbed from sound of frequency ω_q is

$$2\pi\omega_q[1-\exp(-\beta\omega_q)]\,|\phi_q|^2 Z^{-1}$$

$$\times \sum_{m,n} \exp(-\beta E_m)\,|\rho_{qnm}|^2\,\delta(\omega_q - E_n + E_m)$$

$$= -2\omega_q\,|\phi_q|^2 \operatorname{Im} G^\rho(\boldsymbol{q}, \omega_q),$$

where

$$G^\rho(\boldsymbol{q}, \omega_q) = -i \operatorname{Tr}\{[\rho_q(t), \rho_{-q}]\}\,\theta(t).$$

Deduce that the ratio of the rate of absorption of energy from a field of energy $\omega_q (\ll \Delta)$ in the superconducting state to that in the normal state is $2f(\Delta)$.

4. The tunnelling matrix element T_{kq} depends essentially only on the components of momentum $\boldsymbol{k}_\perp, \boldsymbol{q}_\perp$, in the plane of the barrier and $\boldsymbol{k}_\perp = \boldsymbol{q}_\perp$. For isotropic superconductors, derive equation (8.111) from (8.110) and show that

$$|T|^2 = \frac{L_R L_L}{(2\pi)^2} \sum_{k_\perp} |T_{kq}|^2 (V_{FR} V_{FL})^{-1},$$

where L_R and L_L are the lengths, respectively, the right-hand and left-hand superconductors in the direction perpendicular to the barrier and V_{FR}, V_{FL} the Fermi velocities.

Chapter 9

Superfluidity

§9.1 Introduction

When liquid ^{4}He is cooled below 2.17 K it undergoes a phase transition to a new state in which it flows through narrow pores without a measurable viscosity. The new state is the superfluid state of the liquid. Any theory of the behaviour of the liquid must explain this basic property which resembles superconductivity discussed in the last chapter. Other striking properties of superfluid ^{4}He which need to be explained are listed below.

(1) The specific heat near the transition temperature has a characteristic lambda shape as shown in Fig. 9.1.

(2) Although superfluid, the liquid does impede the rotation of flat plates suspended within it. This has suggested that the liquid has two components, a normal and a superfluid one, the former providing viscosity and impeding the rotation of plates, the latter flowing freely through narrow pores. From experiments on the rotation of plates, as well as other experiments, it is deduced that the densities of the two components ρ_n and ρ_s as functions of temperature are as shown in Fig. 9.2.

(3) From neutron scattering experiments it has been possible to deduce the excitation spectrum of quasi-particles in the liquid. The result is as shown in Fig. 9.3. Landau (1941, 1947) had earlier proposed from general considerations that the spectrum would have this general shape. It is, however, still an objective of the

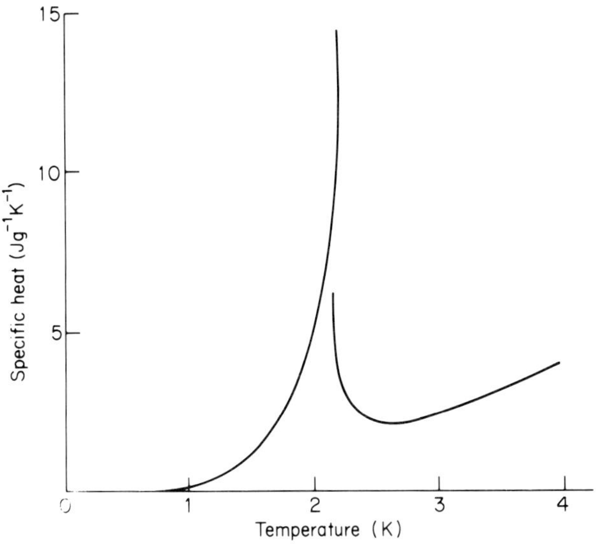

FIG. 9.1. The specific heat of liquid ^{4}He (after Atkins 1959).

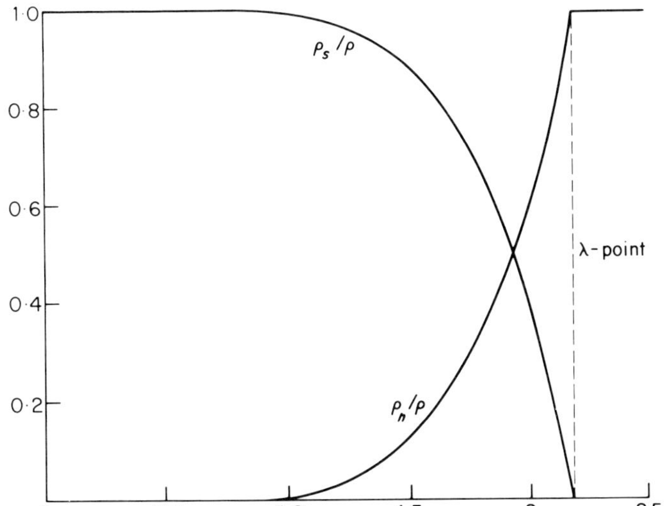

FIG. 9.2. The fraction of normal, ρ_n, and superfluid, ρ_s, densities against temperature as determined from Andronikashvili's experiments (after Atkins 1959).

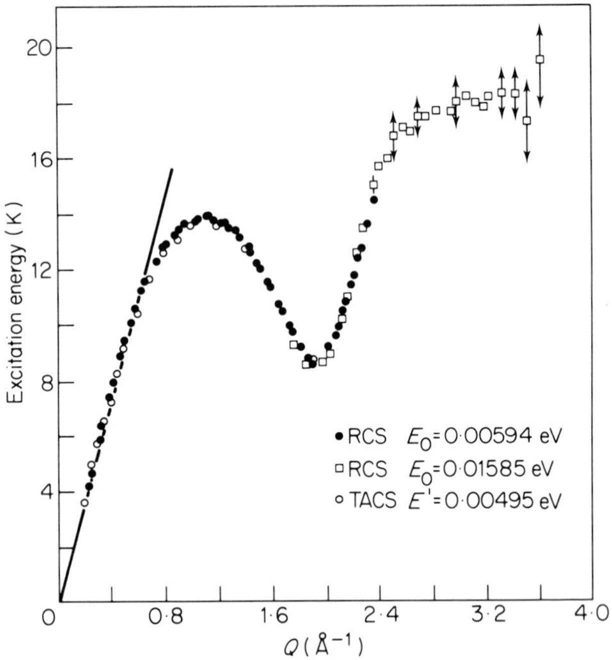

FIG. 9.3. The dispersion curve for He II for the elementary excitations at $T = 1.1$ K and $P \simeq 0$. Data obtained by rotating crystal spectrometer (RCS) and triple-axis crystal spectrometer (TACS) as indicated (reproduced from Cowley and Woods (1971) with permission from the National Research Council of Canada).

theory to explain the shape quantitatively. The part of the spectrum for small momentum $\boldsymbol{k}$ is the phonon part and as $\boldsymbol{k}$ tends to zero the slope is the velocity of sound in the liquid. The part of the spectrum near the minimum is called the roton part. It is an important feature of the spectrum that the only low-energy excitations are phonons.

(4) Experiments by Vinen (1961) and by Rayfield and Reif (1964) suggest that the circulation in the liquid is quantized in conventional units of h/m where m is the mass of an isolated ^{4}He atom.

There are evidently similarities with the properties of superconductors, notably the lack of resistance to flow and the macroscopic quantization, and these led London (1954) to the conclusion that both the superfluid and the superconductor are in macroscopic quantum states in which momentum is ordered. In the case of superconductivity we have seen this view vindicated in the microscopic theory in which paired electrons have the same momentum. In the case of liquid ^{4}He, London suggested that the ordering arose naturally from the Bose statistics obeyed by the helium atoms. In fact, as we show in the following section, a free Bose gas undergoes a transition, known as a Bose–Einstein condensation, to a state where the zero-momentum single-particle state is macroscopically occupied. Most attempts at a theory of superfluidity assume that it is related to this condensation. The experimental evidence for it is not yet unequivocal.

§9.2 Bose–Einstein condensation

Consider a gas of non-interacting bosons. The distribution function $n(\boldsymbol{k})$ for the occupation of the state of momentum $\boldsymbol{k}$ in a grand canonical ensemble at temperature T, chemical potential μ is

$$n(\boldsymbol{k}) = [\exp(\beta\epsilon_k) - 1]^{-1} \tag{9.1}$$

where

$$\epsilon_k = (k^2/2m) - \mu. \tag{9.2}$$

The chemical potential has to be determined from the condition

$$\sum_{\boldsymbol{k}} n(\boldsymbol{k}) = \frac{\mathscr{V}}{(2\pi)^3}\int d^3k\, n(\boldsymbol{k}) = N, \tag{9.3}$$

where N is the mean number of bosons in the system. At high temperatures μ is negative and, as the temperature is lowered μ increases. The maximum value of μ for which the integral in (9.3) has a meaning is $\mu = 0$ and the integral attains *its* maximum value for this value of μ. With $\mu = 0$, equation (9.3) becomes

$$2.612\,\mathscr{V}(m/2\pi\beta)^{3/2} = N. \tag{9.4}$$

This equation determines the temperature at which μ goes to zero for a given density of the gas. Whatever the density there is always a temperature T_c at which this occurs. Inverting (9.4) we find

$$T_c = (2\pi/m)(N/2.612\mathscr{V})^{2/3}. \tag{9.5}$$

What happens as the temperature decreases below T_c? At the absolute zero the answer is quite clear. The ground state of a non-interacting Bose gas is one where all the particles occupy the single-particle state $\boldsymbol{k} = 0$. In order to achieve this we must have $\mu = 0$. At higher temperatures we still expect a macroscopic occupation of the state $\boldsymbol{k} = 0$ and, hence $\mu = 0$. In the sum in equation (9.3) we should, therefore, split off the $\boldsymbol{k} = 0$ state and rewrite the equation for $T < T_c$ as

$$N_0 + \frac{\mathscr{V}}{(2\pi)^3}\int d^3k\,n(k) = N_0 + 2.612\,\mathscr{V}\left(\frac{m}{2\pi\beta}\right)^{3/2} = N. \tag{9.6}$$

This equation determines the macroscopic number of bosons, N_0, occupying the $\boldsymbol{k}=0$ state for any temperature below T_c. Thus the non-interacting Bose gas shows a transition to a state with the macroscopic occupation of the $\boldsymbol{k}=0$ single-particle state as the temperature falls below T_c. If one uses the mass of ^{4}He and the density of the liquid in equation (9.5) one finds that $T_c = 3.1$ K. This is sufficiently close to the transition temperature of superfluid helium to encourage the belief that the two are related.

§9.3 Microscopic theory of liquid helium

Present attempts at a theory of liquid helium start by assuming that only two-body forces between helium atoms are important. The usual potentials chosen are either a suitable Lennard-Jones potential or a hard-sphere one. The results of present theories do not warrant more sophisticated potentials. The Hamiltonian can then be written

$$H = \sum_{\boldsymbol{k}} \frac{k^2}{2m} c_{\boldsymbol{k}}^{+} c_{\boldsymbol{k}} + \tfrac{1}{2} \sum_{\boldsymbol{k},\boldsymbol{k}',\boldsymbol{q}} V(\boldsymbol{q}) c_{\boldsymbol{k}}^{+} c_{\boldsymbol{k}'+\boldsymbol{q}}^{+} c_{\boldsymbol{k}'} c_{\boldsymbol{k}+\boldsymbol{q}}. \tag{9.7}$$

Strictly, for the Lennard-Jones or hard-sphere potential the Fourier transform does not exist and one cannot write the Hamiltonian (9.7). Nevertheless, we can approximate these potentials by ones for which the transform exists and can take an appropriate limit at the end of the calculation. We discuss this point further in §9.4.

Now we showed in §3.2 how to obtain a perturbation expansion for a single-particle Green's function. If, however, there is the possibility of a Bose–Einstein condensation into the state, $\boldsymbol{k}=0$, that expansion is not very useful. For the exact Green's function $\bar{G}(\boldsymbol{k}=0, \zeta)$ is then of order N. This means that great care has to be taken in summing diagrams to obtain the correct dependence on N. For example one can start with an approximation in which $\bar{G}_0(\boldsymbol{k}=0, \zeta)$ is of the order of N by taking μ as part of the perturbation. Then the contribution to $\bar{G}(\boldsymbol{k}, \zeta)$ of the diagram in Fig. 9.4 is of order N and not of order 1. One can similarly find higher-order diagrams which depend on higher powers of N. One then has a major problem to determine which set of diagrams provides a satisfactory approximation to $\bar{G}$. Alternatively one can take the zero-order approximation μ_0 to μ to be negative. Then $\bar{G}_0(\boldsymbol{k}=0, \zeta)$ is of the order unity. In this case, an infinite sum of diagrams is required

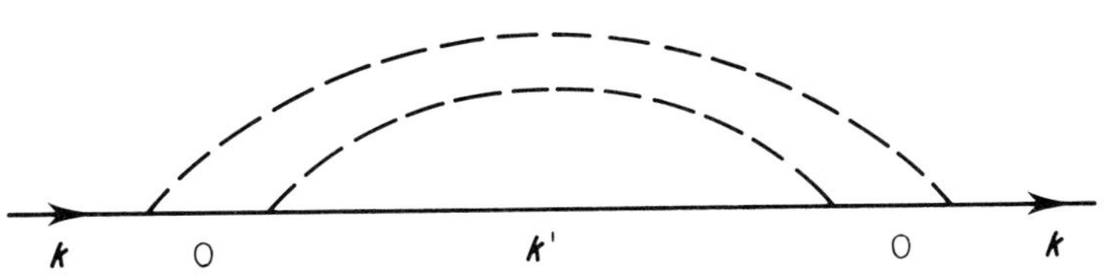

FIG. 9.4. A diagram which makes a contribution of order N to the one-particle Green's function.

to make $\bar{G}(0, \zeta)$ of order N. One is then faced with the difficult problem of ensuring that one gets the N-dependence of other diagrams correct.

To avoid this difficulty, it is usual to treat the single-particle state $k = 0$ separately from the others. Two alternative methods have been proposed for doing this. The first is due to Bogoliubov (1947) and was formalized by Hugenholtz and Pines (1959). The second was due to Beliaev (1958). We adopt the former approach here. If there is a Bose–Einstein condensation

$$\langle c_0^+ c_0 \rangle = N_0, \qquad \langle c_0 c_0^+ \rangle = N_0 + 1. \tag{9.8}$$

We therefore write

$$c_0 = \sqrt{N_0}\,\exp(i\phi) + \zeta_0, \qquad c_0^+ = \sqrt{N_0}\,\exp(-i\phi) + \zeta_0^+, \tag{9.9}$$

so that

$$[\zeta_0, \zeta_0^+] = 1. \tag{9.10}$$

Because of equation (9.8), if we neglect ζ_0 and ζ_0^+ in the Hamiltonian, we will make an error of relative order of magnitude N_0^{-1} in the free energy and the excitation energies. If we do this we no longer conserve particle number in the Hamiltonian but this can be rectified by introducing a chemical potential μ for the particles. This means that in equation (9.7) for the Hamiltonian we replace the kinetic energy $k^2/2m$ by

$$k^2/2m - \mu, \tag{9.11}$$

and choose μ in such a way that

$$\left\langle \sum_{k \neq 0} c_k^+ c_k \right\rangle = N - N_0. \tag{9.12}$$

For each value of N_0 we can find a different free energy of the system. The observed free energy will be the minimum of these.

The Hamiltonian still contains the phase ϕ of c_0. As in the case of superconductivity however, the phase can be eliminated by means of a unitary transformation. We therefore take the phase to be zero in thermodynamic equilibrium.

With the foregoing approximation the Hamiltonian can be written,

$$H = A + \sum_k [(k^2/2m) - \mu + N_0 V(0) + N_0 V(k)] c_k^+ c_k + H_1 + H_2 + H_3, \tag{9.13}$$

where the terms in c_0 are omitted from all sums; A is a constant given by

$$A = \tfrac{1}{2} N_0^2 V(0) - \mu N_0. \tag{9.14}$$

and

$$H_1 = \tfrac{1}{2} N_0 \sum_k V(\boldsymbol{k}) c_k^+ c_{-k}^+ + \text{h.c.},$$

$$H_2 = N_0^{1/2} \sum_{k,q} V(\boldsymbol{q}) c_k^+ c_q^+ c_{k+q} + \text{h.c.},$$

$$H_3 = \tfrac{1}{2} \sum_{k,k',q} V(\boldsymbol{q}) c_k^+ c_{k'+q}^+ c_{k'} c_{k+q}; \tag{9.15}$$

in the last expressions h.c. stands for the hermitian conjugate of the previous term.

Now H_1 is quadratic in the creation and annihilation operators and can be included in the unperturbed Hamiltonian which we take to be

$$H_0 = A + \sum_k \epsilon_k c_k^+ c_k + \tfrac{1}{2} N_0 \sum_k V(\boldsymbol{k}) [c_k^+ c_{-k}^+ + c_{-k} c_k], \tag{9.16}$$

where

$$\epsilon_k = (k^2/2m) - \mu + N_0 V(0) + N_0 V(\boldsymbol{k}). \tag{9.17}$$

The temperature equations of motion for the unperturbed system are

$$dc_k/d\tau = -\epsilon_k c_k - N_0 V(\boldsymbol{k}) c_{-k}^+,$$
$$dc_{-k}^+/d\tau = \epsilon_k c_{-k}^+ + N_0 V(k) c_k. \tag{9.18}$$

We can introduce a matrix notation, similar to Nambu's in the case of superconductivity, to put these equations in a succinct form. Let

$$\alpha_k = \begin{pmatrix} c_k \\ c_{-k}^+ \end{pmatrix}.$$

Then

$$\tau_3 d\alpha_k/d\tau = -\epsilon_k \alpha_k - N_0 V(\boldsymbol{k}) \tau_1 \alpha_k. \tag{9.19}$$

Note that because of the different statistics and hence the different commutation relations, the form of (9.19) is not quite the same as for the superconductor.

If we define the matrix Green's function by

$$G_0(k,\tau) = -\langle[\alpha_k(\tau)\alpha_k^+(0)]\rangle, \tag{9.20}$$

this Green's function satisfies

$$\left[\tau_3 \frac{d}{d\tau} + \epsilon_k + N_0 V(k)\tau_1\right] G_0(k,\tau) = -\delta(\tau) \tag{9.21}$$

and so has the Fourier transform

$$\begin{aligned}\bar{G}_0(k,\zeta) &= [i\tau_3\zeta - \epsilon_k - N_0 V(k)\tau_1]^{-1} \\ &= \frac{i\zeta\tau_3 + \epsilon_k - N_0 V(k)\tau_1}{(i\zeta)^2 - \epsilon_k^2 + N_0^2 V(k)^2}.\end{aligned} \tag{9.22}$$

The use of the Green's functions (9.20) brings some simplicity to the notation but it requires care because the Green's functions carry redundant information. To see this let us write (9.20) in terms of its components. Then

$$G_0(k,\tilde{\tau}) = -\begin{pmatrix}\langle T[c_k(\tau)c_k^+(0)]\rangle & \langle T[c_k(\tau)c_{-k}(0)]\rangle \\ \langle T[c_{-k}^+(\tau)c_k^+(0)]\rangle & \langle T[c_{-k}^+(\tau)c_{-k}(0)]\rangle\end{pmatrix}. \tag{9.23}$$

Hence

$$G_0(-k,-\tau) = -\begin{pmatrix}\langle T[c_{-k}(0)c_{-k}^+(\tau)]\rangle & \langle T[c_{-k}(0)c_k(\tau)]\rangle \\ \langle T[c_k^+(0)c_{-k}^+(\tau)]\rangle & \langle T[c_k^+(0)c_k(\tau)]\rangle\end{pmatrix} \tag{9.24}$$

and

$$G_{011}(k,\tau) = G_{022}(-k,-\tau);$$

$$G_{012}(k,\tau) = G_{012}(-k,-\tau); \qquad G_{021}(k,\tau) = G_{021}(-k,-\tau).$$

It is possible to avoid this redundancy by restricting the values of k to half of k-space. This however, complicates the interaction and it is more convenient to keep the redundancy.

In order to treat H_2 and H_3 by perturbation theory we need to write them in terms of α_k. Now both involve the number density

$$\rho_q = \sum_k c_k^+ c_{k+q} = \sum_k{}' \left(c_k^+ c_{k+q} + c_{-k-q}^+ c_{-k}\right)$$

where the primed sum denotes a sum over half of k-space. Hence,

$$\rho_q = \sum_k{}' \alpha_k^+ \alpha_{k+q} \equiv \tfrac{1}{2}\sum_k \alpha_k^+ \alpha_{k+q} \equiv \tfrac{1}{2}\sum_k \alpha_{-k-q}^+ \alpha_{-k}. \tag{9.25}$$

The redundancy here shows up in the several possible ways of writing ρ_q. Whichever form is used one must remember that

$$\alpha_k^+ \alpha_{k+q} = \alpha_{-k-q}^+ \alpha_{-k}. \tag{9.26}$$

It follows that H_3 can be written in several forms one of which is

$$H_3 = \tfrac{1}{8} \sum_{k,k',q} V(q)(\alpha_k^+ \alpha_{k+q} \alpha_{k'+q}^+ \alpha_{k'}). \tag{9.27}$$

The interaction H_2 depends on $c_q^+ + c_{-q}$ which can be written as

$$c_q^+ + c_{-q} = \alpha_q^+ U \equiv U^+ \alpha_{-q}, \tag{9.28}$$

where

$$U^+ = (1, 1). \tag{9.29}$$

Thus H_2 can be written in several forms one of which is

$$H_2 = \tfrac{1}{2} N_0^{1/2} \sum_{k,q} V(q) \alpha_k^+ \alpha_{k+q} \alpha_q^+ U \tag{9.30}$$

We can now treat H_2 and H_3 by perturbation theory. The interaction H_3 is exactly of the type we have discussed previously and is represented in diagrams as illustrated in Fig. 9.5. In accord with equation (9.27) the unit matrix acts at each vertex. Also the factor $\frac{1}{8}$ is cancelled as follows. Each factor ρ_q in the H_3 acts at either vertex in Fig. 9.5. This yields a factor 2. Further, because of the redundancy discussed after equation (9.25), there is a further factor 2 at each vertex. Hence, the interaction of Fig. 9.5 gives rise to a factor $V(q)$.

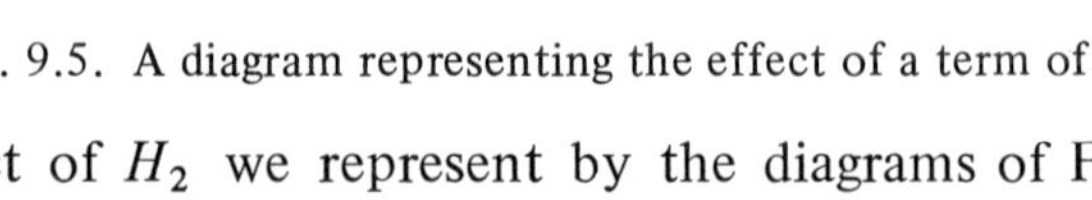

FIG. 9.5. A diagram representing the effect of a term of H_1.

The effect of H_2 we represent by the diagrams of Fig. 9.6. The wavy line with the ingoing arrow represents the matrix U^+ and that with the outgoing arrow U. Both lines carry zero momentum and in each case the column vector is acted upon by the matrix represented by the solid line attached to the same vertex. For example, in terms of components, Fig. 9.7 represents a contribution to the single-particle Green's function $\bar{G}_{\alpha\beta}(k, \zeta)$ of amount

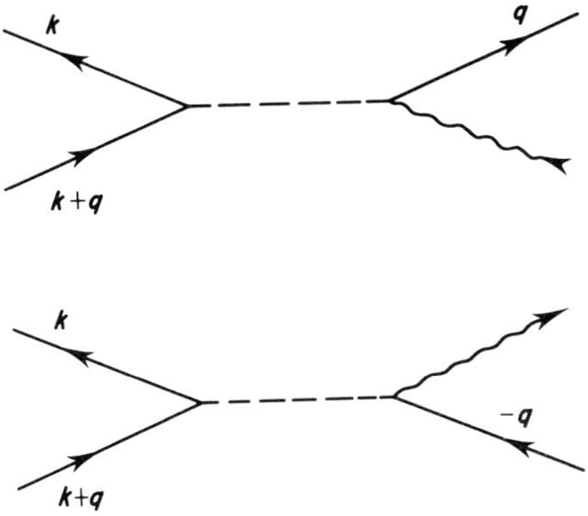

FIG. 9.6. Diagrams representing the effects of H_2. The wavy lines represent the effects of U (on the left) and of U^+ (on the right).

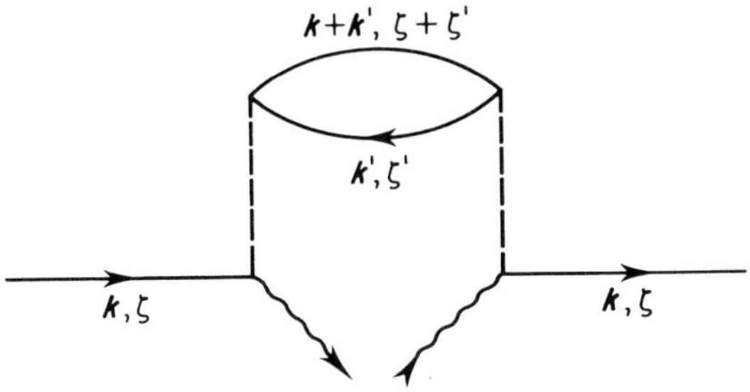

FIG. 9.7. A self-energy contribution to the propagator.

$$N_0\beta^{-1} \sum_{k'\zeta'} \sum_{\gamma\delta\epsilon\rho=1}^{2} \bar{G}_{0\alpha\gamma}(k,\zeta)U_\gamma \bar{G}_{\delta\epsilon}(k'\zeta)\bar{G}_{\epsilon\beta}(k'+k,\zeta+\zeta') \times U_\rho^+ G_{\rho\beta}(k,\zeta). \tag{9.31}$$

If we note that

$$U_\gamma U_\rho^+ = (1+\tau_1)_{\gamma\rho}$$

We can write the contribution of Fig. 9.7 to $\bar{G}(k,\zeta)$ as

$$N_0\beta^{-1} \sum_{k',\zeta'} \bar{G}_0(k,\zeta)(1+\tau_1)\bar{G}_0(k,\zeta)\,\mathrm{Tr}\,\bar{G}(k'\zeta')\bar{G}(k'+k,\zeta+\zeta'). \tag{9.32}$$

In any diagram a continuous boson line will now either:

(i) form a closed loop;
(ii) have two wavy lines at its ends; or
(iii) be an external line with a wavy line at the internal end.

The contribution in case (i) will, as before, be the trace of the matrices which come from its component parts. In case (ii), if M is the matrix coming from the components parts, the contribution is $U^{+}MU$.

Finally the contribution from an incoming external line is of the form [cf. equation (9.31)] M_1U while that of an outgoing line is $U^{+}M_2$. Thus the contribution of the diagram of Fig. 9.8 to the single-particle Green's function is

$$\bar{G}_0(k, \zeta)UM(k, \zeta)U^{+}\bar{G}(k, \zeta)UM(k, \zeta)U^{+}\bar{G}(k, \zeta), \qquad (9.33)$$

where

$$M(k, \zeta) = N_0\beta^{-1} \sum_{k', \zeta'} \mathrm{Tr}\, \bar{G}_0(k', \zeta')\bar{G}_0(k + k', \zeta + \zeta'). \quad (9.34)$$

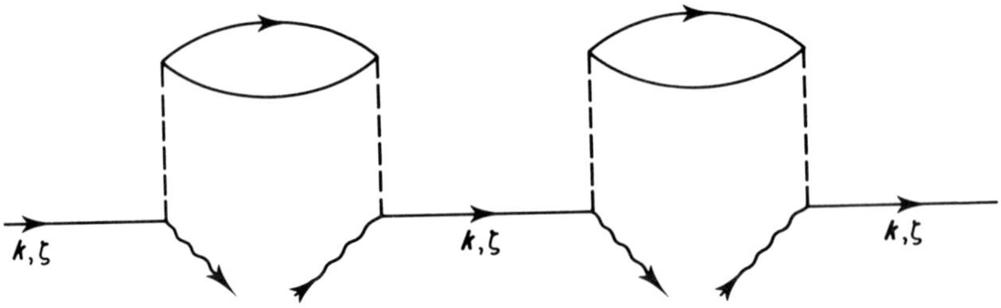

FIG. 9.8. The diagram whose contribution is given in equation (9.33).

We can define irreducible self-energy parts as before as parts of the diagram which cannot be divided into two by cutting a single boson line only and we can define the sum of the contributions of these irreducible self-energy parts to be the self-energy $\Sigma'(k, \zeta)$. Because the Green's function is a matrix the self-energy is also a matrix. For example the central part of Fig. 9.7 (cf. also Fig. 9.8) is an irreducible self-energy part and its contribution to $\Sigma'(k, \zeta)$ is

$$UM(k, \zeta)U^{+} = M(k, \zeta)\,(1 + \tau_1),$$

with $M(k, \zeta)$ given by equation (9.34). The relationship between the single-particle Green's function and the self-energy is given by

$$\begin{aligned}\bar{G}(k, \zeta)^{-1} &= \bar{G}_0(k, \zeta)^{-1} - \Sigma'(k, \zeta)\\ &= i\tau_3\zeta - [(k^2/2m) - \mu + N_0V(0) + N_0V(k)]\\ &\quad + N_0V(k)\tau_1 - \Sigma'(k, \zeta). \qquad (9.35)\end{aligned}$$

It is common practice to include in the self-energy all contributions which depend on the potential and to write

$$\bar{G}(k, \zeta)^{-1} = i\tau_3\zeta - [(k^2/2m) - \mu] - \Sigma(k, \zeta) \tag{9.36}$$

with

$$\Sigma(k, \zeta) = N_0 V(0) + N_0 V(k) + N_0 V(k)\tau_1 + \Sigma'(k, \zeta). \tag{9.37}$$

This form of the self-energy is arrived at naturally if all terms in the Hamiltonian which depend on the interaction are treated by perturbation theory.

Perturbation theory provides a systematic way of calculating properties of the system. The results of the calculation yield G in terms of μ and N_0 which have then to be determined from equation (9.12) and by minimization of the free energy with respect to N_0. The latter condition is hard to apply in practice and can be replaced by a condition first derived by Hugenholtz and Pines. We provide a derivation of this result in the next section. Although perturbation theory provides a way of calculating the properties of liquid ^{4}He, this approach has not yielded good quantitative results because the coupling strength is not small compared with unity. Nevertheless the method has been useful qualitatively.

§9.4 Theorem of Hugenholtz and Pines

In this section we shall follow a method due to Hohenberg and Martin (1965). This exploits the fact that the ground state breaks a symmetry of the Hamiltonian and is, therefore, degenerate. As in the case of superconductivity the symmetry broken is invariant under a change of phase of the field operators. In order to follow through the effect of change of phase we revert to the original form of the Hamiltonian in coordinate space,

$$H = \int d^3r \psi^+(-(\nabla^2/2m) - \mu)\psi + \tfrac{1}{2}\int d^3r V(r - r')\psi^+(r)\psi^+(r')\psi(r')\psi(r).$$

To generate states with a definite phase we add to the Hamiltonian a term

$$H' = \int [\nu(r)\psi^+(r) + \nu^+(r)\psi(r)]\, d^3r,$$

where $\nu(r)$ is a hypothetical classical field which generates states in which $\langle\psi(r)\rangle$ is not zero. This field is analogous to the field used in the dicussion of superconductivity in §8.1. Above the transition temperature $\langle\psi(r)\rangle$ tends to zero as ν tends to zero but below this temperature $\langle\psi(r)\rangle$ tends to a finite limit which, in the last section, we took to be $\sqrt{N_0/\mathscr{V}}$.

Now ν has a definite phase. If we change the phase of ν from ϕ to $\delta\phi$ we can absorb this by making the canonical transformation

$$\psi \rightarrow \exp(i\delta\phi)\psi, \qquad \psi^+ \rightarrow \psi^+ \exp(-i\delta\phi). \tag{9.38}$$

The Hamiltonian is then restored to its original form. Thus, under the transformation,

$$\langle\psi\rangle \rightarrow \langle\psi\rangle \exp(i\delta\phi), \qquad \langle\psi^+\rangle \rightarrow \langle\psi^+\rangle \exp(-i\delta\phi), \tag{9.39}$$

However, the change in $\langle\psi\rangle$ can also be calculated from the Green's functions of the original Hamiltonian. It is convenient to use the matrix notation again and put

$$\Psi(r) = \begin{pmatrix} \psi(r) \\ \psi^+(r) \end{pmatrix}, \qquad V(r) = \begin{pmatrix} \nu(r) \\ \nu^*(r) \end{pmatrix}. \tag{9.40}$$

Then (apart from unimportant additive constants)

$$\begin{aligned} \psi^+(r)\psi(r) &= \tfrac{1}{2}\Psi^+\Psi \\ \nu(r)\psi^+(r) + \nu^*\psi(r) &= V^+\Psi = \Psi^+V. \end{aligned} \tag{9.41}$$

If

$$\nu \rightarrow \nu\exp(i\delta\phi), \qquad \psi \rightarrow \psi\exp(i\delta\phi),$$

then

$$V \rightarrow \exp(i\delta\phi\tau_3)V, \qquad \Psi \rightarrow \exp(i\delta\phi\tau_3)\Psi,$$

and

$$\delta V = i\delta\phi\tau_3 V, \qquad \delta\Psi \rightarrow i\delta\phi\tau_3\Psi, \qquad \delta\langle\Psi\rangle = i\delta\phi\tau_3\langle\Psi\rangle. \tag{9.42}$$

The change in V leads to an extra term in the Hamiltonian

$$\int \Psi(r)^+\delta V(r)d^3r = i\delta\phi\int\Psi^*(r)\tau_3 V(r)dr. \tag{9.43}$$

But we know from the general theory (cf. §1.2) that the change in $\langle\Psi(r)\rangle$ that this produces is given by

$$\delta\langle\Psi(r)\rangle = \int G(r,r',0)\delta V(r')d^3r' + \beta\langle\Psi(r)\rangle\int d^3r'\Psi^+(r')\delta V(r'), \tag{9.44}$$

where $G(r, r', 0)$ is the zero-frequency Fourier component of $G(r, \tau, r', \tau')$. From equation (9.42) this leads to

$$\tau_3\langle\Psi(r)\rangle = \int G(r,r',0)\tau_3 V(r')d^3r' + \beta\langle\Psi(r)\rangle\int d^3r'\Psi^+(r')\tau_3 V(r').$$

For a real uniform field V, $\langle\Psi\rangle$ is real and uniform and

$$\tau_3\langle\Psi\rangle = G(0,0)\tau_3 V, \tag{9.45}$$

where

$$G(0,0) = G(k,\zeta)|_{k=0,\zeta=0}$$

Hence

$$G(0,0)^{-1}\tau_3\langle\Psi\rangle = \tau_3 V. \tag{9.46}$$

In the limit $V \to 0$, $\langle\Psi\rangle \to \langle\psi\rangle U$, and if $\langle\psi\rangle$ is not zero we must have

$$G(0,0)^{-1}\tau_3 U = 0.$$

With equation (9.40) this implies that

$$[\mu - \Sigma(0,0)]\tau_3 \begin{pmatrix}1\\1\end{pmatrix} = 0.$$

This in turn yields two equations involving the components of $\Sigma(0,0)$, namely

$$\mu - \Sigma_{11}(0,0) + \Sigma_{12}(0,0) = 0 \tag{9.47}$$

and

$$-\mu - \Sigma_{21}(0,0) + \Sigma_{22}(0,0) = 0. \tag{9.48}$$

However, because of the redundancy of the notation the different components of Σ are not independent. Indeed it is not difficult to show (Hohenberg and Martin 1965) that

$$\begin{aligned}
\Sigma_{12}(k,\zeta) &= \Sigma_{12}(k,\zeta)^* = \Sigma_{21}(k,\zeta),\\
\Sigma_{11}(k,\zeta) &= \sigma_1(k,\zeta) + i\sigma_2(k,\zeta),\\
\Sigma_{22}(k,\zeta) &= \sigma_1(k,\zeta) - i\sigma_2(k,\zeta),\\
\sigma_1(\zeta) &= \sigma_1(-\zeta) = \sigma_1^*(\zeta),\\
\sigma_2(\zeta) &= -\sigma_2(-\zeta) = -\sigma_2^*(\zeta).
\end{aligned} \tag{9.49}$$

Hence

$$\sigma_2(0) = 0, \quad \Sigma_{12}(0,0) = \Sigma_{21}(0,0), \quad \Sigma_{11}(0,0) = \Sigma_{22}(0,0), \tag{9.50}$$

and equations (9.47) and (9.48) are identical and provide one relation between μ and N_0. This result is the theorem of Hugenholtz and Pines.

Note that a suitable starting point for finding the properties of the system is to put

$$\psi(\mathbf{r}) = \langle\psi(\mathbf{r})\rangle + \phi(\mathbf{r})$$

where the operators $\phi(\mathbf{r})$ have the same commutation relations as $\psi(\mathbf{r})$. Then one can expand in terms of the operators $\phi(\mathbf{r})$ choosing $\langle\psi(\mathbf{r})\rangle$ at the end of the calculation in such a way that $\langle\phi(\mathbf{r})\rangle$ is zero. This is essentially the method we employed in the last section with

$$\langle\psi(r)\rangle = \sqrt{N_0/\mathscr{V}}.$$

However, this approach can also be used for non-uniform systems where the initial choice for $\psi(r)$ is not obvious (Gross 1961, Ginzburg and Pitaevski 1961). In the case of a uniform system we can regard N_0 as given and use equations (9.48) and (9.12) to determine μ and N.

§9.5 Low-density Bose gas

When the density of the gas is low the interaction between the particles is weak and we need retain only the lowest-order of perturbation theory. To first order, Σ' can be neglected in equation (9.37) and

$$\Sigma(k, \zeta) = [N_0 V(0) + N_0 V(k)] + N_0 V(k)\tau_1.$$

Hence,

$$\Sigma_{11}(k) = \Sigma_{22}(k) = N_0 V(0) + N_0 V(k),$$

$$\Sigma_{12}(k) = \Sigma_{21}(k) = N_0 V(k),$$

and equation (9.47) yields

$$\mu = 2N_0 V(0) - N_0 V(0) = N_0 V(0). \tag{9.51}$$

The single-particle Green's function is

$$G(k, \omega) = \frac{\omega\tau_3 + \epsilon_k - N_0 V(k)\tau_1}{\omega^2 - \epsilon_k^2 + N_0 V(k)^2} \tag{9.52}$$

with

$$\epsilon_k = (k^2/2m) + N_0 V(k).$$

The excitation energies of the system are the poles of this function and are given by

$$E_k = [\epsilon_k^2 - N_0^2 V(k)^2]^{1/2}$$

$$= \left[\frac{k^2}{2m}\left(\frac{k^2}{2m} + 2N_0 V(k)\right)\right]^{1/2}.$$

At long wavelengths this becomes the dispersion equation for sound waves

$$E_k = ck \tag{9.53}$$

with sound velocity

$$c = \sqrt{N_0 V(0)/m}. \tag{9.54}$$

It can be shown (Bogoliubov 1947) that, to lowest order in the density, this is the macroscopic velocity of sound in the gas. It can also be shown (Hohenberg and Martin 1965, Hugenholtz and Pines 1959) that, provided that there is a Bose–Einstein condensation, equation (9.53) holds exactly as k tends to zero and that c is the macroscopic velocity of sound. Hohenberg and Martin give a prescription for generating approximations from which these results follow. These results agree with the experimental dispersion curve for liquid helium at long wavelengths (see Fig. 9.3), although the macroscopic velocity of sound is not given by equation (9.54).

The depletion of the zero-momentum state for the low-density gas is given by

$$N' = N - N_0 = \sum_{\mathbf{k}} \langle c_k^+ c_k \rangle$$
$$= -\beta^{-1} \sum_{\substack{\zeta, k \\ \delta \to 0_+}} \frac{(i\zeta + \epsilon_k) \exp(i\zeta\delta)}{(i\zeta)^2 - \epsilon_k^2 + N_0^2 V(k)^2}. \qquad (9.55)$$

This can be regarded as an equation for N_0 in terms of N. For given N, N_0 will tend to zero at a temperature given by

$$N = \sum_{k} [\exp(\beta k^2/2m) - 1]^{-1}. \qquad (9.56)$$

This is the same as equation (9.3) and yields the same transition temperature T_c. Above T_c there is no Bose–Einstein condensation, below it there is.

At the absolute zero of temperature equation (9.55) becomes

$$N' = \frac{\mathscr{V}}{(2\pi)^4} \int_{-\infty}^{\infty} d\zeta \int d^3k \, \frac{(i\zeta + \epsilon_k) \exp(i\zeta\delta)}{\zeta^2 + \epsilon_k^2 - N_0^2 V(\mathbf{k})^2}$$
$$= \frac{\mathscr{V}}{2(2\pi)^3} \int d^3k \, \frac{\epsilon_k - [\epsilon_k^2 - N_0^2 V(\mathbf{k})^2]^{1/2}}{[\epsilon_k^2 - N_0^2 V(\mathbf{k})^2]^{1/2}}.$$

The main contribution to the integral comes from the region

$$k \lesssim [2mN_0 V(0)]^{1/2}.$$

Provided that $V(\mathbf{k})$ varies slowly for such low momenta, $V(\mathbf{k})$ in the integral can be replaced by $V(0)$ and

$$\frac{N'}{N} = \frac{N_0}{N} \frac{[N_0 \mathscr{V}^2 V(0)^3 m^3]^{1/2}}{3\pi^2} \approx \frac{[N\mathscr{V}^2 V(0)^3 m^3]^{1/2}}{3\pi^2}. \qquad (9.57)$$

Hence the assumption of low density amounts to the approximation

$$[N\mathscr{V}^2 V(0)^3 m^3]^{1/2} \ll 1. \qquad (9.58)$$

Further terms in the expansion in this parameter have been obtained by Beliaev (1958) and by Hugenholtz and Pines (1959).

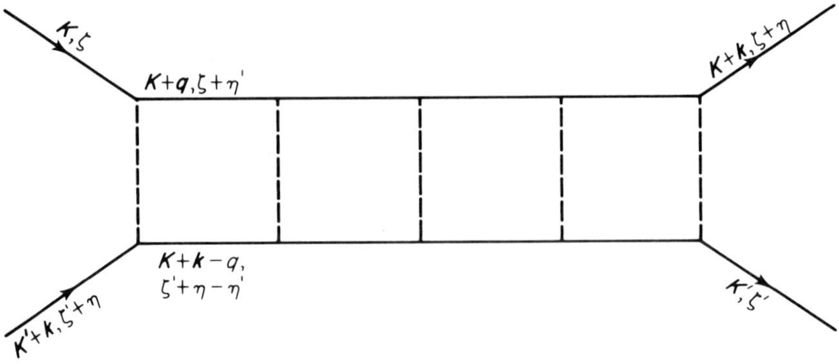

FIG. 9.9. Ladder diagrams which contribute to the t-matrix for particle–particle scattering.

As we pointed out in writing down the Hamiltonian in §9.3, there is one important respect in which this analysis is deficient even for a low-density Bose gas. For a Lennard-Jones potential or hard-sphere interaction, the Fourier transform $V(\boldsymbol{k})$ does not exist. Now in the low-density approximation $V(\boldsymbol{k})$ has been used for the probability amplitude for the scattering of two particles with transfer of momentum $\boldsymbol{k}$. However, this is only the Born approximation to this probability amplitude. With a singular potential this is a poor approximation because any two particles will tend to keep apart. We should, therefore, replace the Born approximation for the scattering of two particles by the full scattering amplitude as illustrated in Fig. 9.9 where all contributions to the ladder are taken into account. Then $V(\boldsymbol{k})$ is replaced by the t-matrix, $t(\boldsymbol{K}, \boldsymbol{K}', \boldsymbol{k}, \zeta, \zeta', \eta)$ which is a function of the three momenta $\boldsymbol{K}, \boldsymbol{K}', \boldsymbol{k}$ and satisfies the equation

$$t(\boldsymbol{K}, \boldsymbol{K}', \boldsymbol{k}, \zeta, \zeta', \eta) = V(\boldsymbol{k}) + \beta^{-1} \sum_{\boldsymbol{q}, \eta'} G(\boldsymbol{k} + \boldsymbol{q}, \zeta + \eta')$$
$$\times G(\boldsymbol{K}' + \boldsymbol{k} - \boldsymbol{q}, \zeta' + \eta' - \eta)\, t(\boldsymbol{K} + \boldsymbol{q}, \boldsymbol{K}' + \boldsymbol{k} - \boldsymbol{q}, \boldsymbol{k}, \zeta + \eta', \zeta' + \eta', \eta). \tag{9.59}$$

This equation can be written in real space even when $V(\boldsymbol{k})$ does not exist; the Fourier transform of the t-matrix always exists. Now we have seen that in the low-density limit, the important particles are those with small momentum and energy. This means that we require the t-matrix in the limit that all its arguments are zero. In that limit it can be shown (Beliaev 1958) that

$$t = 4\pi a/m\mathscr{V}, \tag{9.60}$$

where a is the exact scattering length for the scattering of two free helium atoms from one state of low momentum to another state of low momentum. To obtain the low-density properties of the helium gas, we therefore replace $V(0)$ by t in the preceding discussion. This approximation is discussed in more detail in the papers of Beliaev and Hugenholtz and Pines (*loc. cit.*). The approximation should be valid for

$$K \equiv \rho^{1/2}(4\pi a)^{3/2} \ll 1.$$

where ρ is the number density.

In fact, for the density found in the liquid, $K \approx 12.5$, and the real liquid is very far from this low-density limit. Although the expansion in terms of K has been continued to the next few terms, the results are not applicable for the observed value of K and perturbation theory has not yet proved to be quantitatively useful in the study of superfluid liquid helium. More quantitatively useful results have been obtained from computer and variational calculations (Kalos *et al.* 1974, and McMillan 1965). These suggest that $N_0/N \approx 0.1$ at absolute zero. Attempts have been made to obtain N_0/N from the neutron scattering data for $S(q, \omega)$ at high momentum transfer, assuming the impulse approximation [Mook 1974]. These suggest that $N_0/N \sim 0.03$. However, it is by no means certain that sufficiently high energies have been used to ensure the validity of the impulse approximation.

§9.6 The superfluidity and quantization of circulation

Although perturbation theory does not yet provide satisfactory quantitative results for liquid helium, it does provide an excellent conceptual framework within which the superfluid properties can be understood. Furthermore, this framework is analogous to that within which superconductivity is understood. Different microscopic states of the superfluid are described by different values of the complex order parameter

$$\langle\psi(\boldsymbol{r})\rangle = U(\boldsymbol{r}) \exp [i\chi(\boldsymbol{r})], \tag{9.61}$$

where $U(\boldsymbol{r})$ and $\chi(\boldsymbol{r})$ are real functions. The phase $\chi(\boldsymbol{r})$ is usually fixed by the external constraints on the superfluid, the boundary conditions, while for each value of $\chi(\boldsymbol{r})$, $U(\boldsymbol{r})$ is determined by the condition that the free energy must be a minimum. For states of linear flow of the superfluid, the superfluid momentum is

$$\boldsymbol{p}_s = \nabla\chi(\boldsymbol{r}).$$

In such a macroscopic state, the single-particle state of helium atoms with momentum $\boldsymbol{p}_s$ is macroscopically occupied. Provided that $U(\boldsymbol{r})$ is not zero this is a metastable state. For low superfluid momentum, the mass transport current density $\boldsymbol{j}$ is proportional to $\boldsymbol{p}_s$. We have

$$\boldsymbol{j} = n_s \boldsymbol{p}_s, \tag{9.62}$$

where n_s is defined to be the number density of superfluid atoms. At absolute zero, for pure liquid helium, all the atoms have the extra momentum $\boldsymbol{p}_s$ and

$$n_s = n \qquad (T = 0).$$

In fact, if the spectrum is as given in Fig. 9.3, Landau (1947) showed that for sufficiently small superfluid momentum, the current cannot diminish as a result of the scattering of particles out of the condensed phase. At higher temperatures some excitations are already present and these can be scattered backwards to reduce the free energy and so the current. Thus

$$n_s < n, \qquad T > 0.$$

The observed dependence of n_s on T is shown in Fig. 9.2. Note that the theoretical argument does not equate n_s with n_0 the number density in the $\boldsymbol{k} = 0$ state and experiment seems to show (cf. last section) that n_s and n_0 are very different; at $T = 0$, $n_s = n$ while $n_0 \lesssim 0.1n$.

Equation (9.62) is often written in terms of a superfluid velocity $\boldsymbol{v}_s$ defined by

$$\boldsymbol{v}_s = \boldsymbol{p}_s/m, \tag{9.63}$$

where m is the mass of a free ^{4}He atom. If $\rho_s = mn_s$ is defined to be the superfluid mass density we have

$$\boldsymbol{j} = \rho_s \boldsymbol{v}_s. \tag{9.64}$$

It is possible to have states of flow in which not all of the current is due to the flow of superfluid. We can then write the total current as

$$\boldsymbol{j} = \rho_s \boldsymbol{v}_s + \rho_n \boldsymbol{v}_n \tag{9.65}$$

where ρ_n is defined by

$$\rho_n = \rho - \rho_s \tag{9.66}$$

and is called the density of normal fluid. The velocity of the normal fluid, $\boldsymbol{v}_n$, is defined by equation (9.65). These equations together with those for the conservation of energy and momentum and an equation for the superfluid velocity describe the hydrodynamics of superfluid helium. For a full account of this, the reader is referred to the books by Tilley and Tilley (1974) and by Putterman (1974).

In analogy with superconductivity, in a multiply-connected region, $\boldsymbol{p}_s$ is quantized and, consequently, so is the circulation. If we consider the flow between two cylinders, from the definition of equation (9.61), the phase $\chi(\boldsymbol{r})$ must change by an integer times 2π when one follows one complete loop about the inner cylinder, that is

$$[\chi] = 2n\pi, \qquad n = 0, \pm 1, \pm 2, \ldots . \tag{9.67}$$

Hence, along this loop

$$\oint \boldsymbol{p}_s \cdot d\boldsymbol{s} = \oint \nabla\chi \cdot d\boldsymbol{s} = [\chi] = 2n\pi.$$

If this is written in terms of the velocity, we have

$$\oint \boldsymbol{v}_s \cdot d\boldsymbol{s} = (2\pi/m)n. \tag{9.68}$$

Hence, the circulation is quantized in units of $2\pi/m$. In conventional units this is

$$2\pi\hbar/m = \mathrm{h}/m = 9.98 \times 10^{-8}\ \mathrm{m}^2\ \mathrm{s}^{-1}. \tag{9.69}$$

In the limit in which the radius of the inner cylinder becomes vanishingly small, we have a vortex with circulation given by (9.68). Vortices play an important part in the flow of liquid helium and in its dissipation. The quantization of circulation has been confirmed in experiments by Vinen (1961) and, most clearly, in experiments by Rayfield and Reif (1964) on the motion of vortex rings and by Packard and Sanders (1969).

Again in analogy with superconductivity, there should be Josephson effects in liquid helium. However, these are much more difficult to observe, partly because the coherence length is only of the order of a few Ångstrom units. For a fuller discussion of these effects in liquid helium see Anderson (1966), Richards and Anderson (1965), Anderson and Richards (1965) and Leiderer and Pobell (1973).

References

General

Armitage, J. G. M. and Farquhar, I. E. (eds.) (1975). "The Helium Liquids". Academic Press, London and New York.

Atkins, K. R. (1959). "Liquid Helium". Cambridge University Press, London.

Tilley, D. R. and Tilley, J. (1974). "Superfluidity and Superconductivity". Van Nostrand Reinhold, New York.

Wilks, J. (1967). "The Properties of Liquid and Solid Helium". Clarendon, Oxford.

Special

Anderson, P. W. (1966). "Quantum Fluids" (Ed. D. F. Brewer), p. 146. North Holland, Amsterdam.
Anderson, P. W. and Richards, P. L. (1975). *Phys. Rev.* **B11**, 2702.
Beliaev, S. (1958). *Zh. Eksper. Teor. Fiz.* **34**, 417 (*Sov. Phys. JETP* **7**, 289).
Bogoliubov, N. (1947). *J. Phys., USSR* **11**, 23.
Cowley, R. A. and Woods, A. D. B. (1971). *Can. J. Phys.* **49**, 177.
Ginzburg, V. L. and Pitaevskii, L. P. (1958). *Zh. Eksper. Teor. Fiz.* **34**, 1240 (*Sov. Phys. JETP* **7**, 858).
Gross, E. P. (1961). *Nuovo Cimenuto* **20**, 454.
Hohenberg, P. C. and Martin, P. C. (1965). *Ann. Phys. NY* **34**, 291.
Hugenholtz, N. and Pines, D. (1959). *Phys. Rev.* **116**, 489.
Kalos, M. H., Leversque, D. and Verlet, L. (1974). *Phys. Rev.* **A9**, 2178.
Landau, L. (1941). *J. Phys. Moscow* **5**, 71 (reprinted in Khalatnikov (N) (1965). "Introduction to the Theory of Superfluidity", p. 184. Benjamin, New York).
Landau, L. (1947). *J. Phys. Moscow* **11**, 91 (reprinted in Khalatnikov (N) (1965). "Introduction to the Theory of Superfludity", p. 205. Benjamin, New York).
Leiderer, P. and Pobell, F. (1973). *Phys. Rev.* **A7**, 1130.
London, F. (1954). "Superfluids", Vols. I and II. Wiley, New York.
McMillan, W. L. (1965). *Phys. Rev.* **138**, A442.
Mook, H. A. (1974). *Phys. Rev. Lett.* **32**, 1167.
Packard, R. E. and Sanders, T. M. (1969). *Phys. Rev. Lett.* **22**, 823.
Putterman, S. J. (1974). "Superfluid Hydrodynamics". North Holland, Amsterdam.
Rayfield, G. W. and Reif, F. (1964). *Phys. Rev.* **136**, 1194.
Richards, P. L. and Anderson, P. W. (1965). *Phys. Rev. Lett.* **14**, 540.
Vinen, W. F. (1961). *Proc. Roy. Soc.* **A260**, 218.

Problems

1. Show that in the model discussed in §9.5 the condensate fraction N_0/N tends to zero at the condensation temperature of the free Bose gas.

2. Consider an interacting Bose gas when particles have condensed into the single-particle state with momentum $\boldsymbol{q}$. Then

$$a_q \approx (N_0)^{1/2}, \qquad a_q^+ \approx (N_0)^{1/2}.$$

Keeping terms in the self-energy only up to first order in the interaction, find the single-particle Green's function. Hence show that the excitation energies of the long-wavelength quasi-particles are

$$E(\boldsymbol{k}) = ck - \boldsymbol{q}\,.\,\boldsymbol{k}/m$$

which is always positive if $|\boldsymbol{q}| < mc$. Show that the mass current density in this state is not zero. (This state is a superfluid state of the low-density Bose gas.)

Chapter 10

Magnetism

§ 10.1 Introduction

Magnetic phenomena are so widespread and varied that in this book we cannot hope to provide a comprehensive review. Indeed, our aim is quite limited, namely to provide some examples of magnetic problems where Green's functions are of help. We do this for their own interest and also to bring out the important fact that because of the commutation relations of spin operators, some difficulties arise in the treatment of spin Green's functions which are not seen in the cases of particle Green's functions which we have already treated.

We shall concentrate our attention on ordered magnetic states particularly ferromagnetism, antiferromagnetism and ferrimagnetism. A ferromagnet has a spontaneous magnetic moment below a transition temperature, the Curie temperature T_c. Below the Curie temperature it is believed that the spins tend to align so that at the absolute zero the alignment is complete as shown in Fig. 10.1 for an insulator. A typical plot of spontaneous magnetic moment against temperature is shown in Fig. 10.2. From neutron scattering and other experiments it is clear that ferromagnets possess low-energy excitations with wave characteristics. These are called spin waves, or, in the quantized form, magnons.

Antiferromagnets show a phase transition at a temperature, known as the Néel temperature, below which they are magnetically ordered without possessing a spontaneous magnetic moment. The ordering

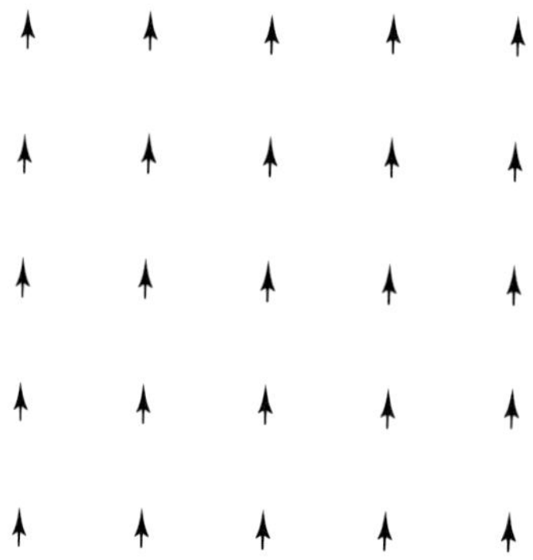

FIG. 10.1. The alignment of spins in the ground state of a ferromagnet.

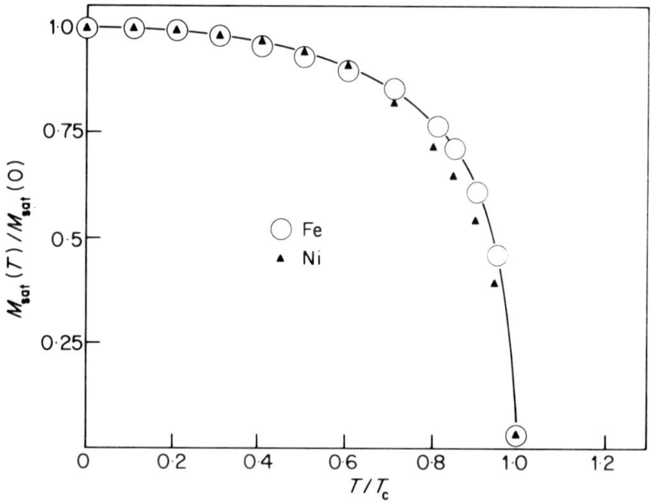

FIG. 10.2. The relative magnetization $M_{sat}(T)/M_{sat}(0)$ as a function of reduced temperature T/T_c. The data for Fe and Ni plot on the same curve (after Wert and Thomson 1970).

at the absolute zero is believed to be close to that shown in Fig. 10.3. The spins tend to be aligned on each of two sublattices but the net magnetic moments on each lattice exactly cancel. Below the transition temperature the magnetic susceptibility is very anisotropic, as illustrated in Fig. 10.4.

A ferrimagnet is similar to an antiferromagnet in that, below a transition temperature, the spins tend to align on two (or more) sublattices but, in the case of the ferrimagnet, the net moments on

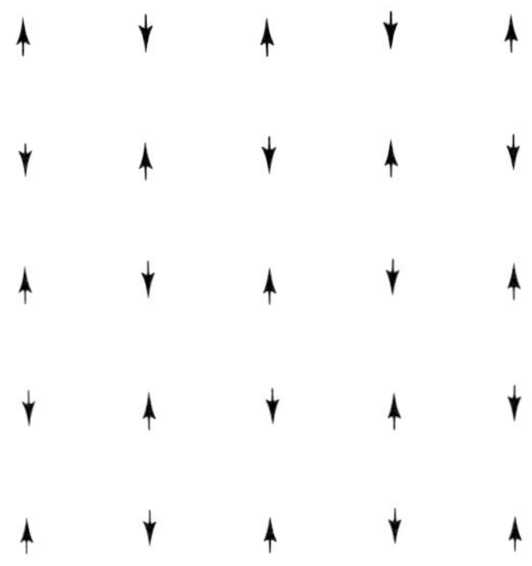

FIG. 10.3. A representation of the alignment of spins in the ground state of an antiferromagnet. Since the state illustrated is not an exact eigenstate of the Hamiltonian, this representation is approximate.

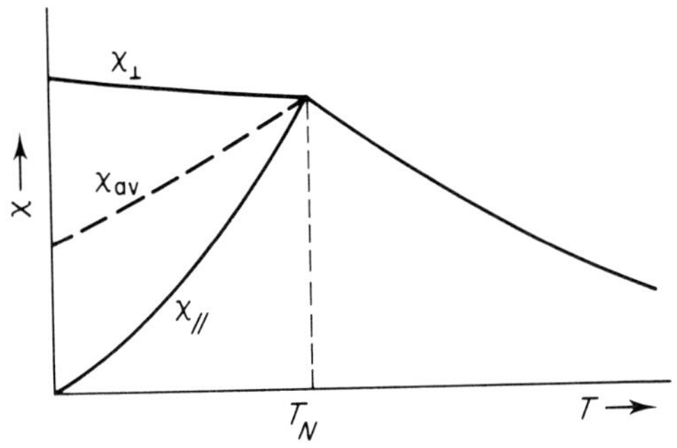

FIG. 10.4. The magnetic susceptibility of an antiferromagnet, MnF_2. The longitudinal, transverse and average susceptibilities are shown (after Sinha 1973).

the sublattices do not cancel and a spontaneous moment remains. Ferrimagnets, too, possess spin waves.

§10.2 Molecular field theory

The first problem in understanding magnetism is to pick out the terms in the Hamiltonian which are most important in giving rise to magnetism. This problem can be partially circumvented by studying model Hamiltonians which are simpler than the original and which can apparently describe the phenomena. Two models which have proved of great value are those due to Heisenberg and to Hubbard. The former is particularly useful for the study of magnetic insulators and the latter for metals. We shall take these as our starting points. For the relationship between these models and the true Hamiltonian, we refer the reader to the literature mentioned at the end of the chapter.

We begin with Heisenberg's Hamiltonian which assumes that the magnetic properties can be described completely in terms of spins $\boldsymbol{S}_i$ situated on the different lattice sites labelled (i). There will be an interaction between the spins tending to align them. The simplest Hamiltonian which contains such an interaction is

$$H = -g\mu_{\mathrm{B}} \sum_i \boldsymbol{H} \cdot \boldsymbol{S}_i - \sum_{i,j} J_{ij} \boldsymbol{S}_i \cdot \boldsymbol{S}_j . \tag{10.1}$$

Here J_{ij} is the strength of the coupling between spins on sites "i" and "j"; it is generally assumed to fall off rapidly with separation. Also included in the Hamiltonian is the effect of an external magnetic field $\boldsymbol{H}$. We have assumed that the spins are identical. It is not difficult to see that if J_{ij} is positive everywhere, the ground state of the Hamiltonian is the one in which all the spins point in the direction of the magnetic field.

We remind the reader that the spins satisfy the commutation relations

$$[\boldsymbol{S}_i, \boldsymbol{S}_j] = 0, \qquad i \neq j, \tag{10.2}$$

$$\boldsymbol{S}_i \times \boldsymbol{S}_i = i\boldsymbol{S}_i. \tag{10.3}$$

Further, if

$$S_i^+ = S_{ix} + iS_{iy}, \qquad S_i^- = S_{ix} - iS_{iy}, \tag{10.4}$$

then

$$[S_{iz}, S_j^+] = S_i^+ \delta_{ij}, \qquad [S_{iz}, S_j^-] = -S_i^- \delta_{ij}, \tag{10.5}$$

and for a particle of spin S,

$$\boldsymbol{S}^2 \equiv S_x^2 + S_y^2 + S_z^2 = S(S+1). \tag{10.6}$$

The eigenstates of S_z are $|s\rangle$ where

$$S_z|s\rangle = s|s\rangle \tag{10.7}$$

and where s takes on the values

$$s = -S, -S+1, \ldots, S-1, S.$$

We also have the relations

$$S^+|s\rangle \propto |s+1\rangle, \qquad S^+|S\rangle = 0 \tag{10.8}$$

and

$$S^-|s\rangle \propto |s-1\rangle, \qquad S^-|-S\rangle = 0. \tag{10.9}$$

If we choose the positive z-direction to be that of the magnetic field, the state with all spins pointing in the direction of the field is

$$|0\rangle = \prod_i |S\rangle_i \tag{10.10}$$

where $|s\rangle_i$ is the state of the ith spin. Since

$$S_i^+ |0\rangle = 0, \qquad \text{for all } i,$$

we find that

$$H|0\rangle = \left[-g\mu_B SH - S^2 \sum_{i,j} J_{ij}\right] |0\rangle, \tag{10.11}$$

and $|0\rangle$ is an eigenvector of H. It is not difficult to see that the eigenvalue is the lowest value the Hamiltonian can attain if all J_{ij} are positive. In this case $|0\rangle$ is the ground state.

Unfortunately, it is not possible to find all the excited states and the thermodynamic potential in this exact way; approximations must be used. Since at the absolute zero each spin is pointing in the z-direction, it is reasonable to assume that at a finite temperature each spin has a finite average value in this direction and that the finite total magnetic moment derives from this spin. If one assumes that fluctuations of spin about the average value are so small that quadratic terms in the fluctuations can be neglected, the Hamiltonian can be linearized and the thermodynamic potential can be obtained self-consistently. If this programme is carried out

$$H \simeq -g\mu_B H \sum_i S_{iz} - 2 \sum_{i,j} J_{ij} \langle S_{iz}\rangle S_{jz} + \sum_{i,j} J_{ij} \langle S_{iz}\rangle \langle S_{jz}\rangle$$

$$= -g\mu_B H_{\text{eff}} \sum_i S_{iz} + \sum_{i,j} J_{ij} \langle S_{iz}\rangle \langle S_{jz}\rangle, \tag{10.12}$$

where

$$g\mu_B H_{\text{eff}} = g\mu_B H + 2 \sum_i J_{ij} \langle S_{iz}\rangle. \tag{10.13}$$

Each spin sits in an effective magnetic field comprising the external field and the average fields due to the other spins. For a uniform crystal H_{eff} is independent of position. The average spin has to be chosen self-consistently from

$$\langle S_z\rangle = \langle S_{iz}\rangle = \frac{\mathrm{Tr}\,[\exp(-\beta H) S_{iz}]}{\mathrm{Tr}\,\exp(-\beta H)}. \tag{10.14}$$

Equations (10.13) and (10.14) constitute a self-consistent approximation for the solution. It was introduced by Weiss and is often called the molecular field approximation. It is analogous to the approximations made to treat superconductivity and superfluidity discussed in Chapters 8 and 9 and preceded them.

Since $\langle S_{iz} \rangle$ is independent of i, equation (10.14) can be rewritten

$$\langle S_z \rangle = \frac{\sum_{s=-S}^{S} s \exp [\beta g \mu_B H_{\text{eff}} s]}{\sum_{s=-S}^{S} \exp [\beta g \mu_B H_{\text{eff}} s]}. \tag{10.15}$$

For the case $S = 1/2$, this can be written

$$\langle S_z \rangle = \tfrac{1}{2} \tanh [\tfrac{1}{2} \beta g \mu_B H_{\text{eff}}]. \tag{10.16}$$

A spontaneous magnetic moment exists if this equation has a non-zero solution when the external field is zero. In that case the equation becomes

$$\langle S_z \rangle = \tfrac{1}{2} \tanh (\beta J_0 \langle S_z \rangle), \tag{10.17}$$

where

$$J_0 = \sum_j J_{ij}. \tag{10.18}$$

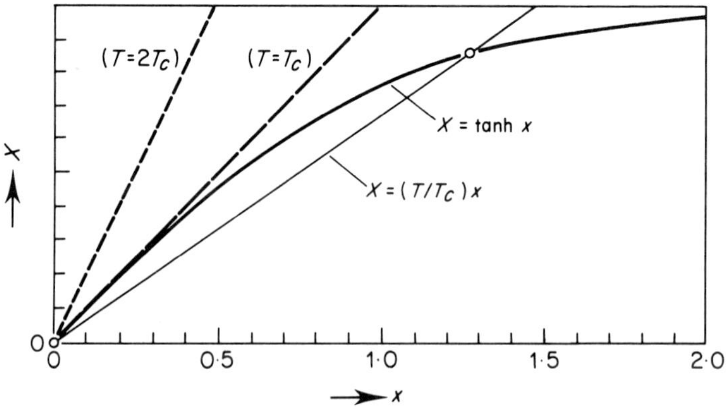

FIG. 10.5. Plots of x and $X = \tanh(\beta J_0 x)$, illustrating the solution of equation (10.17).

A plot of the two sides of equation (10.17) is shown in Fig. 10.5, from which it can be seen that the equation possesses a non-trivial solution as long as

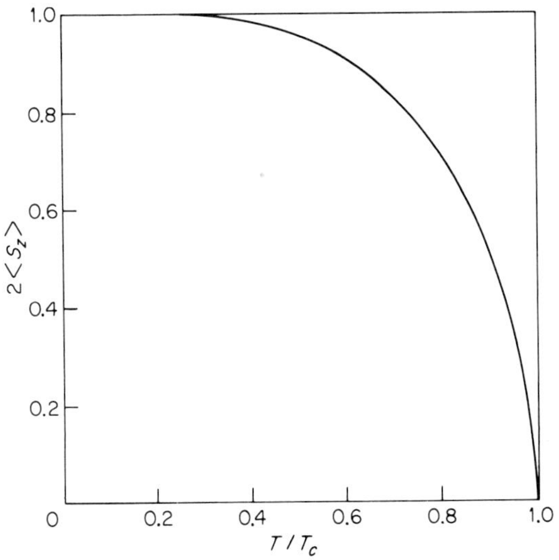

FIG. 10.6. The ratio of the spontaneous magnetic moment to the saturation moment, $2\langle S_z\rangle$ against T as given by molecular field theory. for a spin $-\frac{1}{2}$ system.

$$\tfrac{1}{2}J_0\beta \geqslant 1, \qquad (10.19)$$

that is

$$T \leqslant T_c,$$

where

$$T_c = \tfrac{1}{2}J_0/k_{\mathrm{B}}. \qquad (10.20)$$

Hence there is a critical temperature T_c above which the solid is paramagnetic and below which it possesses a spontaneous magnetic moment. The magnetic moment per unit volume, M_z, is given by

$$M_z = \rho g \mu_{\mathrm{B}} \langle S_z \rangle,$$

where ρ is the number density of the spins. A plot of M versus T for the approximation is seen in Fig. 10.6. It shows the correct qualitative features and is exact at $T = 0$. However, it is quantitatively wrong at higher temperatures. At T_c, $\langle S_z\rangle$ vanishes; near T_c, $\langle S_z\rangle$ is small and equation (10.17) can be expanded to yield

$$\langle S_z\rangle = (T_c/T)\langle S_z\rangle - \tfrac{1}{6}(T_c/T)^3\langle S_z\rangle^3.$$

Hence

$$\langle S_z\rangle \simeq [6(T_c - T)/T_c]^{1/2}.$$

This gives the characteristic temperature dependence,

$$M \propto [(T_c - T)/T_c]^{1/2}, \qquad (10.21)$$

of the order parameter of a molecular field theory near a critical point. We look more closely at behaviour near a critical point in Chapter 12. For other values of spin S one obtains qualitatively similar results, including equation (10.21).

The molecular field approximation is sufficiently general to provide a model of an antiferromagnet and of a ferrimagnet. For simplicity, suppose that all the spins are half and that only the interaction between nearest neighbours is important. Let us look for a solution in which the average spins on neighbouring sites point in opposite directions. Take the positive z-direction to be in the direction of one set of spins, the A-spins; the other spins we call the B-spins. If we again ignore in the Hamiltonian terms quadratic in fluctuations, we have a molecular field theory. If there is no external magnetic field, the effective field on a B-spin is given by

$$g\mu_B H_{\text{eff}\,B} = 2ZJ\langle S_A\rangle,$$

where Z is the number of nearest neighbours of a given spin the coordination number, and J the strength of the interaction between neighbouring spins. Similarly, the effective field acting on an A-spin is given by

$$g\mu_B H_{\text{eff}\,A} = 2ZJ\langle S_B\rangle = -2ZJ\langle S_A\rangle. \qquad (10.22)$$

Then the self-consistent equation for $\langle S_A\rangle$ is

$$\langle S_A\rangle = \tfrac{1}{2}\tanh\left[-\beta ZJ\langle S_A\rangle\right]. \qquad (10.23)$$

This has a non-trivial solution if $J < 0$ and

$$T < T_c = \tfrac{1}{2}|J|/k_B. \qquad (10.24)$$

The magnetic moment per unit volume on the A lattice is

$$M_A = \tfrac{1}{2}\rho g\mu_B\langle S_A\rangle.$$

Qualitatively, this behaves like $M(T)$ in Fig. 10.6 as a function of temperature. Since alternate spins point in opposite directions, the total magnetic moment is zero. In this case, the molecular field approximation is not exact at absolute zero. This is because the state with alternate "up" and "down" spins is not an exact eigenstate of the Hamiltonian.

If the spins and/or the magnetic moments on adjacent sites are different, the average spins on adjacent sites will not cancel in the molecular field approximation and a ferrimagnetic state results.

§10.3 Green's function approach

In the molecular field approximation we neglect quadratic terms in S^+, S^- in comparison with S_z^2. Since, however, we have the commutation relation

$$[S^+, S^-] = 2S_z \tag{10.25}$$

this cannot be an entirely satisfactory approximation. We also have

$$S_x^2 + S_y^2 + S_z^2 = S(S+1). \tag{10.26}$$

Hence, as S_z diminishes in magnitude S_x and S_y must grow. The molecular field approximation should, therefore, become increasingly less satisfactory as one considers higher temperatures.

These comments suggest that we should obtain an improved approximation if we ignore the fluctuations only in the z-component of spin. In the absence of an external magnetic field we therefore replace the Hamiltonian by

$$H = -g\mu_{\mathrm{B}} H_{\mathrm{eff}} \sum_i S_{iz} + \sum_{i,j} J_{ij}\langle S_{iz}\rangle\langle S_{jz}\rangle - \tfrac{1}{2} \sum J_{ij}(S_i^+ S_j^- + S_i^- S_j^+), \tag{10.27}$$

where

$$g\mu_{\mathrm{B}} H_{\mathrm{eff}} = 2 \sum_j J_{ij}\langle S_{jz}\rangle = 2\langle S_z\rangle \sum_j J_{ij}. \tag{10.28}$$

We also replace the commutation relation (10.25) by the approximate one

$$[S^+, S^-] = 2\langle S_z\rangle. \tag{10.29}$$

The equation of motion for $\tilde{S}_i^+(\tau)$ is then

$$\partial \tilde{S}_i^+/\partial\tau = -g\mu_{\mathrm{B}} H_{\mathrm{eff}} \tilde{S}_i^+ + 2 \sum_j J_{ij} \tilde{S}_j^+ \langle S_z\rangle. \tag{10.30}$$

Hence, the equation for the temperature Green's function

$$G(i, j; \tau) = -\langle T(\tilde{S}_i^+(\tau) S_j^-(0)\rangle \tag{10.31}$$

is

$$\left(\frac{\partial}{\partial\tau} + g\mu_{\mathrm{B}} H_{\mathrm{eff}}\right) G(i, j; \tau) = 2 \sum_j J_{il} G(l, j, \tau)\langle S_z\rangle - \delta(\tau)\langle [S_i^+, S_j^-]\rangle. \tag{10.32}$$

The usual "time" Fourier transform satisfies

$$(-i\zeta + g\mu_{\mathrm{B}} H_{\mathrm{eff}})\bar{G}(i, j; \zeta) = 2 \sum_l J_{il}\bar{G}(l, j; \zeta)\langle S_z\rangle - 2\delta_{ij}\langle S_z\rangle,$$
$$\zeta = 2n\pi/\beta. \tag{10.33}$$

Because of the translational invariance of the system, this equation can be solved if the spatial Fourier transform,

$$\bar{G}(i,j;\zeta) = N^{-1} \sum_k \exp[ik\cdot(R_i - R_j)]\,\bar{G}(k,\zeta), \quad (10.34)$$

is introduced. Then

$$(-i\zeta + g\mu_B H_{\text{eff}})\bar{G}(k,\zeta) = 2J(k)\langle S_z\rangle \bar{G}(k,\zeta) - 2\langle S_z\rangle,$$

where

$$J(k) = \sum_i \exp[-ik\cdot(R_i - R_j)]\,J_{ij} \quad (10.35)$$

and

$$\bar{G}(k,\zeta) = 2\langle S_z\rangle/[i\zeta - g\mu_B H_{\text{eff}} + 2J(k)\langle S_z\rangle]. \quad (10.36)$$

The same result can be obtained by deriving the exact equation for $\bar{G}(i,j;\tau)$ and linearizing it by replacing the operator S_z by $\langle S_z\rangle$ wherever it occurs. This is the conventional method.

The next necessary step is to find the self-consistent equation for $\langle S_z\rangle$ by relating it to $\bar{G}$. In general, there is not a unique way of doing this. One satisfactory method has proved to be that used by Bogoliubov and Tyablikov (1959) for spins of magnitude 1/2 and generalized by Tahir-Kheli and ter Haar (1962) for higher spins. We restrict the discussion to the case of spin 1/2 and refer the reader to the paper of Tahir-Kheli and ter Haar as well as the review by Tahir-Kheli (1976) for other cases. For spin 1/2 particles, the operators at any site satisfy two equations,

$$S^2 = S_z^2 + \tfrac{1}{2}(S^+S^- + S^-S^+) = S(S+1) = \tfrac{3}{4}, \quad (10.37)$$

and

$$S_z^2 = \tfrac{1}{4}. \quad (10.38)$$

Hence, using the exact commutation relation for S^+, S^- one finds

$$S^+S^- = \tfrac{3}{4} - \tfrac{1}{4} + S_z = \tfrac{1}{2} + S_z$$

and

$$\langle S_i^+ S_i^-\rangle = -G(i,i,0_+) = \tfrac{1}{2} + \langle S_z\rangle. \quad (10.39)$$

But

$$G(i,i,0_+) = \frac{1}{\beta N}\sum_k \sum_\zeta \frac{2\langle S_z\rangle \exp(-i\zeta 0_+)}{i\zeta - g\mu_B H_{\text{eff}} + 2J(k)\langle S_z\rangle}$$

$$= -\frac{2}{N}\langle S_z\rangle \sum_k [n(k)+1] = -2\langle S_z\rangle\left[\frac{1}{N}\sum_k n(k) + 1\right], \quad (10.40)$$

where

$$n(k) = \{\exp[\beta E(k)] - 1\}^{-1} \quad (10.41)$$

and

$$E(\boldsymbol{k}) = g\mu_{\mathrm{B}}H_{\mathrm{eff}} - 2J(\boldsymbol{k})\langle S_z\rangle = 2\langle S_z\rangle[J(0) - J(\boldsymbol{k})]. \tag{10.42}$$

From equations (10.39) and (10.40) one finds the self-consistent equation

$$\langle S_z\rangle = (1/N)\sum_{\boldsymbol{k}} n(\boldsymbol{k}) + \tfrac{1}{2}, \tag{10.43}$$

where $n(\boldsymbol{k})$ depends on $\langle S_z\rangle$ through equation (10.41) and (10.42).

We see that

$$G(\boldsymbol{k}, \omega) = 2\langle S_z\rangle/[\omega - E(\boldsymbol{k})]. \tag{10.44}$$

Consequently the excitation energies are $E(\boldsymbol{k})$. For a lattice with a centre of symmetry, as $\boldsymbol{k} \to 0$, we find that

$$E(\boldsymbol{k}) \propto k^2 \tag{10.45}$$

and $E(\boldsymbol{k})$ tends to zero in the limit. The existence of excitations with infinitesimal excitation energy is a general consequence of the breaking of the symmetry – in this case rotational symmetry. The operators which create the excitations are of the form

$$\sum_{l} S_i^- \exp(i\boldsymbol{k} \,.\, \boldsymbol{R}_i) \tag{10.46}$$

and represent waves in which spins are flipped with varying amplitude through the lattice. It can be shown that these are exact eigenstates in the Heisenberg ferromagnet at absolute zero provided that $E(\boldsymbol{k}) > 0$.

Equation (10.43) has been solved numerically (Izyumov and Noskova, 1964, Hass and Jarrett 1964, Tyablikov 1965) as well as by expansions at low temperatures and near T_c. At low temperatures one obtains for nearest neighbour interaction the expansion (for $S = 1/2$)

$$\langle S_z\rangle = [\tfrac{1}{2} - a_0 - a_1x^{5/2} - a_2x^{7/2} \ldots - a_0^2x^3 - 4a_0a_1x^4 \ldots], \tag{10.47}$$

where

$$\tau^{-1} = \beta J(0), \qquad x = 3\tau/2\pi\nu, \tag{10.48}$$

$$a_0 = \zeta(3/2), \quad a_1 = \tfrac{3}{4}\pi\nu\zeta(5/2), \quad a_2 = \pi^2 w\nu^2\zeta(7/2), \tag{10.49}$$

$$\begin{aligned} \nu &= 1, & w &= 33/32, & &\text{simple cubic lattice,} \\ \nu &= \tfrac{3}{4}2^{2/3}, & w &= 281/288, & &\text{body-centred cubic lattice,} \\ \nu &= 2^{1/3}, & w &= 15/16, & &\text{face-centred cubic lattice,} \end{aligned} \tag{10.50}$$

and $\zeta(x)$ is the Riemann zeta function. Equation (10.47) differs from an exact low-temperature expansion due to Dyson (1956) only in the terms in τ^3 and τ^4. In Dyson's solution, there is no term in τ^3 and the term in τ^4 has a different coefficient. These differences can be traced back to the fact that interactions between spin waves are neglected in the present method but included in Dyson's. There is a consequent change in the spin wave energies.

It is possible to improve upon these results by continuing the hierarchy of exact Green's functions equations and decoupling at a later stage. If this is done consistently at the next stage, Dyson's results are obtained up to terms in T^4 (Ortenburger 1964).

At temperatures approaching T_c from below, this theory, like the molecular field theory described in the last section, yields the spontaneous magnetic moment,

$$M \propto ((T_c/T) - 1)^{1/2}.$$

The same approach can be used for the study of spin waves in antiferromagnets and ferrimagnets. In these cases the ground state is not known exactly so the approximation is even more open to question. Nevertheless, the results obtained have proved to be useful, especially qualitatively. For an isotropic antiferromagnetic in the absence of an external field there are two degenerate branches of the spin wave spectrum with (Nagmiya *et al.* 1955)

$$E(\mathbf{k}) = S(J(0)^2 - J(\mathbf{k})^2)^{1/2}.$$

As $\mathbf{k} \to 0$, this is the same spectrum as for phonons

$$E(\mathbf{k}) \propto k. \tag{10.51}$$

Green's functions have also proved useful in the study of spin waves in the presence of magnetic impurities, in alloys and in two-magnon problems. Some of these effects are discussed in the review article of Thorpe (1978) who gives further references. Green's functions have also proved particularly useful in the study of excitations in finite Heisenberg magnets (Cottam 1976).

§10.4 Hubbard model

Heisenberg's model describes spins attached to definite sites and is, therefore, useful for the study of magnetic insulators. For metals, where the spins are free to move, a different sort of model is needed, one based on their band structure. A simple band model containing no interaction between the electrons leads to the filling up of the band with equal numbers of electrons with opposite spins and, therefore, no spontaneous magnetic moment. A model which is still comparatively simple and seems to have within it the possibility of ferromagnetism was put forward by Hubbard (1963, 1964). In addition to the band energy it contains a repulsive interaction between electrons of opposite spin on the same lattice site. Thus

$$H = \sum_{i,j,\sigma} \epsilon_{ij} a_{i\sigma}^{+} a_{j\sigma} + \tfrac{1}{2} I \sum_{i\sigma} n_{i\sigma} n_{i-\sigma}, \tag{10.52}$$

where $a_{i\sigma}^{+}$ creates an electron (in a Wannier state) at site "i" with spin σ and

$$n_{i\sigma} = a_{i\sigma}^{+} a_{i\sigma}. \tag{10.53}$$

The first approximation to try with the Hamiltonian (10.52) is a molecular field one. This entails the assumption that the fluctuation of $n_{i\sigma}$ about its mean value $\langle n_{i\sigma}\rangle$ is small. Then

$$H = \sum_{i,j,\sigma} \epsilon_{ij} a_{i\sigma}^{+} a_{j\sigma} + I \sum_{i,\sigma} n_{i\sigma} \langle n_{i-\sigma}\rangle - \tfrac{1}{2} I \sum_{i,\sigma} \langle n_{i\sigma}\rangle \langle n_{i-\sigma}\rangle. \tag{10.54}$$

This approximation is identical with that of Hartree–Fock (§3.6). The equation of motion one then obtains is

$$\frac{\partial a_{i\sigma}(\tau)}{\partial \tau} = -\sum_{j} \epsilon_{ij} a_{j\sigma} - I\langle n_{-\sigma}\rangle a_{i\sigma}, \tag{10.55}$$

where we have assumed a translationally invariant solution with

$$\langle n_{i\sigma}\rangle = n_{\sigma}, \tag{10.56}$$

independent of position. If, as usual,

$$G(i,j;\sigma\colon \tau) = -\langle T a_{i\sigma}(\tau) a_{j\sigma}(0)^{+}\rangle, \tag{10.57}$$

then

$$\left[\frac{\partial}{\partial \tau} + I\langle n_{\sigma}\rangle\right] G(i,j;\sigma:\tau) = -\delta(\tau) - \sum_{l} \epsilon_{il} G(l,j,\sigma:\tau). \tag{10.58}$$

This can be solved by introducing the "time" and space transform of the Green's function. Then

$$[-i\zeta + I\langle n_{-\sigma}\rangle]\, \bar{G}(k,\zeta,\sigma) = -1 - \epsilon_k \bar{G}(k,\zeta,\sigma), \tag{10.59}$$

where ϵ_k is the band energy in the absence of interaction,

$$\epsilon_k = N^{-1} \sum_{i} \epsilon_{ij} \exp[i\boldsymbol{k}\cdot(\boldsymbol{R}_i - \boldsymbol{R}_j)]. \tag{10.60}$$

Hence,

$$\bar{G}(k, \zeta, \sigma) = [i\zeta - I\langle n_{-\sigma}\rangle - \epsilon_k]^{-1}, \tag{10.61}$$

and the up-spins have energy $\epsilon_k + I\langle n_\downarrow\rangle$ while the down-spins have energy $\epsilon_k + I\langle n_\uparrow\rangle$; the different spins are influenced only by the average fields due to the other spins, a characteristic of the Hartree approximation.

The self-consistent equations are now

$$\langle n_\uparrow\rangle = \beta^{-1} \sum_\zeta N^{-1} \sum_k \bar{G}(k, \zeta, \sigma) = \int d\epsilon N(\epsilon) f[\epsilon + I\langle n_\downarrow\rangle], \tag{10.62}$$

$$\langle n_\downarrow\rangle = \int d\epsilon N(\epsilon) f[\epsilon + I\langle n_\uparrow\rangle], \tag{10.63}$$

$$n = \langle n_\uparrow\rangle + \langle n_\downarrow\rangle, \tag{10.64}$$

where $N(\epsilon)$ is the density of states of the unperturbed band and $f(x)$ is the Fermi–Dirac distribution function. The latter depends on the chemical potential, μ, because ϵ_k is measured from μ. Equation (10.62), (10.63) and (10.64) are sufficient for the determination of μ, $\langle n_\uparrow\rangle$, $\langle n_\downarrow\rangle$. They always possess a solution with $\langle n_\uparrow\rangle = \langle n_\downarrow\rangle$ corresponding to a non-magnetic state.

To see under what conditions ferromagnetism might arise we consider a band with a smoothly varying density at a temperature well below its degeneracy temperature so that

$$f(\epsilon) \approx \theta(-\epsilon)$$

and we look for solutions with a small amount of ferromagnetism. We therefore put

$$\langle n_\uparrow\rangle = \tfrac{1}{2}n + x, \qquad \langle n_\downarrow\rangle = \tfrac{1}{2}n - x, \tag{10.65}$$

and look for solutions with x small. It is convenient to put

$$\epsilon = \xi - \mu, \tag{10.66}$$

where ξ is the band energy measured from the bottom of the band and μ is the chemical potential. Then equations (10.62) and (10.63) lead to

$$\tfrac{1}{2}n = \int_0^{\mu - \frac{1}{2}In} d\xi N(\xi) - \tfrac{1}{2} I^2 x^2 N'(\mu - \tfrac{1}{2}In) \tag{10.67}$$

and

$$x[1 - IN(\mu - \tfrac{1}{2}In)] = \tfrac{1}{6} I^3 x^3 N''(\mu - \tfrac{1}{2}In). \tag{10.68}$$

It follows from equation (10.68) that if there exist energies in the band for which

$$IN(\xi_0) = 1, \qquad N''(\xi_0) \neq 0, \tag{10.69}$$

it is possible to choose $(\mu - \frac{1}{2}In)$ near to ξ_0 and find a non-zero solution to equation (10.68) for x. If $N''(\xi_0) < 0$ then we must choose

$$IN(\mu - \tfrac{1}{2}In) > 1, \tag{10.70}$$

and if $N''(\xi_0) > 0$ we must have the opposite inequality. Having fixed $(\mu - \frac{1}{2}In)$ equation (10.68) determines x and equation (10.67) then determines n. Thus, under the conditions (10.69), an itinerant ferromagnetic state can exist according to this theory. In fact, the Hartree–Fock approximation, by ignoring local correlations, tends to overestimate the occurrence of ferromagnetism.

In order to include correlations, Hubbard continued the hierarchy of Green's function equations one stage further. We follow his procedure and use the original Hamiltonian (10.52). Then the equation of motion for one operator becomes

$$\frac{\partial a_{i\sigma}(\tau)}{\partial \tau} = -\sum_j \epsilon_{ij} a_{j\sigma} - I n_{i-\sigma}(\tau) a_{i\sigma}(\tau) \tag{10.71}$$

and

$$\frac{\partial}{\partial \tau} G(i, j, \sigma; \tau) = -\delta(\tau)\delta_{ij} - \sum_l \epsilon_{il} G(l, j, \sigma; \tau)$$
$$+ I\langle T n_{i-\sigma}(\tau) a_{i\sigma}(\tau) a^{+}_{j\sigma}(0)\rangle. \tag{10.72}$$

This differs from equation (10.58) through the appearance of the operator $n_{i-\sigma}(\tau)$ is the last term. If this operator is replaced by its expectation value, equation (10.58) is retrieved.

We now write down the equation of motion of the last Green's function in equation (10.72). This is

$$\frac{\partial}{\partial \tau}\langle T n_{i-\sigma}(\tau) a_{i\sigma}(\tau) a^{+}_{j\sigma}(0)\rangle = -\epsilon_{ii}\langle T n_{i-\sigma}(\tau) a_{i\sigma}(\tau) a_{j\sigma}(0)\rangle$$
$$-\sum_{l\neq i} \epsilon_{il}\langle T n_{i-\sigma}(\tau) a_{l\sigma}(\tau) a^{+}_{j\sigma}(0)\rangle$$
$$+\sum_{l\neq i} \epsilon_{ij}\,[\langle T a^{+}_{l-\sigma}(\tau) a_{i-\sigma}(\tau) a_{i\sigma}(\tau) a_{j\sigma}(0)\rangle$$
$$-\langle T a^{+}_{i-\sigma}(\tau) a_{l-\sigma}(\tau) a_{i\sigma}(\tau) a_{j\sigma}(0)\rangle]$$
$$-I\langle T n_{i-\sigma}(\tau) a_{i\sigma}(\tau) a_{j\sigma}(0)\rangle + \delta_{ij}\langle n_{i-\sigma}\rangle\delta(\tau). \tag{10.73}$$

In the penultimate term we have used the operator equation

$$n_{i-\sigma}(\tau)^2 = n_{i-\sigma}(\tau).$$

The last term derives from the θ-functions implicit in the use of the T operator and the commutation relation

$$[n_{i-\sigma}a_{i\sigma}, a_{j\sigma}^{+}] = n_{i-\sigma}\delta_{ij}.$$

On the assumption that correlations between electrons on different sites are less important than those on the same site, the third term in equation (10.73) is neglected. We also assume that in the second term the fluctuations in $n_{i-\sigma}$ can be ignored and the operator replaced by its expectation value. Both these terms are zero when ϵ_{ij} is diagonal and no hopping takes place. The result we obtain is therefore exact in this case. For the more interesting case when there is hopping, the result is approximate. From the method of approximation we expect the result to be best when hopping is weak and this will be the case for narrow energy bands for which the approximation was originally devised.

With these approximations equation (10.73) reduces to

$$\left(\frac{\partial}{\partial\tau} + \epsilon_{ii} + I\right) \langle Tn_{i-\sigma}(\tau)a_{i\sigma}(\tau)a_{j\sigma}(0)\rangle$$
$$= \langle n_{-\sigma}\rangle\delta_{ij}\delta(\tau) - \langle n_{-\sigma}\rangle \sum_{l\neq i} \epsilon_{il}\langle Ta_{l\sigma}(\tau)a_{j\sigma}^{+}(0)\rangle. \qquad (10.74)$$

It we introduce the "time" Fourier transform with

$$\langle Tn_{i-\sigma}(\tau)a_{i\sigma}(\tau)a_{j\sigma}^{+}(0)\rangle = \beta^{-1} \sum_{\zeta} \exp(-i\zeta\tau)G_2(i,j,\sigma;\zeta) \qquad (10.75)$$

equation (10.74) becomes

$$(-i\zeta + \epsilon_{ii} + I)G_2(i,j,\sigma;\zeta) = \langle n_{-\sigma}\rangle \left[\delta_{ij} + \sum_{l\neq i} \epsilon_{il}\bar{G}(l,j,\sigma;\zeta)\right]$$

Hence, if this is substituted into equation (10.72),

$$(-i\zeta)\bar{G}(i,j,\sigma:\zeta) = -\delta_{ij} - \sum_{l} \epsilon_{il}\bar{G}(l,j,\sigma:\zeta) \qquad (10.76)$$
$$+ \frac{I\langle n_{-\sigma}\rangle \left[\delta_{ij} - \epsilon_{ii}\bar{G}(i,j,\sigma;\zeta) + \sum_{l} \epsilon_{il}\bar{G}(l,j,\sigma,\zeta)\right]}{(-i\zeta + \epsilon_{ii} + I)}.$$

If we now introduce the space Fourier transform of $\bar{G}$ we obtain

$$\left[-i\zeta + \epsilon_{k}\left(1 - \frac{I\langle n_{-\sigma}\rangle}{-i\zeta + \epsilon_0 + I}\right) + \frac{\epsilon_0 I\langle n_{-\sigma}\rangle}{-i\zeta + \epsilon_0 + I}\right] \bar{G}(\boldsymbol{k},\sigma;\zeta)$$

$$= -1 + \frac{I\langle n_{-\sigma}\rangle}{i\zeta + \epsilon_0 + I}, \tag{10.77}$$

where

$$\epsilon_0 = \epsilon_i.$$

Hence,

$$\bar{G}(\boldsymbol{k}, \sigma; \zeta) = \frac{i\zeta - \epsilon_0 - I(1 - \langle n_{-\sigma}\rangle)}{(-i\zeta + \epsilon_k)(-i\zeta + \epsilon_0 + I) + I\langle n_{-\sigma}\rangle(\epsilon_0 - \epsilon_k)} \tag{10.78}$$

and

$$G(\boldsymbol{k}, \sigma; \omega) = \frac{\omega - \epsilon_0 - I(1 - \langle n_{-\sigma}\rangle)}{(\omega - \epsilon_k)(\omega - \epsilon_0 - I) + I\langle n_{-\sigma}\rangle(\epsilon_0 - \epsilon_k)}. \tag{10.79}$$

Because the denominator is quadratic in ω, for given $\langle n_{-\sigma}\rangle$ and $\boldsymbol{k}$ the Green's function has two poles $E_{k\sigma}^{(1)}$ and $E_{k\sigma}^{(2)}$. Since the denominator is positive for ω large and real and negative when

$$\omega = \epsilon_0 + I(1 - n)$$

the poles of G are real and satisfy the inequality

$$E_{k\sigma}^{(1)} < \epsilon_0 + I(1 - \langle n_{-\sigma}\rangle) < E_{k\sigma}^{(2)}. \tag{10.80}$$

Consequently, in this approximation, the excitations are quasi-particles which lie in two disparate bands.

In the case of no hopping, $\epsilon_k = \epsilon_0$, the quasi-particle energies are ϵ_0 and $\epsilon_0 + I$ corresponding, respectively, to the energy required to put one electron on a site and the energy required to put a second electron at each site. The effect of hopping is to spread out these energy levels into two separate bands.

To complete the solutions we have to add the self-consistency conditions which determine $\langle n_\sigma\rangle$ and the chemical potential μ. It is again useful to make the chemical potential explicit and this can be done formally by changing ω to $\omega + \mu$ in equation (10.79). Since the Green's function is of the form

$$\bar{G}(\boldsymbol{k}, \sigma; \zeta) = -\frac{\partial D(\zeta, \epsilon_k)/\partial \epsilon_k}{D(\zeta, \epsilon_k)}, \tag{10.81}$$

it is possible to write

$$\bar{G}(\boldsymbol{k}, \sigma; \zeta) = \frac{dE_{k\sigma}^{(1)}/d\epsilon_k}{-i\zeta - E_{k\sigma}^{(1)} + \mu} + \frac{dE_{k\sigma}^{(2)}/d\epsilon_k}{-i\zeta - E_{k\sigma}^{(2)} + \mu}. \tag{10.82}$$

Then at low temperatures

$$\begin{aligned}\langle n_\uparrow \rangle &= \beta^{-1} \sum_{\zeta} N^{-1} \sum_{k} \bar{G}(k, \uparrow; \zeta) \\ &= \int d\epsilon\, N(\epsilon) \frac{dE_{k\uparrow}^{(1)}}{d\epsilon_k} \theta(\mu - E_{k\uparrow}^{(1)}) + \int d\epsilon\, N(\epsilon) \theta(\mu - E_{k\uparrow}^{(2)}) \frac{dE_{k\uparrow}^{(2)}}{d\epsilon_k} \\ &= \int_{-\infty}^{\mu} dE\, N[\epsilon_\uparrow(E)], \end{aligned} \tag{10.83}$$

where we have changed the variable of integration from ϵ to the quasi-particle energy E. The function $\epsilon_\uparrow(E)$ relates the unperturbed energy ϵ to the perturbed energy E and is given implicitly by

$$E^2 - E(\epsilon_\uparrow + \epsilon_0 + I) + I\langle n_\downarrow \rangle(\epsilon_0 - \epsilon_\uparrow) + \epsilon_\uparrow(\epsilon_0 + I) = 0.$$

Hence,

$$\epsilon_\uparrow = \frac{E^2 - E(\epsilon_0 + I) + I\langle n_\downarrow \rangle \epsilon_0}{E + I(\langle n_\downarrow \rangle - 1) - \epsilon_0}. \tag{10.84}$$

The integration in equation (10.83) is taken over the two bands of allowed energies E up to μ. In a similar way

$$\langle n_\downarrow \rangle = \int_{-\infty}^{\mu} dE\, N[\epsilon_\downarrow(E)], \tag{10.85}$$

and we also have

$$\langle n_\uparrow \rangle + \langle n_\downarrow \rangle = n. \tag{10.86}$$

Equations (10.83) to (10.86) determine $\langle n_\uparrow \rangle$, $\langle n_\downarrow \rangle$, μ as functions of the total density of electrons n.

As in the Hartree–Fock approximation, the equations always possess a non-magnetic solution

$$\langle n_\downarrow \rangle = \langle n_\uparrow \rangle = \tfrac{1}{2} n.$$

If there is exactly one electron per atom, the lowest band will be filled and the upper band empty and, as a result of the interaction, the metal becomes an insulator. This transition is often referred to as a Mott–Hubbard transition. According to the solution given here the transition will take place in a half-filled band, however weak the interaction I. Improved approximations (see below) show that actually the transition will take place only if I exceeds a critical value I_c.

For a small amount of ferromagnetism we require the first-order change in equation (10.85) due to a small change in $\langle n_\uparrow \rangle$ from $\frac{1}{2}n$, to be zero. This implies

$$1 = -\int_{\infty}^{\mu} dE \left. \frac{\partial N[\epsilon_{\uparrow}(E)]}{\partial \langle n_{\downarrow} \rangle} \right|_{\langle n_{\downarrow} \rangle = \frac{1}{2}n}. \tag{10.87}$$

Without a specific model this result is not very transparent. For a square band with

$$N(\epsilon) = W^{-1} \quad \text{for} \quad -\tfrac{1}{2}W < \epsilon < \tfrac{1}{2}W, \quad \epsilon_0 = 0, \tag{10.88}$$

one finds that equation (10.87), with μ in the lower band, becomes

$$1 = \tfrac{1}{2}I[I^2 + \tfrac{1}{4}W^2 + IW(1-n)]^{-1/2}. \tag{10.89}$$

If μ is in the lower band $n < 1$ and equation (10.89) cannot be satisfied for any I and W. Thus although the Hartree–Fock approximation suggests that ferromagnetism can result from a square band (or, strictly speaking, a band that is close to a square band since we require $N''(\epsilon) \neq 0$), the new and, we believe, improved approximation suggests otherwise. Because of the symmetry between electron and hole states one arrives at the same conclusion if the chemical potential lies in the upper band.

As Hubbard has pointed out, band shapes can be found for which a ferromagnetic state (or at least equation (10.87)) can be realized. One possibility is

$$N(\epsilon) = \begin{cases} \delta^{-1} & \text{for } \epsilon_0 - \tfrac{1}{2}W < E < \epsilon_0 - \tfrac{1}{2}W + \tfrac{1}{2}\delta \\ & \text{and for } \epsilon_0 + \tfrac{1}{2}W - \tfrac{1}{2}\delta < E < \epsilon_0 + \tfrac{1}{2}W, \\ 0, & \text{otherwise.} \end{cases} \tag{10.90}$$

For this structure equation (10.87) becomes

$$\delta < \tfrac{1}{2}WI[(\tfrac{1}{2}I + \tfrac{1}{2}W)^2 - \tfrac{1}{4}nWI]^{-1/2}, \tag{10.91}$$

and this can be satisfied for sufficiently small δ. The model density of states (10.90) resembles that of a density of states with two peaks, a common feature of ferromagnetic materials. It must be remembered, however, that the real metals usually possess degenerate d and f bands, not non-degenerate ones. Any deductions from the model should, therefore, be treated with caution.

Hubbard's model has received a great deal of attention since Hubbard's first discussion described here. The main features to have been included are spin-disorder scattering and resonance broadening by Hubbard himself (1964) who used methods to be described in the next chapter. The former concentrates on the scattering of an electron of one spin by the disordered state of the other spins in the metal. Resonance broadening includes spin-flip scattering and the scattering of a σ-spin into a $-\sigma$-spin hole. These improved calculations reveal that for suitable values of the parameters, the model with a half-filled band will exhibit a Mott–Hubbard transition only if I exceeds a critical value. For further details of this work, the reader is referred to the Conference Proceedings reported in *Rev. Mod. Phys.* (Vol. **40**, 1968) as well as to papers by Hubbard (1964), Doniach (1969), Bartel and Jarrett (1974) and Economu and White (1977) where further references may be found. An exact solution for the problem in one dimension has been given by Lieb and Wu (1968), Shiba and Pincus (1972) and Shiba (1972).

References

General

Doniach, S. (1969). *Adv. in Phys.* **20**, 1.
Mattis, D. C. (1965). "The Theory of Magnetism". Harper and Row, New York.
Rado, G. T. and Suhl, H. (eds.) (1963–73). "Magnetism", Vols. 1–5. Academic Press, New York and London.
Tahir-Kheli, R. A. (1976). "Phase Transitions and Critical Phenomena" (Eds. C. Domb and M. S. Green), Vol. 5b, p. 259. Academic Press, London and New York.
"Proc. Int. Conf. on the Metal–Non-Metal Transition San Francisco, 1968". *Rev. Mod. Phys.* **40**, 673–844.

Special

Bartel, L. C. and Jarrett, H. S. (1974). *Phys. Rev.* **B10**, 946.
Bogoliubov, N. and Tyablikov, S. V. (1959). *Doklady Akad. Nauk SSSR* **126**, 53 (*Sov. Phys. – Doklady* **4**, 604).
Cottam, M. G. (1976). *J. Phys. C. Solid State Phys.* **9**, 2121.
Dyson, F. J. (1956). *Phys. Rev.* **102**, 1217, 1230.
Economu, E. N. and White, C. T. (1977). *Phys. Rev. Lett.* **38**, 289.
Haas, C. W. and Jarrett, H. S. (1964). *Phys. Rev.* **135A**, 1089.
Hubbard, J. (1963). *Proc. Roy. Soc.* (*London*). **A276**, 238.
Hubbard, J. (1964). *Proc. Roy. Soc.* (*London*). **A277**, 237; **A281**, 401.
Izyumov, Y. A. and Noskova, M. M. (1964). *Fiz. Met. Metalloved.* **18**, 20.
Lieb, E. H. and Wu, F. Y. (1968). *Phys. Rev. Lett.* **20**, 1445.
Nagmiya, T., Yosida, K. and Kubo, R. (1955). *Adv. in Phys.* **4**, 6.
Ortenburger, I. (1964). *Phys. Rev.* **136A**, 1374.
Shiba, H. (1972). *Prog. Theor. Phys.* **48**, 2171.
Shiba, H. and Pincus, P. A. (1972). *Phys. Rev.* **B5**, 1966.
Sinha, K. P. (1973). "Electrons in Crystalline Solids". International Atomic Energy Agency, Vienna.
Tahir-Kheli, R. A. and ter Haar, D. (1962). *Phys. Rev.* **127**, 88, 95.
Thorpe, M. F. (1978). "Correlation Functions and Quasi-Particle Interactions in Condensed Matter" (Ed. J. Woods Halley), p. 261. Plenum, New York.

Tyablikov, S. V. (1965). "Metody kvantovoi teoru magnetisma". Nauka, Moscow (Translation: "Methods in the Quantum Theory of Magnetism". Plenum, New York, 1967].
Wert, C. A., and Thomson, R. M. (1970). "Physics of Solids", 2nd edn. McGraw-Hill, New York.

Problems

1. By following the analysis given in §10.3 find the spin–spin Green's function for the spins on one sub-lattice of an antiferromagnet with a centre of symmetry in terms of the average spin on the lattice. Deduce that at long wavelengths the quasi-particle spectrum is phonon-like. (There is no need to find the average spin on a sub-lattice.)
2. Deduce equation (10.89) from equation (10.87) and the model (10.88).
3. Consider the scattering of electrons in a metal by a single magnetic impurity at the origin by taking as a model interaction

$$H' = -\frac{J}{2} \sum_{k,k'} c_k^+ \boldsymbol{\sigma} \cdot \boldsymbol{S} c_{k'}.$$

Show that the equation of motion for the single-particle Green's function $G(\boldsymbol{k}, \boldsymbol{k}', \tau)$ involves the function

$$\Gamma(\boldsymbol{k}, \boldsymbol{k}', \tau) = -\langle T[c_{k'\uparrow}(\tau) S_z(\tau) + c_{k\downarrow}(\tau) S_-(\tau)]\, c_{k\uparrow}^+\rangle.$$

Show that the equation of motion for $\Gamma(\boldsymbol{k}, \boldsymbol{k}', \tau)$ involves the quantities $\langle T[c_{k'\gamma}^+(\tau) c_{p\delta}(\tau) c_{q\beta}(\tau)(\boldsymbol{\sigma}_{\gamma\delta} \times \boldsymbol{S}(\tau)) \cdot \boldsymbol{\sigma}_{\alpha\beta}]\, c_{k\alpha}^+\rangle$. To obtain a closed set of equations, assume that the product of operators $c_{k'\gamma}^+(\tau) c_{q\beta}(\tau)$ can be replaced by their unperturbed expectation value $f_{k'} \delta_{qk'} \delta_{\gamma\beta}$. (This approximation should be valid in the weak-coupling limits.) Solve the resulting equations to third order in J and show that

$$\Sigma(\boldsymbol{k}, \omega) = -(J^2/4)\Gamma(\omega)[1 - J\,\Lambda(\omega)],$$

where

$$\Gamma(\omega) = -S(S+1)\sum_k \frac{1}{\omega - \epsilon_k}, \qquad \Lambda(\omega) = \sum_k \frac{f_k - \frac{1}{2}}{\omega - \epsilon_k}.$$

For a band with a constant density of states for

$$-D < \epsilon < D \qquad (D \gg k_B T),$$

show that

$$\Lambda(0) \propto \ln (D/k_B T).$$

There is then a temperature (the Kondo temperature) at which the lifetime becomes infinite in this approximation. This effect (the Kondo effect) leads to an explanation of the resistance minimum in magnetic alloys.

Chapter 11

Disordered Systems

§11.1 Introduction

We have already considered one problem involving disorder, that of the effect of impurities on the transport properties of metals (Chapter 4). We there pointed out that since appropriate averages of the Green's functions are directly related to measurable quantities they are particularly useful for solving that particular problem. For the same reason they are useful in solving a large variety of problems involving disorder. Indeed, at the present time, this is the best analytical method for solving such problems except for the method of the renormalization group (Chapter 12) when that is applicable. We should not ignore the fact, however, that computer studies of disordered systems have added a great deal to our knowledge.

The problem of the scattering electrons in metals by impurities contains one parameter $(k_F l)^{-1}$ the smallness of which in many realistic situations could be exploited to obtain a useful solution. In this solution electron states with a definite momentum had a finite lifetime. At the same time the density of states in energy of the electrons was unchanged. These features were important for the understanding of the transport properties.

Most problems involving disorder, however, do not contain a small parameter like $(k_F l)^{-1}$ and the previous method has at least to be generalized. Even in metal alloys $(k_F l)^{-1}$ is not small and in semiconductors there is no comparable parameter. One result of this is that the approximations used are often not controlled. We cannot always say that this approximation will be valid for a certain range of parameters. Often we expect the approximation to produce certain features of exact results but not all of them.

One feature of disordered systems is that whatever the physical background of the system, the problem can be stated in the same mathematical form and comparable quantities will show the same features. Thus densities of states of the following excitations have many features in common:

(i) electrons in alloys;

(ii) electrons in amorphous semiconductors and in semiconductors containing high concentrations of impurities;

(iii) phonons in disordered and amorphous solids;

(iv) magnons in disordered magnets and antiferromagnets.

This gives rise to a certain economy. The same methods are useful for all of these problems and results obtained for one of the problems throw light on the others.

In all cases we expect that states with a definite momentum will have a finite lifetime or, more accurately, the average spectral distribution function $A(\boldsymbol{k}, E)$ will have a finite width. In general, too, we expect that the density of states will be different from that for a perfect lattice. Indeed it is known from theory (see next section) and from experiments on samples with low concentrations of impurities that individual impurities can give rise to energy levels outside the normal bands. As the concentration of impurities increases or the state of disorder increases, these isolated levels can overlap and form bands which can eventually merge with the original band. Also, as the concentration increases, one can find levels which arise from clusters of impurities. Thus the density of states can be quite complicated.

The possible structure of the spectrum was first revealed in computer calculations by Dean (1961, 1972) of the density of states of phonons in one-dimensional chains of atoms with uniform force constants and two isotopic masses which were varied randomly. Results that Dean has obtained for one- and two-dimensional lattices are shown in Figs. 11.1 and 11.2. It has been possible to identify much of the structure with vibrations localized at particular arrangements of light (L) and heavy (H) atoms. It will be realised that quite a sophisticated analytical theory is needed to reveal all of this structure.

A further question of importance is whether all the states are localized. This is of particular importance for alloys and disordered

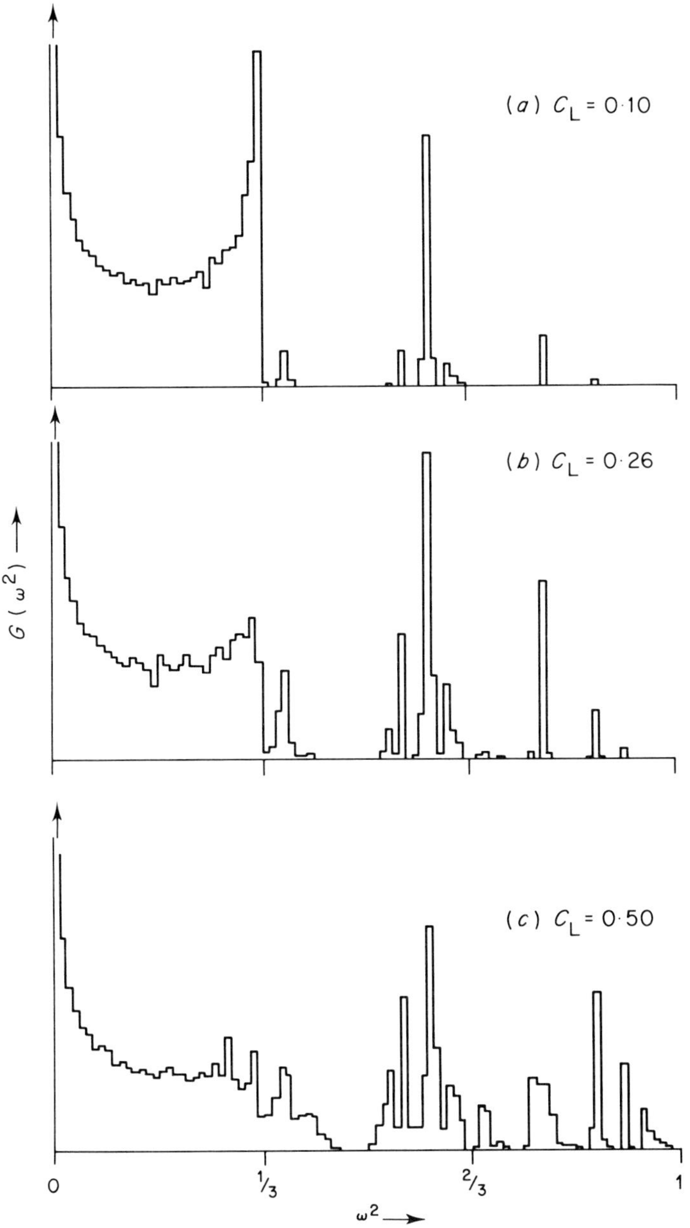

FIG. 11.1. Computed squared frequency spectra for disordered chains of length 8000 atoms of mass ratio 3 : 1 with equal nearest neighbour force constants and different concentrations, c_L, of the light atoms (after Dean 1972).

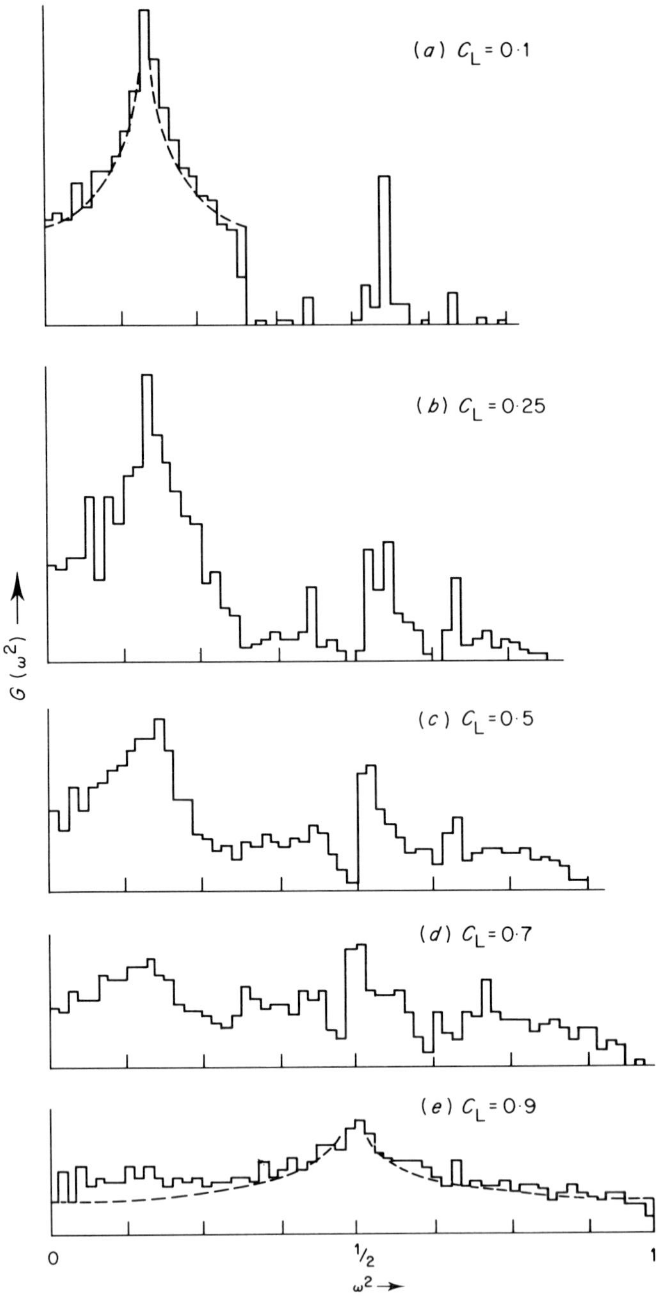

FIG. 11.2. Squared frequency spectra for the disordered two-component simple quadratic lattice of size 56 × 16 for which the mass ratio is 3 : 1 and the nearest neighbour force constants are equal. The broken line in (*a*) is the spectrum of the monatomic heavy lattice, and that in (*e*) is the spectrum of the monatomic light lattice (after Dean 1972).

semiconductors where one wishes to know which states will conduct electricity. We have mentioned that low concentrations of impurities in semiconductors produce localized levels outside the bands. As the concentration increases these levels merge with the band in such a way that near the upper and lower tails the states are localized and near the centre of the band the states are delocalized. The position of the change from localized to delocalized states is of importance as well as is the mobility of electrons near this energy. The methods we discuss in this chapter have provided useful information on these points.

For strongly disordered systems the question arises as to whether disorder can lead to complete localization. A fundamental, remarkable (and at the time controversial) paper by Anderson (1958) led to the conclusion that in some circumstances disorder could lead to complete localization. He studied a problem in which particles could hop from site to site with matrix elements of order V (falling off faster than r^{-3} with distance) and random site energies spread over a range W. He found that for

$$(W/V) > (W/V)_{\text{crit}} \tag{11.1}$$

where the right-hand side is a critical value for the ratio, "all" states are localized. Anderson's model has been studied since in greater detail by others [see, for example, Thouless and co-workers (1971, 1972), Herbert and Jones (1971) and Economou and Cohen (1972)]. The conclusion is essentially the same although the actual value of the critical ratio is still a matter of debate.

Although different disordered systems have much in common, the possible kinds of disorder fall into three distinct classes which are of fundamental importance for the way in which we treat them from the outset. Examples of the three classes are shown in Figs. 11.3(*a*), (*b*) and (*c*). Fig. 11.3(*a*) illustrates substitutional disorder. The atoms in the disordered lattice still form a lattice but the kinds of atoms at the different sites are distributed randomly. Disordered alloys, lattices containing isotopes as well as many magnetic systems are close to this model. Fig. 11.3(*b*) illustrates structural disorder. Here the original lattice is distorted by the disorder but there is a 1–1 correspondence between the disordered sites and the original ones. Fig. 11.3(*c*) illustrates topological disorder. There is not even a 1–1 correspondence between the sites of the disordered state and those of the ordered one. Structural and topological disorder are much more difficult to categorize than substitutional disorder and most analytic work has been carried out for the latter. In this chapter, we confine the discussion to substitutional disorder.

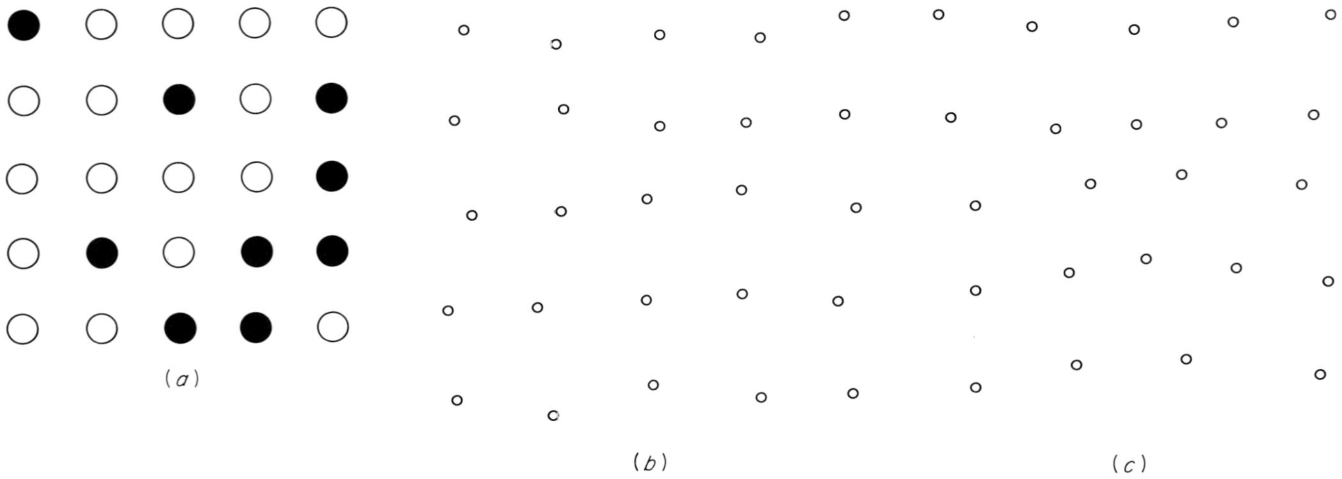

FIG.11.3. Different types of disorder; (*a*) is substitutional (*b*) is structural and (*c*) is topological.

We relate the effect of substitutional disorder to the properties of the underlying lattice. The pure crystal has a band structure. In our discussion, we assume that the effect of substitution is to modify the original bands separately. This is a simplifying but inessential assumption. Without substitution the Hamiltonian is

$$H = \sum \epsilon_k a_k^+ a_k, \qquad (11.2)$$

where ϵ_k are the band energies and a_k^+ creates the appropriate particle or excitation. Because the substitution takes place at individual sites it is convenient to write this with Wannier functions as a basis. Then

$$H = \sum_{i,j} W_{ij} a_i^+ a_j, \qquad (11.3)$$

where

$$a_i^+ = (1/N^{1/2}) \sum_k \exp(-i\boldsymbol{k} \cdot \boldsymbol{R}_i) a_k^+ \qquad (11.4)$$

and a_i^+ creates an excitation at site i. The energies W_{ij} are the matrix elements for hopping from site j to site i and are given by

$$W_{ij} = (1/N) \sum_k \epsilon_k \exp[i\boldsymbol{k} \cdot (\boldsymbol{R}_i - \boldsymbol{R}_j)]. \qquad (11.5)$$

If we now suppose that we introduce a substitution at site ν we shall have an increased potential $V(\nu)$ at the site. This can produce hopping between neighbouring sites with matrix elements $V_{ij}(\nu)$. With many such substitutions the Hamiltonian becomes

$$H = \sum_{i,j} W_{ij} a_i^+ a_j + \sum_{\nu,i,j} V_{ij}(\nu) a_i^+ a_j. \qquad (11.6)$$

If the band is narrow, the Wannier functions are very localized and the largest matrix element of $V(\nu)$ is expected to be $V_{\nu\nu}(\nu)$ which we denote by V_ν. If we ignore, as we shall often do, the other matrix elements of V, the Hamiltonian reduces to

$$H = \sum_{i,j} W_{ij} a_i^+ a_j + \sum_{\nu} V_\nu a_\nu^+ a_\nu. \qquad (11.7)$$

The retarded Green's function

$$G(i, j, t) = -i\langle [a_i(t), a_j^+(0)]_\epsilon \rangle \, \theta(t-t') \qquad (11.8)$$

then satisfies

$$\sum_l \left[\delta_{il}\left(\frac{i\partial}{\partial t} - V_i\right) - W_{il}\right] \cdot G(l, j, t) = \delta(t)\delta_{i,j}. \tag{11.9}$$

Its time Fourier transform, therefore, satisfies

$$\sum_l [(\omega - V_i)\delta_{il} - W_{il}]\, G(l, j, \omega) = \delta_{i,j}. \tag{11.10}$$

Because this is a one-particle problem, there is no difficulty in satisfying the time boundary conditions by finding first the Green's function which is analytic in the lower and upper half-planes of ω. Other Green's functions can be found from this as special cases.

Now the approximate solutions of equation (11.10) are usually related to that for a single impurity. We therefore begin by solving equation (11.10) when only one impurity is present.

§ 11.2 One impurity in a lattice

There is no loss of generality if we take the impurity to be situated at the origin. Then

$$V_i = V\delta_{0i}, \tag{11.11}$$

and

$$\sum_l (\delta_{il}\,\omega - W_{il})G(l, j; \omega) = \delta_{i,j} + \delta_{i,0}\, VG(0, j; \omega). \tag{11.12}$$

Now the Green's function $G_0(l, j; \omega)$ which satisfies the unperturbed equation

$$\sum_l (\delta_{il}\,\omega - W_{il})G_0(l, j; \omega) = \delta_{i,j} \tag{11.13}$$

is known. As we have done before (§ 3.1) we can use this to write equation (11.12) as

$$G(i, j; \omega) = G_0(0, j; \omega) + G_0(i, 0; \omega)VG(0, j; \omega). \tag{11.14}$$

If we take i at the origin we find

$$G(0, j; \omega) = G_0(0, j; \omega) + G_0(0, 0; \omega)VG(0, j; \omega)$$

whence

$$G(0, j; \omega) = \frac{G_0(0, j; \omega)}{1 - VG_0(0, 0; \omega)}. \tag{11.15}$$

If this is now substituted into the right-hand side of equation (11.14) we see that

$$G(i, j; \omega) = G_0(i, j; \omega) + \frac{G_0(i, 0, \omega)VG_0(0, j; \omega)}{1 - VG_0(0, 0; \omega)}. \tag{11.16}$$

The Green's function has poles where the denominator vanishes, that is, where

$$1 = VG_0(0, 0; \omega) = (V/N) \sum_k (\omega - \epsilon_k)^{-1}. \qquad (11.17)$$

This is an equation of the Nth degree in ω. A plot of the sum as a function of ω is shown schematically in Fig. 11.4. It can be seen that equation (11.17) has a solution between every two consecutive values of ϵ_k. It also possesses one solution, the impurity level outside the band. For V positive it is above the band, for V negative below.

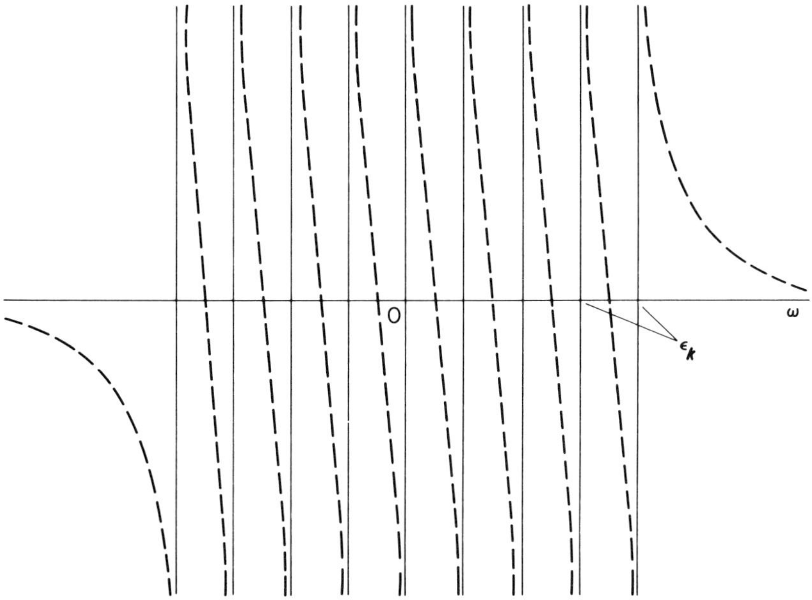

FIG. 11.4. A graph illustrating the solution of equation (11.17). The dashed curves are a schematic plot of the right-hand side of the equation. Wherever they meet the unit ordinate there is a solution of the equation.

While correct, this method of analysis does not show whether the level outside the band is separated by an energy O(1), which is observable, or O(1/N) which is not. To decide this, we look for solutions of equation (11.7) where ω is separated from the band by an energy of order 1. Then, in the usual way, the sum in equation (11.17) can be replaced by an integral and

$$1 = V \int \frac{N(\epsilon)d\epsilon}{\omega - \epsilon}. \qquad (11.18)$$

Near the band edge $N(\epsilon)$ usually approaches zero. Hence, the integral converges as ω approaches this edge. This means that for an impurity level we must have $V \geqslant V_0$ or $V \leqslant V_0'$ where

$$\frac{1}{V_0} = \int \frac{N(\epsilon)d\epsilon}{\omega_B - \epsilon}, \qquad \frac{1}{V_0'} = \int \frac{N(\epsilon)d\epsilon}{\omega_{B'} - \epsilon}, \tag{11.19}$$

and ω_B, $\omega_{B'}$ are the energies at the upper and lower edges of the band respectively. Schematically the integral of equation (11.18) is shown in Fig. 11.5.

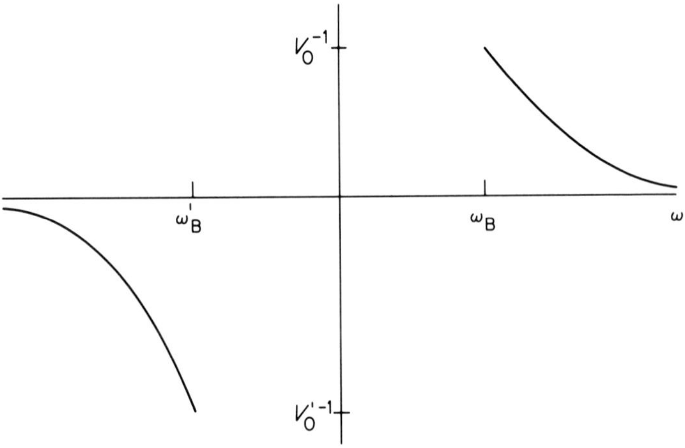

FIG. 11.5. A graph illustrating the solution of equation (11.18). The solid curve is a plot of the integral in the equation beyond the band edges ω_B and $\omega_{B'}$. At the band edges the integral takes on the values V_0^{-1} and $V_0'^{-1}$. For a solution $V_0^{-1} \geqslant V^{-1} \geqslant V_0'^{-1}$.

From equation (11.15) the density of states at the impurity is

$$-\frac{1}{\pi} \operatorname{Im} G(0,0,\omega + i\delta) = -\frac{1}{\pi} \operatorname{Im} \frac{G_0(0,0,\omega+i\delta)}{1 - VG_0(0,0,\omega+i\delta)}. \tag{11.20}$$

If we write

$$G_0(0,0,\omega+i\delta) = P\int \frac{N(\epsilon)d\epsilon}{\omega - \epsilon} - i\pi N(\omega) = a - i\pi N(\omega). \tag{11.21}$$

the density of states at the impurity becomes

$$N(\omega)[(1 - Va)^2 + V^2\pi^2 N(\omega)^2]^{-1}. \tag{11.22}$$

For ω within the band it is reduced by the factor in the denominator from its value in the host crystal. This may represent resonant enhancement if

$$(1 - Va) \equiv 1 - VP \int \frac{N(\epsilon)d\epsilon}{\omega - \epsilon} = 0 \qquad (11.23)$$

for some value of ω in the band. Then $V\pi N(\omega)$ gives the width of the resonance.

This method of studying the effect of a single impurity can be extended to the case where the range of the perturbing potential extends beyond the single-site so that V_{0j} is not zero. In this case the Green's function shows greater structure and more than one impurity level may result. The method can also be extended to the case of a few impurities. Then impurity levels may appear at several sites. If the sites are sufficiently close the levels will interfere with each other. This may lead to displacement of the levels and broadening. When the concentration of the impurities becomes finite all these effects should be present.

§11.3 Formalism for many impurities

The equation we have to solve is (11.10) where V_i varies from site to site, being zero at host sites. The Green's function then depends on the individual values of the V's. As was pointed out in Chapter 4 this is not the Green's function which is measured. Indeed, the measurement is usually of a macroscopic quantity which averages over the positions of the impurities subject to some overall constraints such as the relative concentrations of different atoms. If it is the density of states which is under consideration we need to find the average, $\langle G \rangle$, of G. If it is a transport coefficient that is to be determined (as in §11.7) we need to average the appropriate two-particle Green's function. In this section we confine our attention to $\langle G \rangle$.

The average we have to perform depends on the statistical information available about the disorder. Failing any other information we assume that the disorder is random; this is the only case dealt with here. However, it may be in practice that, in alloys for example, either like atoms tend to be adjacent or unlike atoms tend to be adjacent. Apart from the limitations of other approximations this could be a source of discrepancy between theory and experiment.

If there are two kinds of atoms (or sites), A and B, with concentrations $(1-c)$ and c, respectively, and if V takes on the values

V_A and V_B at these sites then the probability distribution for the potential at site i is

$$P(V_i) = (1-c)\delta(V_i - V_A) + c\delta(V_i - V_B). \tag{11.24}$$

Hence,

$$\langle f(V_i)\rangle = (1-c)f(V_A) + cf(V_B). \tag{11.25}$$

If the A atoms are regarded as the host atoms, V_A is zero. However, as we shall see this is not the only useful way of distributing the potential. If the distribution of the atoms amongst the sites is random the total probability distribution is

$$P(V_1, V_2 \ldots) = \prod_i P(V_i)$$

and

$$\langle G\rangle = \prod_i \int dV_i P(V_i) G. \tag{11.26}$$

The last equation gives the explicit form of the required average.

Equation (11.10), the equation to be solved, can be put in several other forms. First, the unperturbed Green's function G_0, defined by equation (11.14) can be used to invert it to

$$G(i,j;\omega) = G_0(i,j;\omega) + \sum_l G_0(i,l;\omega)V_l G(l,j;\omega). \tag{11.27}$$

It is now quite useful to simplify the notation by thinking of $G(i,j;\omega)$ and $G_0(i,j;\omega)$ as elements of matrices G and G_0. Since ω appears everywhere in the equation we omit writing it and consider it understood. The equation (11.27) can then be written

$$G = G_0 + G_0 VG, \tag{11.28}$$

where V is the diagonal matrix with elements

$$V(i,j) = V_i \delta_{ij}. \tag{11.29}$$

Equation (11.28) can be expanded in powers of V as was done in §3.1 to yield

$$G = G_0 + G_0VG_0 + G_0VG_0VG_0 + \ldots + G_0(VG_0)^n + \ldots \tag{11.30}$$

This can be represented by diagrams in the standard way as illustrated in Fig. 11.6.

It is also convenient to write equation (11.30) as

$$G = G_0 + G_0TG_0, \tag{11.31}$$

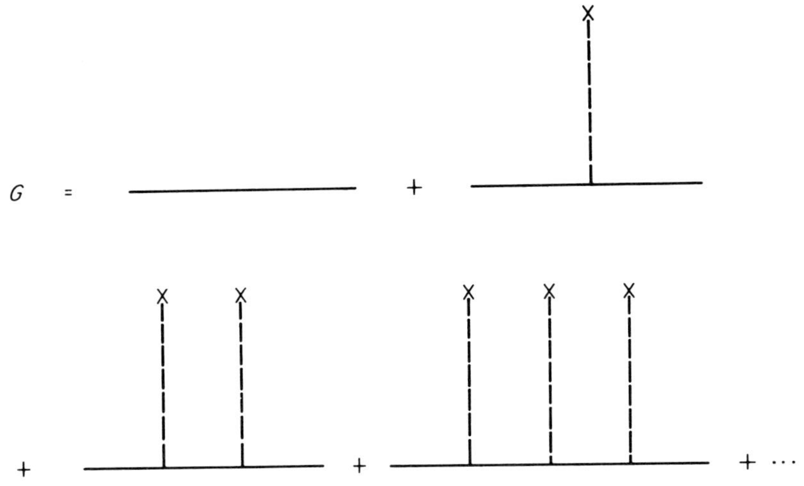

FIG. 11.6. The expansion of the one-particle Green's function in powers of the impurity potential.

where

$$T = V + VG_0V + \ldots + V(G_0V)^n + \ldots \tag{11.32}$$

or, comparing (11.31) and (11.28)

$$T = VGG_0^{-1}. \tag{11.33}$$

The matrix T is called the t-matrix. It contains all the effects of scattering; it is zero if there is no scattering, and V if the scattering is weak. From equation (11.31), we see that

$$\langle G\rangle = G_0 + G_0\langle T\rangle G_0, \tag{11.34}$$

and the required result is given in terms of the average t-matrix.

The problem of the single impurity discussed in the last section can also be discussed in terms of the t-matrix. For this case V contains only one diagonal element V at the impurity site i, say. By inspection of equation (11.16) we see that the t-matrix is

$$t_i = V[1 - \mathrm{Tr}\,(VG_0)]^{-1}. \tag{11.35}$$

It is convenient to define a self-energy operator for the average Green's function through the usual relation,

$$\langle G\rangle = G_0 + G_0\Sigma\langle G\rangle$$

or

$$\langle G\rangle^{-1} = G_0^{-1} - \Sigma.$$

$G_0\langle T\rangle G_0 = G_0\Sigma\langle G\rangle$

$\Sigma^{-1}\langle T\rangle G_0 = \langle G\rangle$

$G_0^{-1}\langle T\rangle^{-1}\Sigma = G_0^{-1} - \Sigma$

$\langle G\rangle = (G_0^{-1} - \Sigma)^{-1} = G_0 + G_0\langle T\rangle G_0$

$(G_0^{-1} - \Sigma)^{-1} = (1 + G_0\langle T\rangle)G_0$

$G_0^{-1}(\langle T\rangle^{-1} - 1) = \Sigma$

$G_0^{-1} - \Sigma = G_0^{-1}(1 + G_0\langle T\rangle)^{-1}$

It is then not difficult to show that, in general, Σ is related to the average t-matrix through the equation

$$\Sigma = \langle T\rangle(1 + G_0\langle T\rangle)^{-1}. \tag{11.36}$$

§ 11.4 The virtual crystal approximation

A simple practical approximation that can be made is to ignore the correlation between V and G in equation (11.28) when taking the statistical average. Then

$$\langle G\rangle = G_0 + G_0\langle V\rangle\langle G\rangle$$

and

$$\Sigma = \langle V\rangle = [(1-c)V_A + cV_B]E, \tag{11.36}$$

where E is the unit operator which we usually denote by 1. The same result is obtained if one averages the V's in equation (11.32) independently, ignoring the coincidences of sites in the products. Then the potential at each site is replaced by an effective potential which is the average of that at the two different kinds of sites.

For the particular case we have been studying of a single-site perturbation, the average potential simply adds a constant to all energies. However, the approximation can be applied straightforwardly to more general cases. The effect is to put the same average potential $\langle V\rangle$ on each site. The potential is then still periodic and the states can still be assigned crystal wave vectors. Hence, the approximation is called the virtual crystal approximation. It is quite useful for finding band energies of disordered systems and provides an interpolation for the density of states as a function of concentration which is exact at the two extremes of pure crystals of types A and B. However, the approximation clearly misses all impurity levels and localized states. Nevertheless, even when it does not provide a complete description of the states it can often be used, as we see below, as the zeroth approximation.

§ 11.5 The average *t*-matrix approximation (ATA)

The virtual crystal approximation misses the impurity levels because it does not properly include the scattering from single impurities. The average t-matrix approximation remedies this deficiency. At least at low concentrations different impurities should not interfere with each other and repeated scatterings from one impurity will be more important than scatterings from different impurities. We therefore write the t-matrix in a form which distinguishes between such scatterings.

To focus on the different impurities we write the operator V as a sum of contributions from the different sites. Thus

$$V = \sum_i v_i \tag{11.37}$$

where v_i has just one non-zero element,

$$v_i(j, k) = V_i \delta_{ji} \delta_{ki}. \tag{11.38}$$

Then

$$T = \sum_i v_i + \sum_i v_i G_0 \sum_j v_j + \sum_{i,j,k} v_i G_0 v_j G_0 v_k + \ldots \tag{11.39}$$

Now we collect terms where consecutive scatterings take place at the same site and write

$$\begin{aligned} T = {} & \sum_i [v_i + v_i G_0 v_i + v_i G_0 v_i G_0 v_i + \ldots] \\ & + \sum_i (v_i + v_i G_0 v_i + \ldots) G_0 \left(\sum_{j \neq i} (v_j + v_j G_0 v_j + \ldots) \right) \\ & + \sum_i (v_i + v_i G_0 v_i + \ldots) G_0 \left(\sum_{j \neq i} (v_j + v_j G_0 v_j + \ldots) \right) G_0 \\ & \times \sum_{k \neq j} (v_k + v_k G_0 v_k + \ldots) \\ & + \ldots\ldots \end{aligned} \tag{11.40}$$

To ensure that there is no double counting of terms and that no terms are missed we carry out this process more formally.

In equation (11.40) the Green's function within the brackets always involves the propagation from one site back to the same site. This is therefore just the diagonal part of G_0 which we shall denote as G_D. In fact because of the translational symmetry of the host lattice all the diagonal elements are equal and G_D is proportional to the unit operator. On the other hand the Green's function between brackets is always related to propagation between different sites. It is, therefore, the off-diagonal part, G_0', of G_0. We therefore write

$$G_0 = G_D + G_0'. \tag{11.41}$$

Now the equation for T, (11.32), can be written

$$T = V[1 - G_0 V]^{-1}. \tag{11.42}$$

Hence

$$T(1 - G_0 V) = V, \tag{11.43}$$

that is,

$$T(1 - G_D V) = V + TG_0' V$$

and

$$T = V(1 - G_D V)^{-1} + TG_0' V(1 - G_D V)^{-1}. \qquad (11.44)$$

If this equation is now iterated, we find that

$$T = V(1 - G_D V)^{-1} \sum_{n=0}^{\infty} [G_0' V(1 - G_D V)]^n. \qquad (11.45)$$

If the factors $(1 - G_D V)^{-1}$ are further iterated one obtains exactly equation (11.40).

Since V and G_D are diagonal so is $V(1 - G_D V)^{-1}$. However, equation (11.35) shows that, in the present notation,

$$t_i = v_i [1 - G_D v_i]^{-1}$$

and

$$V[1 - G_D V]^{-1} = \sum_i t_i. \qquad (11.46)$$

In the average t-matrix approximation (ATA) it is assumed that, because G_0' involves propagation between different sites, T is approximately statistically independent of $V(1 - G_D V)^{-1}$ in the last term of equation (11.44). Thus, when the equation is averaged

$$\langle T \rangle = \left\langle \sum_i t_i \right\rangle + \langle T \rangle G_0' \left\langle \sum_i t_i \right\rangle. \qquad (11.47)$$

But

$$\left\langle \sum_i t_i \right\rangle = (1 - c) t_A + c t_B$$

and

$$\langle T \rangle = [(1 - c) t_A + c t_B][1 - G_0' \{(1 - c) t_A + c t_B\}]^{-1}. \qquad (11.48)$$

From equations (11.36) and (11.48) one derives the self-energy

$$\Sigma = \frac{(1 - c) t_A + c t_B}{1 + G_D[(1 - c) t_A + c t_B]}. \qquad (11.49)$$

There is some freedom of choice in the use of this result depending on the choice made for the unperturbed system. For example, if the pure A crystal is the unperturbed system, t_A is zero and t_B depends on the potential difference between the B and A sites. If the virtual crystal is the unperturbed system then t_A and t_B are related, respectively, to the differences between the potentials at the A and B sides and the crystal potential. The latter approach leads to a result symmetric in the two kinds of sites.

Some results obtained by use of this method are shown in Fig. 11.7.

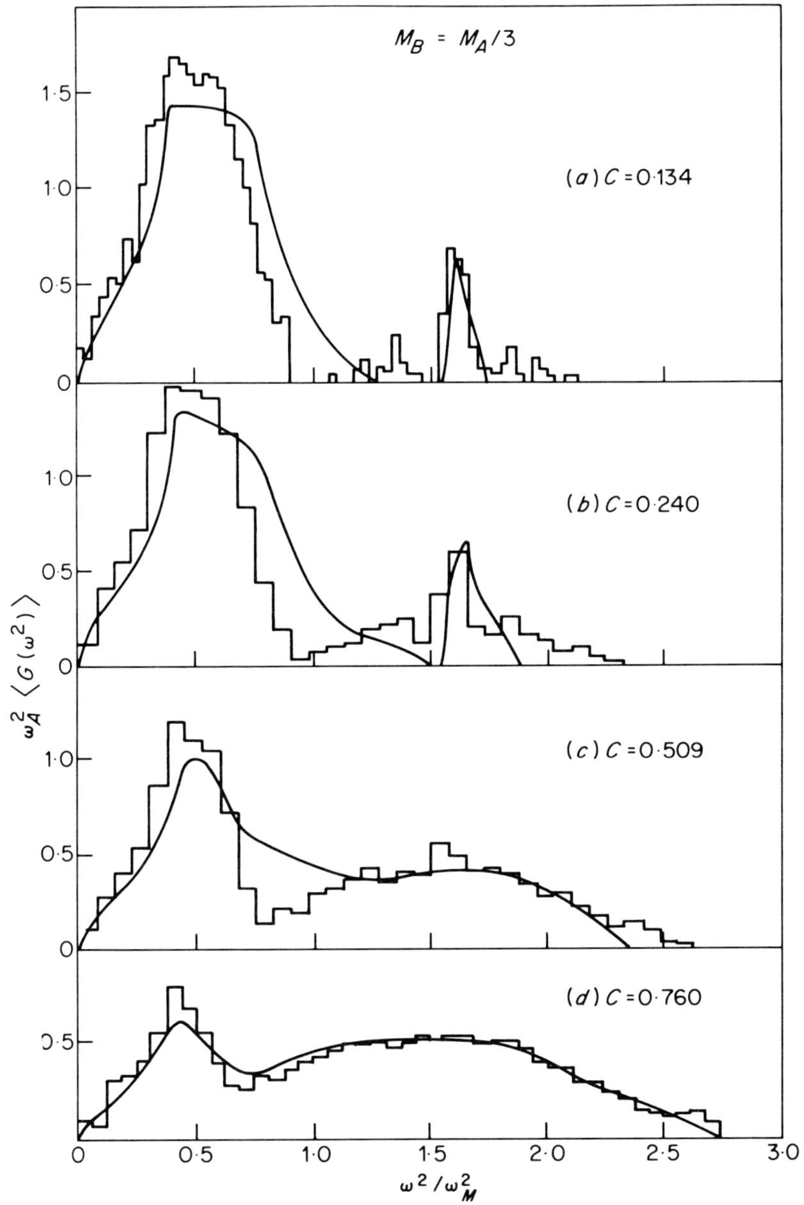

FIG. 11.7. The phonon density of states $G(\omega^2)$ for disordered simple cubic lattices with mass ratio 3 : 1 and different concentrations c of light atoms. The histograms are the result of machine calculations of Payton and Visscher (1967) and the solid curves are obtained in the ATA with the virtual crystal as the reference lattice (after Leath and Goodman 1969). The maximum frequency of the heavy monatomic lattice is denoted by ω_m.

§11.6 The coherent potential approximation (CPA)

In the ATA just described we use the host medium or virtual crystal as the unperturbed system and ensure that we get the single-site scattering correct. Correlations between the scatterings are ignored. The coherent potential approximation (denoted CPA) takes this idea logically one step forward (Taylor 1967, Soven 1967). Instead of using the virtual or host crystal as the zeroth approximation one uses the average or medium Green's function as the zeroth approximation, and ensures that there is no resultant scattering off single impurities. Since the medium Green's function is unknown at the outset the method is a self-consistent one.

The equation for G is

$$G_0^{-1}G = 1 + VG, \tag{11.50}$$

which we write as

$$(G_0^{-1} - \Sigma)G = 1 + (V - \Sigma)G; \tag{11.51}$$

we choose Σ so that

$$G_0^{-1} - \Sigma = \langle G\rangle^{-1} \equiv G_M^{-1} \tag{11.52}$$

Then

$$G = G_M + G_M(V - \Sigma)G \tag{11.53}$$

$$= G_M + G_M T G_M \tag{11.54}$$

where $(V - \Sigma)$ is the perturbation and T is the t-matrix resulting from this perturbation. At each site, we therefore have a potential $(V_A - \Sigma)$ or $(V_B - \Sigma)$ which now depends on energy ω. From equation (11.54) we find

$$G_M = \langle G\rangle = G_M + G_M\langle T\rangle G_M. \tag{11.55}$$

Hence

$$\langle T\rangle = 0. \tag{11.56}$$

The last equation is the self-consistent one for Σ. If we could solve it exactly, the problem would be solved. Except in special cases we cannot do this and we have to replace (11.56) by an equation that is soluble. In the CPA one replaces (11.56) by

$$\langle t_i\rangle = 0. \tag{11.57}$$

This means that we assume that in the medium the average scattering by a single impurity is zero.

It follows that in this approximation, $\Sigma(\omega)$ is diagonal and the self-consistent equation for it, (11.57), becomes

$$(1 - c)t_A + ct_B = 0,$$ Soven eqn.

where

$$t_{A,B}(\omega) = [V_{A,B} - \Sigma(\omega)][1 - G_M(0,0;\omega)\{V_{A,B} - \Sigma(\omega)\}]^{-1}.$$

After some rearrangement the self-consistent equation can be written

$$\Sigma(\omega) - V_A = \frac{c(V_B - V_A)}{1 - G_M(0,0;\omega)[V_B - \Sigma(\omega)]}. \tag{11.58}$$

For a given host band structure this equation determines $\Sigma(\omega)$ for the medium and hence the average band structure for the disordered system.

The CPA tends to the correct results in the pure crystal limits of $c = 0$ and $c = 1$. Further, since Σ depends on V and Σ only in the combination $V - \Sigma$ and G_M, a different choice for the host crystal in the same problem (e.g. pure A, pure B or virtual crystal) is compensated for by a change in Σ which leaves G_M invariant. Thus the approximate result is independent of the choice made for the host crystal. Moreover, the structure of equation (11.58) is such that $\Sigma(\omega)$ always has the correct property of being analytic in the upper and lower halves of the complex ω-plane.

One other advatage of the CPA is that it gives the correct result in what is generally called the split band limit, the limit in which the width of the host band tends to zero. Thus

$$I_{ij} = W_0 \delta_{ij}.$$

For this case the eigenstates are atomic states on the two sorts of atom and the exact solution for $\langle G \rangle$ must be

$$\langle G \rangle = \frac{1-c}{\omega - W_0 - V_A} + \frac{c}{\omega - W_0 - V_B}. \tag{11.59}$$

However,

$$G_M^{-1} = \omega - W_0 - \Sigma(\omega).$$

With this substitution in equation (11.58) one does indeed find after a little algebraic manipulation that equation (11.59) holds for $\langle G \rangle = G_M$. This suggests that the approximation should be good for narrow bands as well as for low concentrations of impure sites. Otherwise, it is not evident from the approximation itself under what circumstances it will be valid. Comparisons with exact calculations, where these have been possible, throw light on this. The article by Elliott *et al.* (1974) reviews such comparisons. We mention a few by way of illustration.

In one dimension where fluctuations are very important and where there is usually a high density of clusters of various kinds, the

CPA is rather poor. In three dimensions where fluctuations are less important, the approximation seems to work quite well. In Fig. 11.8, we show a comparison of the CPA with machine calculations of Payton and Visscher (1967) for the phonon density of states of a simple cubic lattice with a disordered arrangement of isotopic masses. The approximation can be seen to work remarkably well and, as is to be expected, reproduces the impurity bands as well as the main band. The approximation is poor at low concentration of the light atoms where the scattering by clusters of atoms is relatively more important.

Extensive comparisons of the ATA and CPA have been carried out by Schwartz *et al.* (1971) for electronic bands with single-site perturbations. These show the expected features. At a fixed concentration, there is no impurity band for sufficiently small values of $\delta = (V_A - V_B)/W$, where W is the band width. As δ increases an impurity band splits off. For the band model with

$$\rho_0(E) = \begin{cases} (2/W^2)(W^2 - E^2)^{1/2}, & |E| \leqslant W, \\ 0 & |E| > W, \end{cases} \tag{11.60}$$

band splitting takes place at very low values of δ in the ATA and at $\delta = 1.0$ for the CPA. Elliott *et al.* (*loc. cit.*) have pointed out that the known exact result (Thouless 1970) is that band splitting should take place at $\delta = 2.0$. This discrepancy is again probably due to the effect of clusters.

In general, the comparisons show that the CPA does provide a good qualitative description of the effect of disorder and for much of the band structure a good quantitative expression. Not surprisingly, there are regions of the band structure where clusters play an important role and the CPA is poor. Because of the overall success of the approximation, many attempts with varied success have been made to extend the theory to deal with clusters. For a discussion of some of these attempts we refer the reader to the review article of Elliott *et al.* (1974) as well as to papers by Leath (1973), Butler (1973) and Freed and Cohen (1971). There have also been many attempts to widen the theory to deal with extended defects. These also are reviewed in the article by Elliott *et al.*

We have so far not mentioned the light thrown on localization by the CPA when the impurity band overlaps the main band. In Green's function theory this is best discussed in terms of the transport coefficients, to which we turn in the next section. The localized states do not contribute to the transport of electric current, for example.

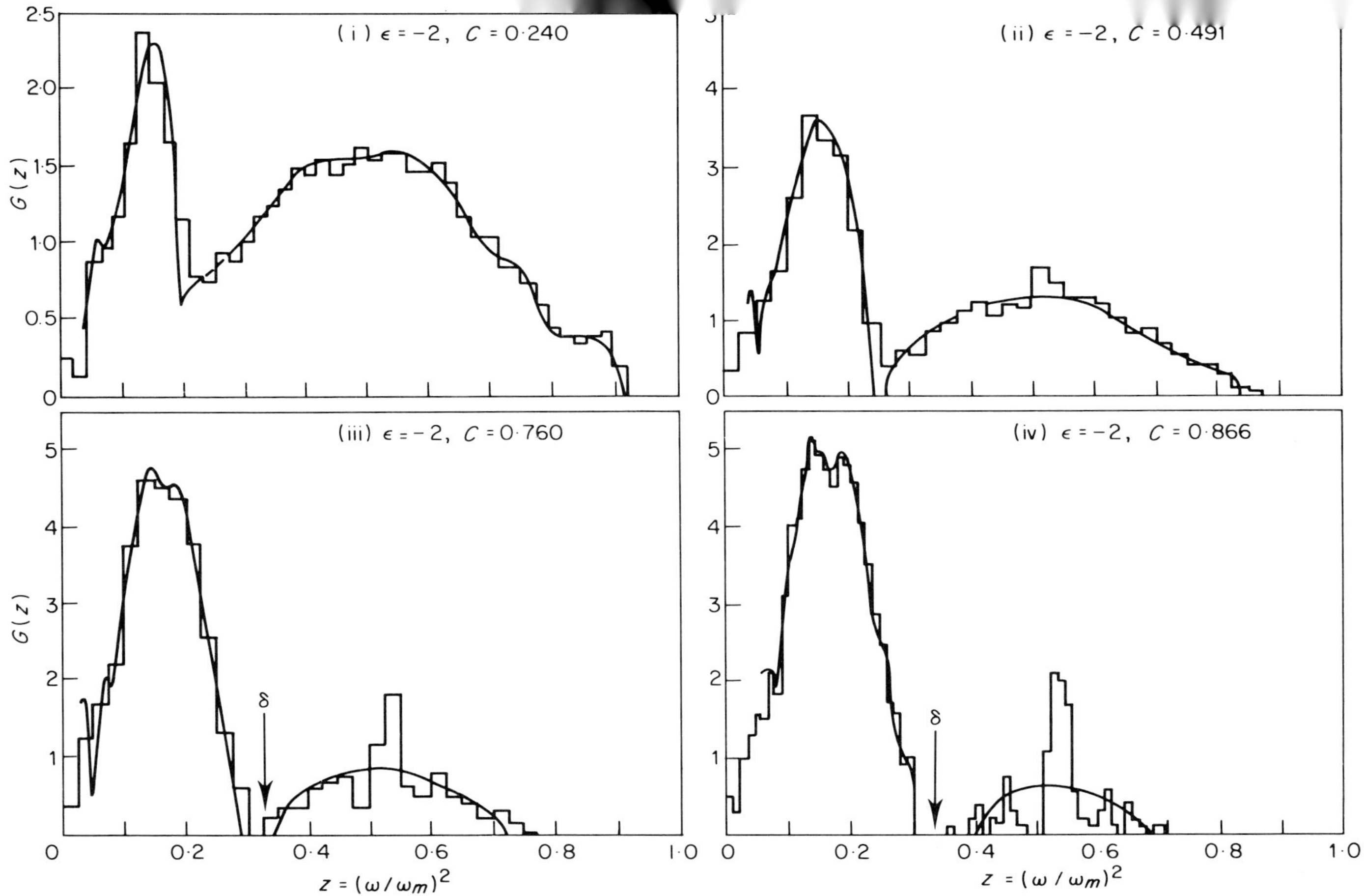

FIG. 11.8. The phonon density of states $G(\omega^2)$, against ω^2/ω_m^2 for disordered simple cubic lattices with mass ratio 3:1 at different concentrations c of light atoms. The histograms are the result of machine calculations of Payton and Visscher (after Taylor 1967).

§ 11.7 Electrical conductivity

We take as an example of a transport coefficient in a disordered system the electrical conductivity of an alloy. Now we have already investigated the conductivity of a metal (Chapter 4) and the methods used there can be taken over into the present case; the only difference is the form of the Green's functions. We saw in Chapter 4 that if the electric field is described by a vector potential $A(r, \omega)$ at one frequency ω, the induced current is given [cf. equation (4.60)] by

$$J_\alpha(\boldsymbol{r}, \omega) = -\frac{ne^2}{m} A_\alpha(\boldsymbol{r}, \omega) - \sum_\beta \int d^3 r' A_\beta(\boldsymbol{r}', \omega) G^j_{R\alpha\beta}(1, 1', \omega), \tag{11.61}$$

where G^j_R is a retarded Green's function related by analytic continuation to the temperature function

$$G^j_{\alpha\beta}(1, 1') = -\frac{e^2}{4m^2}(\nabla_{2'\beta} - \nabla_{1'\beta})(\nabla_{2\alpha} - \nabla_{1\alpha}) \times \langle G_{\sigma\sigma'}(2, 1') G_{\sigma'\sigma}(2', 1)\rangle_{\substack{r'_2, \tau'_2 = r'_1, \tau'_1; \\ r_2, \tau_2 = r_1, \tau_1.}} \tag{11.62}$$

The angular brackets here indicate that the average of the enclosed function over the site potentials are to be taken. To ensure that the sum rules and Ward identity are satisfied it is necessary to ensure that the average of the product is approximated in a manner which is consistent with that for the single-particle Green's functions. From equations (11.61) and (11.62) we were able to derive a satisfactory theory of the conductivity of metals containing moderate concentrations of impurities.

When we turn to the electrical conductivity of alloys and semiconductors differences arise only from the shape of the band or bands and in the form of the Green's functions. The band structure determines the form of the current operator which, for a single band and a uniform field represented by a vector potential $A(t)$, has the form

$$j(\boldsymbol{r}) = e\psi^+(\boldsymbol{r}) v[-i\nabla - eA(t)]\,\psi(\boldsymbol{r}),$$

where $v(\boldsymbol{k})$ is the velocity function given by

$$v(\boldsymbol{k}) = \partial\epsilon(\boldsymbol{k})/\partial\boldsymbol{k}$$

and $\epsilon(\boldsymbol{k})$ is the band energy for the pure host. For the parabolic band

$$v(\boldsymbol{k}) = \boldsymbol{k}/m,$$

and this leads to the expression used in Chapter 4 and expressions (11.61) and (11.62). To obtain the new expressions we have to replace $\boldsymbol{k}/m$ by $v(\boldsymbol{k})$ or $(-i\boldsymbol{\nabla})$ by $v(-i\boldsymbol{\nabla})$. If we ignore the small static paramagnetism, one frequency component induced by the field is given [cf. equation (4.60)] by

$$J_{\alpha}(\omega) = -\sum_{\rho}\int d^3r' A_{\beta}(\omega)[G^j_{R\alpha\beta}(1,1';\omega) - G^j_{R\alpha\beta}(1,1';0)], \tag{11.63}$$

where $G^j_{\alpha\beta}(1, 1'; \omega)$ is now a retarded Green's function related by analytic continuation to the temperature function

$$G^j_{\alpha\beta}(1,1') = \lim_{2,2'\to 1,1'} e^2 v_{\beta}(-i\boldsymbol{\nabla}_{2'})v_{\alpha}(-i\boldsymbol{\nabla}_2)\langle G_{\sigma\sigma'}(2,1')G_{\sigma'\sigma}(2',1)\rangle, \tag{11.64}$$

where the angular brackets again indicate an average of the enclosed function over the site potentials. To ensure that the sum rules and Ward identity are satisfied it is necessary that the average of the product again be approximated in a manner which is consistent with that for the single-particle Green's functions. This is described in §§3.9 and 4.3 and is guaranteed if the CPA is used in the presence of the external field. The result is that in general the average of the product differs from the product of the averages. However, when the perturbation is a single-site perturbation, the scattering is s-like. When the transport coefficient involves vector operators, as does the conductivity, the corrections in the long-wavelength limit vanish by symmetry for s-scattering when the band has a centre of symmetry. This can be seen to be true for the case discussed in §4.3 and is quite general. It is discussed in more detail in Elliott *et al.* (*loc. cit.*).

For this case then, the product in equation (11.62) can be replaced by the product of the averages which were evaluated in the last section. For a macroscopically translationally invariant system one obtains the analogues of equations (4.66) to (4.68) which can be written

$$J_{\alpha}(\omega) = -\sum_{\beta} K_{\alpha\beta}(\omega)A_{\beta}(\omega), \tag{11.65}$$

where

$$K_{\alpha\beta}(\omega) = G^j_{\alpha\beta}(\omega) - G^j_{\alpha\beta}(0) \tag{11.66}$$

and

$$G^j_{\alpha\beta}(\omega) = (2e^2/\beta\mathscr{V})\sum_{\boldsymbol{k},\lambda} v_{\alpha}(\boldsymbol{k})v_{\beta}(\boldsymbol{k})G_M(\boldsymbol{k},\omega'_{\lambda})G_M(\boldsymbol{k},\omega'_{\lambda}+\omega_{\mu}), \tag{11.67}$$

where G_M is the single-particle Green's function calculated by one of the approximations given previously.

The expression (11.67) can still be treated in the same way as the corresponding expression in Chapter 4 [equation (4.68)] and it yields for the retarded function

$$G^{j}_{R\alpha\beta}(\omega) = \frac{2e^2}{\mathscr{V}} \sum_{\boldsymbol{k}} v_\alpha(\boldsymbol{k}) v_\beta(\boldsymbol{k}) \int dx\, dy \frac{A(\boldsymbol{k},x)A(\boldsymbol{k},y)[f(x)-f(y)]}{x-y+\omega+i\delta}, \tag{11.68}$$

where $A(\boldsymbol{k}, x)$ is the spectral distribution function of $G_M(\boldsymbol{k}, \omega)$; it is real and given by

$$A(\boldsymbol{k},x) = -(1/\pi)\,\mathrm{Im}\, G(\boldsymbol{k}, x+i\delta). \tag{11.69}$$

The function $f(x)$ is the Fermi distribution. Equations (11.65), (11.66) and (11.68) provide an explicit set from which the frequency-dependent conductivity, $\sigma_{\alpha\beta}(\omega)$, can be calculated. Since

$$E = i\omega A, \tag{11.70}$$

$$\sigma_{\alpha\beta}(\omega) = iK_{\alpha\beta}(\omega)/\omega. \tag{11.71}$$

In the limit $\omega \to 0$, one obtains the d.c. conductivity for which a slightly simpler expression can be given. Note that, apart from the denominator the integrand is antisymmetric in x and y. Hence

$$K_{\alpha\beta}(\omega) = (e^2/\mathscr{V}) \sum_{\boldsymbol{k}} v_\alpha(\boldsymbol{k}) v_\beta(\boldsymbol{k}) \int dx\, dy\, A(\boldsymbol{k},x)A(\boldsymbol{k},y)[f(x)-f(y)] \times \left\{ \frac{1}{x-y+\omega+i\delta} - \frac{1}{y-x+\omega+i\delta} - \frac{2P}{x-y} \right\}. \tag{11.72}$$

If this expression is expanded to the terms linear in ω, only the imaginary part of the first-order term contributes. Hence

$$K_{\alpha\beta}(\omega) = -2i\frac{\pi e^2}{\mathscr{V}}\omega \sum_{\boldsymbol{k}} v_\alpha(\boldsymbol{k}) v_\beta(\boldsymbol{k}) \times \int dx\, dy\, A(\boldsymbol{k},x)A(\boldsymbol{k},y)[f(x)-f(y)] \frac{\partial \delta(x-y)}{\partial x}. \tag{11.73}$$

Integration by parts now leads to

$$\sigma_{\alpha\beta}(\omega) = -\frac{2\pi e^2}{\mathscr{V}} \sum_{\boldsymbol{k}} v_\alpha(\boldsymbol{k}) v_\beta(\boldsymbol{k}) \int dx\, A(\boldsymbol{k},x)A(\boldsymbol{k},x) \frac{\partial f(x)}{\partial x}. \tag{11.74}$$

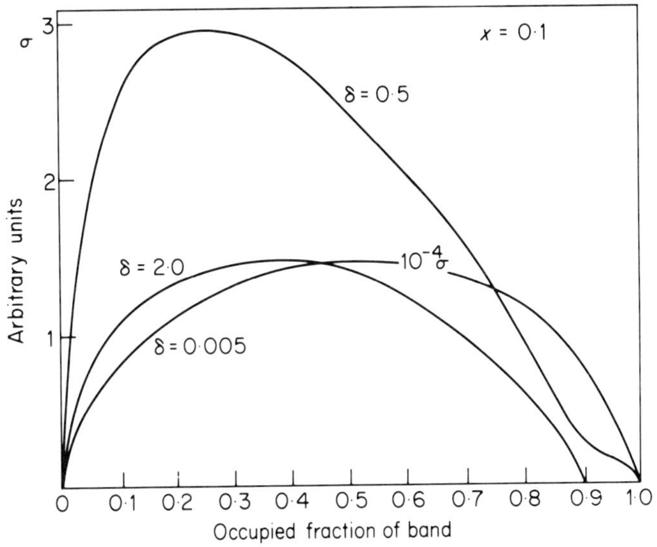

FIG. 11.9. The CPA d.c. electrical conductivity σ at 0 K for the model of equation (11.60) is plotted the fraction of the band occupied by electrons, for the three values of δ, 0.005, 0.5 and 2.0 (after Velicky 1969).

When scattering is weak $A(\boldsymbol{k}, x)$ has the standard Lorentzian form

$$A(\boldsymbol{k}, x) = \frac{1/2\tau}{(x - \epsilon_k - V_B)^2 + (1/2\tau)^2}. \tag{11.75}$$

For a metal with a spherical Fermi surface

$$-\frac{\partial f}{\partial x} \approx \delta(x - \mu)$$

and integration of (11.74) yields the standard result

$$\sigma = (2\pi^2/3)N(0)e^2v^2\tau|_{\epsilon=\mu}. \tag{11.76}$$

For weak scattering we can use the ATA to obtain an expression for τ. From equation (11.4) for Σ and

$$\Sigma = ct_B/[1 + cG_Dt_B],$$

$$\Sigma \approx cV_B\,[1 + (1-c)G_D\,V_B].$$

Hence,

$$\text{Re}\,\Sigma \approx cV_B, \qquad \text{Im}\,\Sigma \equiv 1/2\tau = c(1-c)V_B^2\,\text{Im}\,G_D. \tag{11.77}$$

Consequently, as a function of concentration, the conductivity is symmetric about $c = 0.5$ when the scattering is weak. Equations (11.76) and (11.77) agree with the result of Chapter 4 when $c \approx 0$ and $c \approx 1$, that is for nearly pure crystals.

Velicky (1969) has made calculations of σ based on equation (11.74) for stronger scattering and a non-degenerate electron gas with the band given by equation (11.60). The results of his calculations for $c = 0.1$ and for three values of $\delta = (V_B - V_A)/W$ are plotted versus the occupied fraction of the energy band in Fig. 11.9. The localization of states is indicated by the reduction of the conductivity by two orders of magnitude when $\delta = 2$ and the occupied fraction of the band exceeds 0.9. However, for $\delta = 2$ this is the regime of the split band and one expects the conductivity to be exactly zero. Thus although the CPA shows signs of the localization its predictions are deficient in this regime. A full discussion of the question of localization goes beyond the scope of this book; we therefore refer the interested readers to papers by Anderson (1958), Thouless (1972), Mott and Davis (1971), Economu and Cohen (1972), Licciardello and Economu (1975) and Wegner (1976).

References

General

Economu, E. N. (1979). "Green's Functions in Quantum Physics". Springer-Verlag, Berlin.

Elliott, R. J., Krumhansl, J. A. and Leath, P. L. (1974). *Rev. Mod. Phys.* **46**, 465.

Mott, N. F. and Davis, E. A. (1971). "Electronic Processes in Non-Crystalline Materials". Oxford University Press, London.

Special

Anderson, P. W. (1958). *Phys. Rev.* **109**, 1492.

Butler, W. H. (1973). *J. Phys. C: Solid State Phys.* **6**, 1559.

Dean, P. (1961). *Proc. Roy. Soc. London* **A260**, 263.

Dean, P. (1972). *Rev. Mod. Phys.* **44**, 127.

Economu, E. N. and Cohen, M. H. (1972). *Phys. Rev. B.* **5**, 2931.

Edwards, J. T. and Thouless, D. J. (1972). *J. Phys. C: Solid State Phys.* **5**, 807.

Freed, K. F. and Cohen, M. H. (1971). *Phys. Rev. B.* **3**, 3400.

Herbert, D. and Jones, R. (1971). *J. Phys. C: Solid State Phys.* **4**, 1145.

Leath, P. L. (1973). *J. Phys. C: Solid State Phys.* **6**, 1559.

Leath, P. L. and Goodman, B. (1969). *Phys. Rev.* **181**, 1062.

Licciardello, D. C. and Economu, E. N. (1975). *Phys. Rev. B.* **11**, 3697.
Payton III, D. N. and Visscher, V. M. (1967). *Phys. Rev.* **154**, 802.
Schwartz, L., Brouers, F., Vedyayev, A. V. and Ehrenreich, H. (1971). *Phys. Rev. B.* **4**, 3383.
Soven, P. (1967). *Phys. Rev.* **156**, 809.
Taylor, D. W. (1967). *Phys. Rev.* **156**, 1017.
Thouless, D. J. (1970). *J. Phys. C: Solid State Phys.* **3**, 1559.
Thouless, D. J. (1972). *J. Noncryst. Solids* **8–10**, 461.
Velicky, B. (1969). *Phys. Rev.* **184**, 614.
Wegner, F. J. (1976). *Z. Phys. B.* **25**, 327.

Problems

1. Show that for a band with density of states $N(\epsilon)$,

$$G_D(\omega) = \int_{-\infty}^{\infty} \frac{N(\epsilon)d\epsilon}{\omega - \epsilon}.$$

   Show also that

$$G_M(0, 0; \omega) = G_D[\omega - \Sigma(\omega)].$$

2. Show explicitly that in the coherent potential approximation G_M is the same whether the host crystal is taken to be the pure A-crystal or the virtual crystal.
3. For the band shape given in equation (11.60) show that

$$G_D(\omega) = (2/W^2)[\omega - (\omega^2 - W^2)^{1/2}],$$

   where the square root is in the same half-plane (upper or lower) as ω. Show that for this band the CPA leads to a cubic equation for $\Sigma(\omega)$.

Chapter 12

Critical Behaviour

§ 12.1 Second-order phase transitions

In Chapters 8, 9 and 10, we have discussed a number of phase transitions. In all these cases we have seen that the transition is characterized by an order parameter (e.g. the gap parameter, the condensate wave function and the magnetic moment) which tends smoothly to zero as the temperature tends to the transition temperature from below. Above the transition temperature, the order parameter is zero. Within the mean or molecular field theory, which we used to describe the transition, we found that in each case the order parameter tends to zero according to the law,

$$\text{order parameter} \propto (T_c - T)^{1/2}.$$

All these transitions belong to the class of second-order phase transitions for which Landau (1965) gave a very general theory independent of the particular transition.

Landau assumed that the transition was characterized by a real macroscopic order parameter ϕ_i which could have n components ($i = 1, 2, \ldots, n$). In the case of superconductivity and superfluidity the gap parameter is complex in general and so has two real components. In the case of magnetism the magnetization has three real components. He further assumed that near the transition temperature the Gibbs free energy F could be expanded in powers of ϕ_i where, because of symmetries in the problem, only even powers of ϕ_i would appear. As the second-order term can always be made diagonal, the expansion for the free energy density near T_c is for a homogeneous system of the form,

$$f = a \sum_i \phi_i^2 + \tfrac{1}{2} \sum_{ijkl} b_{ijkl}\phi_i\phi_j\phi_k\phi_l. \tag{12.1}$$

Landau also assumed that the parameters a and b_{ijkl} are analytic functions of $T - T_c$.

At a particular temperature, the equilibrium value of ϕ_i is the one which minimizes the free energy. It therefore satisfies

$$a\phi_i + \sum_{jkl} b_{ijkl}\phi_j\phi_k\phi_l = 0, \tag{12.2}$$

where we have taken advantage of the fact that b_{ijkl} can always be written in a form symmetric in pairs of indices.

Equation (12.2) always possesses the solution

$$\phi_i = 0, \qquad \text{all } i, \tag{12.3}$$

and this is believed to be the solution for $T > T_c$. For $T < T_c$, if (12.2) possesses a solution, it will be such that

$$\phi_i \propto (-a)^{1/2}. \tag{12.4}$$

Since ϕ_i tends to zero as $T \to T_c$ we must have

$$\lim_{T\to T_c} a = 0. \tag{12.5}$$

The simplest form for a which is analytic and satisfies (12.5) is

$$a \propto (T - T_c). \tag{12.6}$$

Except by accident some components of b_{ijkl} will not tend to zero at T_c. Hence, near T_c, b_{ijkl} can be taken to be independent of temperature and

$$\phi_i \propto (T_c - T)^{1/2}. \tag{12.7}$$

This accords with the particular cases we have discussed.

To see how the two solutions (12.3) and (12.7) arise let us specialize the free energy further to

$$f = a|\phi|^2 + \tfrac{1}{2}b|\phi|^4, \tag{12.8}$$

where a and b are real (to make f real) and

$$|\phi|^2 = \sum_{i=1}^{n} \phi_i^2. \tag{12.9}$$

The cases we have discussed do, for reasons of symmetry, conform to (12.8). (Superfluid liquid ^{3}He does not (Anderson and Brinkman 1975).) Then equation (12.2) becomes

$$a\phi_i + b\phi_i|\phi|^2 = 0. \tag{12.10}$$

Thus,

$$\phi_i = 0 \qquad \text{or} \qquad |\phi|^2 = -a/b \tag{12.11}$$

and the corresponding values of f are

$$f = 0 \qquad \text{or} \qquad f = -a^2/2b. \tag{12.12}$$

Since we must have $|\phi|^2 > 0$, two solutions exist only when a/b is negative. Further, the second solution will occur in nature only if the corresponding free energy is negative (i.e. below the free energy when ϕ_i is zero). Hence, for the second solution to occur and so yield a phase transition we must have

$$b > 0 \tag{12.13}$$

and a negative for some temperatures. The form (12.6) ensures this. In the theories of superfluids and magnetism a is negative below the transition temperature and positive above.

Equations (12.6) and (12.12) lead to the temperature dependence of the free energy

$$f \propto -(T_c - T)^2, \qquad T < T_c,$$
$$f = 0 \qquad\qquad T > T_c.$$

Hence, near T_c the specific heat is constant but at T_c it changes discontinuously from one value to another.

It is possible, too, to look at the order parameter correlation function by studying the change in the order parameter due to an external field. This discussion is a generalization of that given in Chapters 8–10. We denote the components of the field (which is a magnetic field in the case of magnetism) by $h_i(\boldsymbol{r})$ and assume a linear coupling. Since we wish to study spatial correlation we allow h_i to depend on position. The gap parameter will then also depend on position. For the study of long-range correlations h_i can vary slowly in space. The order parameter will then vary slowly in space, too. In general, there will then be a contribution to the free energy from the derivatives of the order parameter in space. For slow variations in space we need keep only the first non-vanishing order in the derivatives which, for isotropic systems, is in general quadratic in $\boldsymbol{\nabla}_{\phi_i}$. The total free energy in the field can then be written

$$F = \int dv \left\{ a|\phi|^2 + \tfrac{1}{2}b|\phi|^4 + c|\boldsymbol{\nabla}\phi|^2 - \sum_i h_i\phi_i \right\}, \tag{12.14}$$

where, for simplicity, we have assumed symmetry in the components of ϕ_i. If the phase transition is homogeneous in the absence of the external field, variations of ϕ in space must increase the free energy. Hence we must have $c > 0$. Except by accident c does not vanish at T_c. It can, therefore, be taken as constant for temperatures near T_c.

For F to be a minimum in the presence of the field h_i we must have

$$2a\phi_i + 2b|\phi|^2\phi_i - 2c\nabla^2\phi_i - h_i = 0. \qquad (12.15)$$

To find the first-order change in ϕ_i due to the field we linearize equation (12.15) about the equilibrium value of ϕ_i. Above T_c, $\phi_i = 0$ for all i. Hence, the cubic term in (12.10) can be neglected and

$$2a\phi_i - 2c\nabla^2\phi_i = h_i(\boldsymbol{r}). \qquad (12.16)$$

We can Fourier transform this equation to obtain

$$2(a + cq^2)\phi_i(\boldsymbol{q}) = h_i(\boldsymbol{q}). \qquad (12.17)$$

We then have a wave vector dependent "susceptibility"

$$\chi_i(\boldsymbol{q}) = 2(a + cq^2)^{-1}, \qquad T > T_c.$$

The Fourier transform of $\chi_i(q)$ is the order parameter correlation function $\chi_i(r)$. In three dimensions we find that

$$\chi_i(r) \propto \exp(-r/\zeta)/r, \qquad T > T_c \qquad (12.18)$$

with

$$\zeta = (c/a)^{1/2} \propto (T - T_c)^{-1/2}. \qquad (12.19)$$

The range of the correlation is ζ which is, therefore, called the correlation length. As $T \to T_c$, ζ becomes infinite in accordance with equation (12.19). This long range of the correlations has played an important role in our improved understanding of behaviour near phase transitions. At the transition ζ^{-1} is zero and

$$\chi_i(r) \propto r^{-1}, \qquad T = T_c. \qquad (12.20)$$

Below the temperature T_c, the equilibrium value of ϕ is not zero. However, according to equation (12.11) only $|\phi|^2$ is determined uniquely. There are many different solutions for ϕ_i, consistent with this and each allowed solution represents a different phase below T_c. For example, in the case of the ferromagnet these different phases correspond to different directions of the spontaneous moment. Let us consider the phase with $\phi_1 = (-a/b)^{1/2} = \phi_0$, say, and with $\phi_r(r = 2, \ldots, n) = 0$. There is no loss of generality in this choice because of the symmetry in the free energy. Then, to first order in h, we have

$$|\phi|^2 = \phi_1^2.$$

Linearization of equation (12.15) then results in

$$\begin{aligned}(2a+6b\phi_0^2)\phi_1 - 2c\nabla^2\phi_1 &= h_1,\\ (2a+2b\phi_0^2)\phi_i - 2c\nabla^2\phi_i &= h_i, \qquad i = 2, 3, \ldots, n.\end{aligned} \tag{12.21}$$

Because of the definition of ϕ_0 the first term of the second of these equations vanishes. We then find two susceptibilities, a longitudinal one

$$\chi_{\|}(q) = \chi_{\|}(q) = (-4a + 2cq^2)^{-1}, \tag{12.22}$$

and a transverse one

$$\chi_{\perp}(q) = \chi_{ii}(q) = (2cq^2)^{-1}, \qquad i = 2, 3, \ldots, n. \tag{12.23}$$

In coordinate space, the longitudinal susceptibility again has the form given by equation (12.18) but ζ is given by

$$\zeta = (c/2a)^{1/2} \propto (T_c - T)^{-1/2}, \qquad T < T_c. \tag{12.24}$$

Thus its magnitude is changed but it approaches infinity as $T \to T_c$ with the same power law. The transverse susceptibility in three dimensions follows the simple power law

$$\chi_{\perp}(r) \propto r^{-1}. \tag{12.25}$$

All the results of this section derive from a simple and general theory which is not very sensitive to the details of the phase transition. They include the results of the specific mean field theories described in previous chapters. However, as we see in the next section, the results are not, in fact, in very good agreement with experiment (except in the case of superconductors) nor with the results obtained from exactly soluble models such as the two-dimensional Ising model. Despite this, the observed behaviour does seem to be insensitive to the details of the transition. The explanation of the behaviour seems to lie in the scaling and renormalization group theories which we introduce in later sections.

§ 12.2 Critical exponents

We saw in the last section that Landau's theory of second-order phase transitions leads to certain power laws. Similarly, both experiment and more general theories suggest the existence of power laws although the exponents are not usually those given by Landau's theory; the latter are now called the classical exponents. For this reason it is usual to define and label the exponents and laws in a particular way. As the observable quantities differ, we list the

TABLE 12.1. Summary of definitions of critical-point exponents for fluid systems. Here $\epsilon \equiv T/T_c - 1$

Exponent	Definition	Conditions ϵ	$P - P_c$	$\rho - \rho_c$	Quantity
α'	$C_v \sim (-\epsilon)^{-\alpha'}$	< 0	0	0	Specific heat at constant volume $V = V_c$
α	$C_v \sim \epsilon^{-\alpha}$	> 0	0	0	
β	$\rho_L - \rho_G \sim (-\epsilon)^{\beta}$	< 0	0	$\neq 0$	Liquid–gas density difference (or shape of coexistence curve)
γ'	$K_T \sim (-\epsilon)^{-\gamma'}$	< 0	0	$\neq 0$	Isothermal compressibility
γ	$K_T \sim \epsilon^{-\gamma}$	> 0	0	0	
δ	$P - P_c \sim \lvert\rho_L - \rho_a\rvert^{\delta} \operatorname{sgn}(\rho_L - \rho_a)$	0	$\neq 0$	$\neq 0$	Critical isotherm
ν'	$\xi \sim (-\epsilon)^{-\nu'}$	< 0	0	$\neq 0$	Correlation length
ν	$\xi \sim \epsilon^{-\nu}$	< 0	0	0	
η	$G(r) \sim \lvert r\rvert^{-(d-2+\eta)}$	0	0	0	Pair correlation function ($d =$ dimensionality)

TABLE 12.2. Summary of definitions of critical-point exponents for magnetic systems. Here $\epsilon \equiv T/T_c - 1$

Exponent	Definition	Conditions ϵ	H	M	Quantity
α'	$C_H \sim (-\epsilon)^{-\alpha'}$	< 0	0	0	Specific heat at constant magnetic field
α	$C_H \sim \epsilon^{-\alpha}$	> 0	0	0	
β	$M \sim (-\epsilon)^{\beta}$	< 0	0	$\neq 0$	Zero-field magnetization
γ'	$\chi T \sim (-\epsilon)^{-\gamma'}$	> 0	0	$\neq 0$	Zero-field isothermal susceptibility
γ	$\chi T \sim \epsilon^{-\gamma}$	> 0	0	0	
δ	$H \sim \lvert M\rvert^{\delta} \operatorname{sgn}(M)$	0	$\neq 0$	$\neq 0$	Critical isotherm
ν'	$\xi \sim (-\epsilon)^{-\nu'}$	< 0	0	$\neq 0$	Correlation length
ν	$\xi \sim \epsilon^{-\nu}$	> 0	0	0	
η	$\Gamma(r) \sim \lvert r\rvert^{-(d-2+\eta)}$	0	0	0	Pair correlation function ($d =$ dimensionality)

TABLE 12.3. Values of critical-point exponents for selected systems.

System	$T<T_c$				$T=T_c$		$T>T_c$		
	α'	β	γ'	ν'	δ	η	α	γ	ν
Fluids									
CO_2	~ 0.1	0.34	~ 1.0	–	4.2	–	~ 0.1	1.35	–
Xe	< 0.2	0.35	~ 1.2	0.57	4.4	–	–	1.3	–
Magnets									
Ni	$\alpha'_s = -0.3$	0.42	–	–	4.22	–	0	1.35	–
EuS	$\alpha'_s = -0.15$	0.33	–	–	–	–	0.05	–	–
$CrBr_3$	–	0.368	–	–	4.3	–	–	1.215	–
Soluble Models									
Classical	0 (disc)	$\frac{1}{2}$	1	$\frac{1}{2}$	3	0	0 (disc)	1	$\frac{1}{2}$
Ornstein–Zernike	–	–	–	–	5	0	$\alpha_s = -1$	2	1
$d = 3$, spherical model	–	$\frac{1}{2}$	–	–	5	0	$\alpha_s = -1$	2	1
$d = 2$, Ising model	0 (log)	$\frac{1}{8}$	$\sim \frac{7}{4}$	1	~ 15	$\frac{1}{4}$	0 (log)	$\sim \frac{7}{4}$	1

exponents and their definitions for the liquid–gas transition and the ferromagnetic transition in Tables 12.1 and 12.2, respectively (taken from Stanley (1971)). In this table, the laws express the dominant behaviour as $T \to T_c$, $r \to \infty$, $\rho \to \rho_c$, $H \to 0$. In the liquid–gas transition, the order parameter is $\rho - \rho_c$ where ρ is the density and ρ_c the critical density. The classical theory of a liquid is Van der Waals' theory (Kac *et al.* 1963). In the case of the ferromagnet, the order parameter is the magnetic moment M.

In Table 12.3 (also taken from Stanley (1971)) we display some observed experimental values for the exponents as well as some calculated from soluble models and the classical theory discussed above. It can be seen that the results of the classical theory do not agree well with either observation or the results from soluble models.

Certain relations have been found to obtain between the various critical exponents and these agree with the phenomenological scaling hypothesis (see Kadanoff *et al.* (1967) or Stanley (1971) for a review). According to this hypothesis, near the critical point the Gibbs free energy is a homogeneous function of some power of $t[=(T-T_c)/T_c]$ and the order parameter, or external field. In the case of the ferromagnetic transition this implies

$$G(T,H) = |t|^{2-\alpha} f^{\pm}(H/|t|^{\Delta}), \tag{12.26}$$

where the functions f^+ and f^- exist above and below the transition temperature, respectively, and may not be identical. All the critical exponents can then be derived from the two parameters α and Δ. Only two exponents are independent according to this hypothesis and relations exist between those given in Tables 12.1 and 12.2 as follows:

$$\alpha + 2\beta + \gamma = 2, \qquad \alpha + \beta(\delta + 1) = 2,$$
$$\gamma(\delta + 1) = (2 - \alpha)(\delta - 1), \qquad \gamma = \beta(\delta - 1),$$
$$\delta = \frac{2 - \alpha + \gamma}{2 - \alpha - \gamma}, \qquad \alpha = \alpha', \qquad \gamma = \gamma'. \tag{12.27}$$

One aim of a fundamental theory is to explain these relations as well as the numerical values of the exponents.

§12.3 The sum-over-states and fluctuations

Landau's theory of second-order phase transitions is seductively simple. Why does it not explain the facts? Essentially because it is too simple and not entirely consistent. For example we were able to derive the susceptibility of the system from the free energy and we know from our general theory that this is related to fluctuations of the order parameter. However, the theory itself contains no fluctuations of this order parameter. It is assumed that the system has different states with well defined values of the order parameter and that a well defined free energy exists for each value of the order parameter. The equilibrium state is the one which minimizes the free energy subject to any external constraints. It is necessary to modify this theory to include fluctuations and to do this we need to push it one stage further back and consider, instead of the free energy, the sum-over-states Z.

We still assume that there exist states with well defined values of the order parameter. To construct Z, we assume that each of these states contributes an amount $\exp\{-F[\phi]/k_B T\}$ where $F[\phi]$ is the free energy expression of Landau. The total sum-over-states is obtained by summing over all allowed values of the order parameter ϕ. We write the result as

$$Z = \int \mathscr{D}\,\phi \exp[-F(\phi)/k_B T]. \tag{12.28}$$

According to this theory the observed free energy is

$$\mathscr{F} = -k_B T \ln Z. \tag{12.29}$$

The symbols $\int \mathscr{D}\phi$ in equation (12.28) indicate that we ascribe different values of $\phi(\boldsymbol{r})$ to each point $\boldsymbol{r}$ in space and that, at each point, we integrate over all possible values of $\phi(\boldsymbol{r})$. Since the volume of the system contains a continuous infinity of points, the expression (12.28) involves a continuous infinity of integrals, a formidable prospect. Expressions like this are called functional integrals and we shall see below examples of how to evaluate them. One common method is to expand $\phi(\boldsymbol{r})$ in terms of a complete set of orthonormal basis functions $u_n(\boldsymbol{r})$ according to

$$\phi_i(\boldsymbol{r}) = \sum a_{ni} u_n(\boldsymbol{r}) \tag{12.30}$$

and integrate over the coefficients a_n. The Jacobean for this transformation is unity and

$$Z = \int \mathscr{D}\, a_{ni} \exp[-F'(a_{ni})/k_B T]. \tag{12.31}$$

Then we have to deal with a discrete infinity of integrals.

Since Z depends on many values of $\phi(\boldsymbol{r})$, it allows for fluctuations in $\phi(\boldsymbol{r})$. The mean value of $\phi(\boldsymbol{r})$ is then

$$\langle \phi_i(\boldsymbol{r}) \rangle = Z^{-1} \int \mathscr{D}\, \phi \exp[-F(\phi)/k_B T]\, \phi_i(\boldsymbol{r}), \tag{12.32}$$

while the correlation function $\langle \phi_i(\boldsymbol{r}) \phi_j(\boldsymbol{r}') \rangle$ is given by

$$\langle \phi_i(\boldsymbol{r}) \phi_j(\boldsymbol{r}') \rangle = Z^{-1} \int \mathscr{D}\, \phi \exp[-F(\phi)/k_B T]\, \phi_i(\boldsymbol{r}) \phi_j(\boldsymbol{r}'). \tag{12.33}$$

The susceptibility discussed in §12.1 is now related to these fluctuations by

$$k_B T \chi_{ij}(\boldsymbol{r}) = \langle \phi_i(0) \phi_j(\boldsymbol{r}) \rangle - \langle \phi_i \rangle \langle \phi_j \rangle. \tag{12.34}$$

This is consistent with the general theory.

The free energy $F(\phi)$ is proportional to the volume and is therefore macroscopically large. Hence, we might expect to be able to evaluate the integral by the method of steepest descents. This means evaluating $F(\phi)$ close to its stationary points. These stationary points are just the values ϕ_0 picked out in Landau's theory and $F(\phi) \approx F(\phi_0)$ at these points. If this value of F dominates the integral

$$\mathscr{F} = F(\phi_0) \tag{12.35}$$

and one obtains Landau's theory. Thus Landau's theory should be a good approximation when the method of steepest descents is valid.

As the point is important, let us pursue the evaluation in more detail for the model with $F[\phi]$ given by equation (12.14), but with no external field. Thus

$$F(\phi) = \int dv\,[a|\phi|^2 + \tfrac{1}{2}b|\phi|^4 + c|\nabla\phi|^2]. \tag{12.36}$$

Consider the case of a single-component order parameter and $a < 0$. Then

$$\phi_0 = \pm(-a/b)^{1/2}. \tag{12.37}$$

These two values of ϕ_0 each contribute the same amount to the integral. We shall therefore evaluate Z near

$$\phi_0 = (-a/b)^{1/2}. \tag{12.38}$$

It might be thought that Z should then be multiplied by 2. However, when the method of steepest descents is valid, the system has to pass through values of $F(\phi)$ which are macroscopically different from $F(\phi_0)$ to change from ϕ_0 to $(-\phi_0)$. Since this change would take an astronomical time, we confine the integration the the neighbourhood of $\langle\phi\rangle$. If we were to sum over the contributions from the neighbourhoods of both ϕ_0 and $-\phi_0$ we should find

$$\langle\phi\rangle = 0. \tag{12.39}$$

For the reason given above, one finds in practice that

$$\langle\phi\rangle \approx \phi_0. \tag{12.40}$$

(This result can also be obtained by including a small but finite external field h and allowing h to tend to zero at the end of the calculation.)

If we now put

$$\phi = \phi_0 + \psi$$

and retain only quadratic terms in ψ in F we find that

$$Z = \exp[-F(\phi_0)/k_B T] \int \mathscr{D}\psi \times \exp\left\{-(k_B T)^{-1} \int [(a + 3b\phi_0^2)\psi^2 + c(\nabla\psi)^2]\,dv\right\}. \tag{12.41}$$

To evaluate the integral which is of the Gaussian type we expand ψ in plane waves

$$\psi(\boldsymbol{r}) = \sum_{\boldsymbol{q}} \psi_{\boldsymbol{q}} \exp(i\boldsymbol{q}\cdot\boldsymbol{r})/\mathscr{V}^{1/2}, \tag{12.42}$$

$$\psi_{\boldsymbol{q}} = \psi^*_{-\boldsymbol{q}}.$$

Then

$$Z = \exp[-F(\phi_0)/k_B T] \int \mathscr{D} \psi_q \times \exp\left[-(k_B T)^{-1} \sum_q (-2a + cq^2)\psi_q \psi_q^*\right]$$

$$= \exp[-F(\phi_0)/k_B T] \prod_q{}' \int \mathscr{D} \psi_q \times \exp[-2(k_B T)^{-1}(-2a + cq^2)|\psi_q|^2], \qquad (12.43)$$

where the product is over only half of $\boldsymbol{q}$-space. If

$$\psi_q = a_b + ib_q,$$

$$\int \mathscr{D} \psi_q \exp(-\alpha|\psi_q|^2) = \int da_q db_q \exp[-\alpha(a_q^2 + b_q^2)]$$

$$= 2\pi/\alpha. \qquad (12.44)$$

Hence

$$Z = \exp[-F(\phi_0)/k_B T] \prod_q \pi k_B T/(-2a + cq^2), \qquad (12.45)$$

and

$$\mathscr{F} = F(\phi_0) - k_B T \sum_q \ln[\pi k_B T/(-2a + cq^2)]. \qquad (12.46)$$

We see that because of the large number of degrees of freedom involved the correction term is also proportional to the volume and may be comparable in magnitude to the first.

To compare the correction term to the Landau term we have to evaluate the sum over $\boldsymbol{q}$ which diverges for large $\boldsymbol{q}$. This divergence shows that large values of $\boldsymbol{q}$ contribute significantly to the free energy and that the assumption of a slowly varying order parameter used to obtain our form of free energy is not valid. (This is a point to which we shall return.) If, however, we are concerned only with the dominant temperature dependence near T_c only the small values of q are important. This is clearly illustrated if we consider the specific heat C. For the dominant temperature dependence near T_c we can put $T = T_c$ except in a which is proportional to $T - T_c$. Then

$$C \propto -\frac{\partial^2 \mathscr{F}}{\partial T^2} \propto -\frac{\partial^2 F}{\partial a^2} = \frac{1}{b} + \frac{4k_B T_c}{(2\pi)^3} \int \frac{d^3 q}{(-2a + cq^2)^2}. \qquad (12.47)$$

The integral converges and the important values of q are of the order of $(-a/c)^{1/2}$ which is small very close to T_c. The ratio of the second term to the first is then

$$\sim k_B T_c b/(-ac^3)^{1/2}. \tag{12.48}$$

For values of a for which this ratio is much less than unity, mean field theory dominates. This region requires that $|T-T_c|$ should not be too small. It may be that, since we are not at the extreme limit $T \to T_c$, it is necessary to incorporate more terms in ϕ into the expression $F[\phi]$ so that one does not see the pure classical behaviour. Nevertheless, a mean field theory should provide a satisfactory description for T not too close to T_c. On the other hand, when the ratio (12.48) is of the order of unity, fluctuations are important and the Landau theory of the transition breaks down. This criterion for the importance of fluctuations was first proposed by Ginzburg and is known as the Ginzburg criterion. For bulk superconductors the temperature range where fluctuations are important is so narrow that one sees almost pure mean field theory. For other phase transitions the fluctuations are usually very important.

One can use the same technique to evaluate the correlation function. One then finds from equation (12.34) that

$$\chi_{ij}(\boldsymbol{q}) = (1/k_B T)\langle \psi_{\boldsymbol{q}} \psi_{\boldsymbol{q}}^* \rangle = [2(-2a + cq^2)]^{-1}. \tag{12.49}$$

This agrees as it should with equation (12.22). In this version the theory is self-consistent.

We have now given a consistent basis for Landau's theory. (Although the starting point, equation (12.28), is phenomenological it can actually be derived from microscopic theory for certain theories. In the case of superconductivity such a derivation has been given by Langer (1964).) The correction to Landau's theory derived in section (12.3) has also shown that Landau's theory will fail in the critical region close to T_c. This explains why it does not yield the observed critical exponents. An improved theory to deal with the critical region has been advanced by Wilson (1969, 1971a, b). Although there is still a gap to be bridged between theory and observation, Wilson's approach has stimulated a wealth of ideas and calculations to which we cannot hope to do justice in this chapter. Indeed it is an unfortunate fact that even explaining the ideas without supplying proofs takes up a great deal of space. We therefore refer the reader to the review articles of Wilson and Kogut (1974), Wilson (1975), Fisher (1974), Ma (1974), those in Volume 6 of "Phase Transitions and Critical Phenomena" edited by Domb and Green (1976), and to the books of Pfeuty and Toulouse (1975), Ma (1976) and of Amit (1978). In the next sections, we simply try to whet the reader's appetite for this work and show that Green's functions play some role in the methods.

§ 12.4 Perturbation theory for the correlation function

There are two distinct but equivalent approaches to the renormalization group as it is applied to critical phenomena. The first is closest to Wilson's original ideas and is based on certain scale transformations of the functional in equation (12.28). It has the advantage of intuitive appeal and it reveals the group nature of what is called the renormalization group. However, this approach uses techniques rather far removed from those in this book and would be out of place here. It is described in some detail in the review article of Wilson and Kogut (1974), in the article by Wegner in Domb and Green (1976) and in the book of Pfeuty and Toulouse (1975).

The second approach uses the methods of quantum field theory and is closer to the techniques used in this book. It is discussed in some detail in the book by Amit (1978) and the articles by Brezin, Le Guillou and Zinn-Justin and by Wallace in Domb and Green (1976). This account relies heavily on these sources, especially the last. We provide a very brief outline of these methods without proofs as we consider the problem only as T_c is approached from $T > T_c$. General theorems relate the exponents below T_c to those above it.

We consider only the special form for $F[\phi]$ given by equation (12.4). Since $\phi(\boldsymbol{r})$ is variable and the sum-over-states is related to F through equation (12.28), F is often called the Hamiltonian of the system. Without loss of generality the variables ϕ and $\boldsymbol{r}$ can be changed so that F can be written in the standard form

$$\mathscr{H} \equiv (k_B T)^{-1} F = \int d^d x \left\{ \tfrac{1}{2} |\nabla \phi|^2 + \tfrac{1}{2} r_0 |\phi(x)|^2 + \frac{\bar{u}_0}{4!} |\phi(x)|^4 + \frac{\Lambda^2}{2} |\nabla^2 \phi(x)|^2 \right\}, \tag{12.50}$$

where, near T_c, $\bar{u}_0$ is constant and

$$r_0 \propto T - T_c .$$

The parameter r_0 is then a measure of the temperature. The last term has been introduced to provide a cut-off at large values of q and ensure the convergence of integrals. If only long-wavelengths fluctuations are important near T_c, the physically significant results should be obtained in the limit $\Lambda \to \infty$ and they should not depend on the assumed form of cut-off.

We have made one further generalization in writing down equation (12.50) and that is an extension to any number of dimensions d. For reasons which will appear, this has proved to be a powerful, practical generalization. The number of dimensions is not even assumed to be integral! What the expression then means will be discussed below. In this initial discussion d can be taken to be an integer.

The first point to note is that Z can be expanded in powers of $\bar{u}_0$ and the terms of the expansion related to Feynman diagrams in complete analogy with the perturbation series described in other parts of this book. The zero-order term is

$$Z_0 = \int \mathscr{D}\phi \exp\left\{-\int d^dx\left[\tfrac{1}{2}|\nabla\phi|^2 + \tfrac{1}{2}r_0|\phi(x)|^2 + \tfrac{1}{2}\Lambda^{-2}|\nabla^2\phi(x)|^2\right]\right\}.$$

If, as before, we expand ϕ_i in plane waves according to

$$\phi_i = \mathscr{V}^{-1/2}\sum \exp(iqx)\phi_{iq}, \qquad \phi^*_{iq} = \phi_{i-q},$$

we find that

$$Z_0 = \int \mathscr{D}\phi_q \exp\left[-\sum_{i,q}{}' (r_0 + q^2 + \tfrac{1}{2}\Lambda^{-2}q^4)\phi_{iq}\phi_{i-q}\right],$$

where the sum is over half-q space. The integrals are now all Gaussian and yield

$$Z_0 = \prod_q{}' \left[\frac{2\pi}{r_0 + q^2 + \tfrac{1}{2}\Lambda^{-2}q^4}\right]^n \tag{12.51}$$

where n is the number of components of the order parameter.

In the same way the zero-order approximation to the correlation function

$$G_{i,j}(x,x') = \langle\phi_i(x)\phi_j(x')\rangle$$

$$= Z^{-1}\int \mathscr{D}\phi \exp[-\mathscr{H}]\phi_i(x)\phi_j(x') \tag{12.52}$$

with Fourier transform

$$G_{ij}(q) = \langle\phi_{iq}\phi_{j-q}\rangle, \tag{12.53}$$

is

$$G^0_{ij}(q) = \delta_{ij}/(r_0 + q^2 + \tfrac{1}{2}\Lambda^{-2}q^4)^{-1}. \tag{12.54}$$

We can now consider the expansion for G_{ij} or Z. As we have concentrated on the former in other parts of this book we do the same here. The thermodynamic functions can easily be related to G_{ij}. For example the entropy S per unit volume is given by

$$S \propto -2\partial \mathscr{F}/\partial r_0 = \langle|\phi(x)|^2\rangle = \sum_i G_{ii}(x, x). \qquad (12.55)$$

The specific heat is then obtained by further differentiation with respect to r_0. The free energy itself can be obtained by integrating equation (12.55).

The perturbation expansion for G_{ij} is obtained in a way analogous to the previous expansions obtained in §3.2 and 7.5. To make the analogy clearer let us write the quartic term in the fields in H as

$$\int d^dx \frac{\bar{u}_0}{4!}|\phi(x)|^4 = \int d^dx d^dx' V(x-x')|\phi(x)|^2|\phi(x')|^2 \qquad (12.56)$$

with

$$V(x_1 - x_2) = \frac{\bar{u}_0}{4!}\delta(x_1 - x_2) \equiv V(1, 2).$$

Further, to simplify the notation, let x stand symbolically for the field label i as well as x and let the integral over x include a sum over i.

Then the expansion of G in powers of $\bar{u}_0$ yields

$$\begin{aligned} G(x_1, x_2) &= Z^{-1}\int \mathscr{D}\phi \exp\{-\mathscr{H}_0[\phi]\}\Big\{\phi(x_1)\phi(x_2) \\ &\quad \times \prod_{i=1}^{n} di'di''V(i', i'')\phi(i')^2\phi(i'')^2\Big\} \\ &= (Z_0/Z)\Big\{\prod_{i=1}^{n}\int di'di''V(i', i'')\Big\} \\ &\quad \times G_0^{2n+2}(1, 2, 1', 1', 1'', 1'', \ldots, n', n', n'', n''), \end{aligned} \qquad (12.57)$$

where G_0^n is the n-particle correlation function for the system described by $\mathscr{H}_0$, that is,

$$\int G_0^n(1, 2, \ldots, n) = \langle\phi(1)\phi(2)\ldots\phi(n)\rangle_0. \qquad (12.58)$$

Now these Green's function obey a theorem similar to that for the corresponding phonon function discussed in §7.5.

$$G_0''(1, 2, \ldots, n) = \sum_{\text{all pairs}} G_0(1, 2)G_0(3, 4)\ldots G_0(n-1, n), \qquad (12.59)$$

for n even (the only case which arises in equation (12.57)) and is zero for n odd.

To prove equation (12.59) we look at the small change in G_0^n due to a small change, $\delta\phi$, in ϕ in the integrand. Since we integrate over all functions ϕ, this can make no difference to G_0^n. Similarly it can make no difference to Z_0. Hence

$$\delta(G_0^n Z_0) = 0. \tag{12.60}$$

However,

$$\delta[G_0^n(1, \ldots, n)Z_0] = \int \mathscr{D}\phi\,\delta[\exp\{-\mathscr{H}_0[\phi]\}\phi(x_1)\ldots\phi(x_n)]. \tag{12.61}$$

From the calculus of variations we know that

$$\delta \exp[-\mathscr{H}_0(\phi)] = \int d1'\delta\phi(1')\exp\{-\mathscr{H}_0[\phi]\} \times \{\nabla^2\phi(1') - r_0\phi(1') - \Lambda^{-2}\nabla^4\phi(1')\}.$$

Hence the total change in (12.61) can be written

$$Z_0 \int d1'\delta\phi(1')\left\{(\nabla_{1'}^2 - r_0 - \Lambda^{-2}\nabla_{1'}^4)G_0^{n+1}(1', 1, 2, \ldots, n) + \sum_i \delta(i, i')G_0^{n-1}(1, 2, \ldots, \not{i}, \ldots n)\right\}.$$

Since this is zero for all variations $\delta\phi$ we must have

$$(-\nabla_{1'}^2 + r_0 + \Lambda^{-2}\nabla_{1'}^4)G_0^{n+1}(1', 1, 2, \ldots, n) = \sum \delta(i, i')G_0^{n-1}(1, 2, \ldots, \not{i}, \ldots, n). \tag{12.62}$$

For $n = 1$ this is

$$(-\nabla_{1'}^2 + r_0 + \Lambda^{-2}\nabla_{1'}^4)G_0^2(1', 1) = \delta(1', 1)$$

With the solution with Fourier transform (12.54). We can then proceed as in §3.2 to integrate equation (12.62) and deduce by the method of induction that equation (12.59) holds.

As in §3.2 the contributions to G can be represented by Feynman diagrams constructed as in that section with continuous lines representing $G_0^2(a, b)$ and dashed lines representing the interaction $V(1, 2)$. As before, only the contributions of the topologically distinct diagrams need be included. There is, however, one subtlety connected with the combinatorial algebra. First we note that interchanging x and x' in the interaction term (12.56) does not change the value of

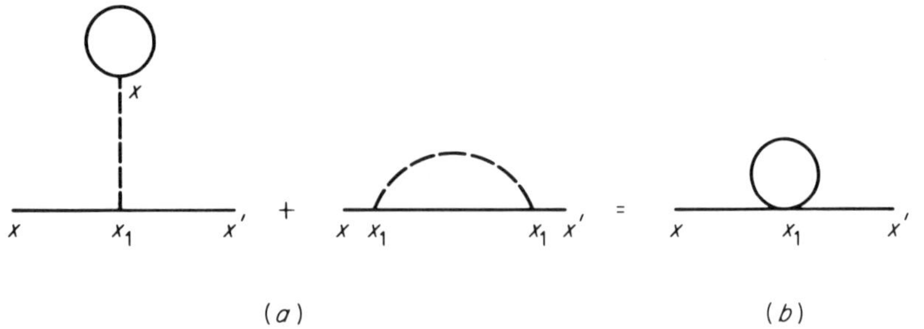

FIG. 12.1. First-order contributions to the one-particle Green's function, or second-order correlation function. The interaction can be shrunk to a point and the two graphs of (a) represented by the single graph (b).

the integral although it interchanges the labelling of vertices in a diagram. Hence when we restrict ourselves to distinct diagrams we should multiply the contribution from each interaction by 2. Similarly if we interchange the two $\phi(x)$'s with each other and the two $\phi(x')$'s with each other in (12.56) we do not change its value although we get other terms in (12.59). Consequently we must multiply the contribution from each interaction by 4 to account for this effect. However if we interchange the ϕ's at each point around a continuous loop we do not get a different contribution to (12.59). Therefore there is an extra factor of $\frac{1}{2}$ from each continuous loop. These conclusions mean that the rules for contributions from the diagrams are the same as given in §3.2 except that the contribution from each dashed line is

$$\bar{u}_0\,\delta(x_1 - x_2)/3$$

and there is an extra factor $(\frac{1}{2})^l$ where l is the number of continuous closed loops.

It is now simple to reintroduce the field label i. Since G_{0ij} is diagonal each continuous line carries the same label. On the backbone this will be the external label. On a closed continuous loop the label is free and must be summed over. This yields a further factor n for each closed loop. We find for example that the first-order contribution to G_{ij} is illustrated in Fig. 12.1 and is found to be

$$G'_{ij}(x,x') = -\delta_{ij}\int \tfrac{1}{6}\bar{u}_0\, G^0_{ii}(x,x_1) \times [n\,G^0_{ii}(x_1,x_1) + 2G^0_{ii}(x_1,x_1)]\,G^0_{ii}(x_1,x')d^dx_1 . \tag{12.63}$$

Because the interaction is a point one, it is possible without ambiguity to illustrate it in the diagrams by shrinking the dashed line to a point. Then the two diagrams in Fig. 12.1(*a*) are represented by the single diagram in Fig. 12.1(*b*). This is in accord with the fact that they make the same contribution apart from a constant factor. The new diagram (*b*) gives a contribution according to the same rules as those of (*a*) apart from the constant factor which represents the number of ways in which the diagram can occur. From its origin we see that this factor is $(n + 2)$. In the general case, the constant factor is often the most difficult to calculate.

As before it is possible to define a proper self-energy $\Sigma(q)$ and express G as

$$G_{ii}(q)^{-1} = [r_0 + q^2 + \tfrac{1}{2}\Lambda^{-2}q^4 - \Sigma(q)]. \tag{12.64}$$

We take the transition temperature to be the highest temperature at which this becomes infinite. For a uniform transition this will first occur for $q = 0$. The true transition temperature is therefore given by the equation

$$r_0(T_c) - \Sigma(0) = 0. \tag{12.65}$$

Consequently, it is shifted slightly from the transition temperature in r_0. Since r_0 does not tend to zero at the real transition temperature, it is usual to use a renormalized parameter r defined at any temperature by

$$r = r_0(T) - \Sigma(0, r). \tag{12.66}$$

Then $r \to 0$ at the real transition temperature and we expect physical quantities near T_c to be proportional to powers of r. In terms of r, we can write

$$G_{ij}(q)^{-1} = \delta_{ij} r + q^2 + \tfrac{1}{2}\Lambda^{-2}q^4 - \Sigma(q) + \Sigma(0). \tag{12.67}$$

From equation (12.63) we can see that the contribution to $\Sigma(0, r)$ which is first order in $\bar{u}_0$ is

$$\begin{aligned}\Sigma(0, r) &= -\frac{\bar{u}_0}{6}\frac{(n+2)}{\mathscr{V}}\sum_q G^0(q) \\ &= -\frac{\bar{u}_0}{6}\frac{(n+2)}{(2\pi)^d}\int\frac{d^d q}{r + q^2 + \frac{1}{2}\Lambda^{-2}q^4}.\end{aligned} \tag{12.68}$$

This leads to a new relationship between r and $T - T_c$ which can be obtained as follows. From equation (12.66) and the definition of T_c we have

$$0 = r_0(T_c) - \Sigma(0, 0).$$

Hence

$$\begin{aligned}t \equiv r_0(T) - r_0(T_c) &= r + \Sigma(0, r) - \Sigma(0, 0) \\ &= r + \frac{\bar{u}_0}{6}\frac{(n+2)}{(2\pi)^d} r \int\frac{d^d q}{(r + q^2 + \frac{1}{2}\Lambda^{-2}q^4)(q^2 + \frac{1}{2}\Lambda^{-2}q^4)}\end{aligned} \tag{12.69}$$

where the parameter t is proportional to $(T-T_c)$. The integral in this equation converges at large q for $d<8$. It is quite sufficient for our purposes to limit d by this inequality. The behaviour of the integral for small r depends on the behaviour of the integrand for small q and on the value of d. For $r=0$ and $q\to 0$, the integrand behaves like q^{-4}. Hence for $r=0$, the integral converges for $d>4$ and diverges for $d\leqslant 4$. We separate these cases.

(i) *$d>4$.*

As $r\to 0$, the integral converges to a constant and

$$t \approx r + C\bar{u}_0 r, \tag{12.70}$$

where C is a constant. We still have $r\propto t$, the effect of the perturbation being to change the constant of proportionality slightly. This result remains true to all orders in u_0 and we have a mean field theory.

(ii) *$2<d\leqslant 4$.*

The integral is singular as $r\to 0$. To evaluate it we have to perform the angular integral which is often denoted by S_d. The remaining factor is

$$\int_0^\infty \frac{q^{d-1}\,dq}{(r+q^2+\frac{1}{2}\Lambda^{-2}q^4)(q^2+\frac{1}{2}\Lambda^{-2}q^4)} \approx \frac{\pi/2}{\sin(\pi\epsilon/2)}\, r^{-\epsilon/2}, \tag{12.71}$$

where we use the conventional notation

$$\epsilon = 4-d. \tag{12.72}$$

Comparing the first and second terms on the right-hand side of equation (12.69) for small values of r we see that the second is greater than the first by a factor proportional to $\bar{u}_0 r^{-\epsilon/2}$. This suggests that this parameter is the real expansion parameter in the problem and one sees that for any finite value of $\bar{u}_0$ it is large for sufficiently small values of r. Perturbation theory breaks down.

For the special case of $d=4$, the expansion parameter is $\bar{u}_0 \ln(\Lambda^2/r)$ which is also large for sufficiently small values of r.

For $d\leqslant 2$, $\Sigma(0,0)$ does not exist and this discussion is not relevant.

From the propagator it is clear that r and Λ^2 have the same dimensions and will occur in the ratio r/Λ^2. To preserve this throughout it is conventional to write

$$\bar{u}_0 = u_0 \Lambda^{\epsilon}.$$

In terms of u_0 the expansion parameter is

$$u_0(\Lambda^2/r)^{\epsilon/2} \tag{12.73}$$

and Λ^2 always occurs in the ratio (Λ^2/r), or, if we are looking at q-dependent functions, (Λ^2/q^2). Then the critical limits $r \to 0, q \to 0$ are equivalent to $\Lambda \to \infty$.

§12.5 The renormalization group and critical phenomena

Since the perturbation expansion diverges in the limit $r \to 0$ for $d \leqslant 4$, an alternative approach is needed. This is provided by considering the alternative equivalent limit of $\Lambda \to \infty$. Now we cannot simply proceed to this limit because certain diagrams will again diverge for q large. For example, the self-energy Σ diverges for $\Lambda \to \infty$ and $d \geqslant 2$. For the ϕ^4 field theory this divergence is well known and well understood. It is found that only limited classes of diagrams diverge and all the diagrams in each class diverge in the same way. For example all self-energy diagrams diverge as Λ^{d-2}. As a result, it is possible to rescale the fields and parameters of the theory in such a way that when expressed in terms of the new parameters, all correlation functions are finite as $\Lambda \to \infty$. The new quantities themselves would diverge as $\Lambda \to \infty$, so that there is no inconsistency, but, when expressed in terms of the new parameters, physical quantities are finite. When a field theory can be made finite by rescaling a finite number of parameters it is called a renormalizable theory. The process of rescaling is referred to as renormalization.

To be specific to the case of the Hamiltonian (12.45), this Hamiltonian can be written in the form

$$\mathscr{H} = \int d^dx \left[\tfrac{1}{2}(\nabla\phi_R)^2 + \tfrac{1}{2} r\phi_R^2 + \tfrac{1}{2}\Lambda^{-2}Z^2(\nabla\phi_R)^4 + \frac{u}{4!}\phi_R^4 \right] + \mathscr{H}_c, \tag{12.74}$$

$$\mathscr{H}_c = \int d^dx \left[\tfrac{1}{2}(Zr_0 - r)\phi_R^2 + \tfrac{1}{2}(Z-1)(\nabla\phi_R)^2 + \frac{u}{4!}(Z_4 - 1)\phi_R^4 \right], \tag{12.75}$$

where

$$\phi = Z^{1/2}\phi_R, \qquad u(Z_4 - 1) = u_0 Z^2 - u. \tag{12.76}$$

The parameters u, Z, r can be chosen in such a way that the terms of $\mathscr{H}_c$ (called the counter terms) just cancel the divergent parts arising from the remainder of $\mathscr{H}$. When this is done, the renormalized correlation functions are finite functions of r and u even in the limit $\Lambda \to \infty$. The cancellation yields u_0, r_0 and Z as functions of u, r and Λ.

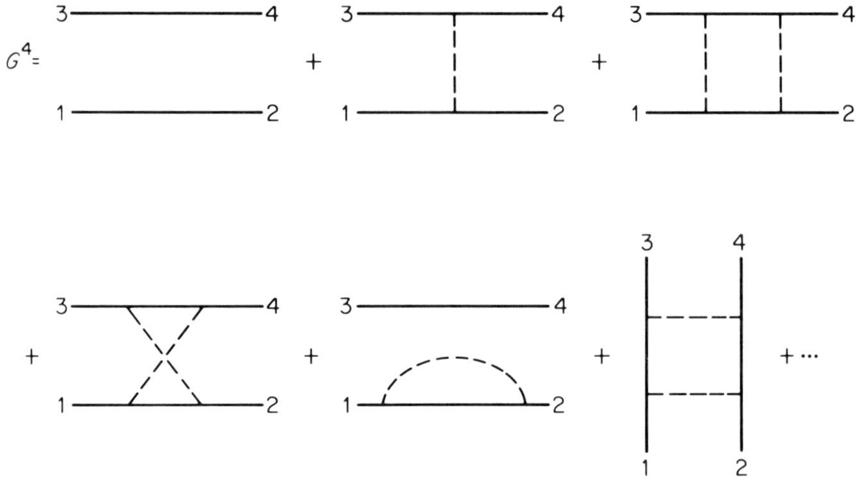

FIG. 12.2. Contributions to the fourth-order correlation function.

The possibility of renormalization has consequences for the correlation functions which, in the limit $\Lambda \to \infty$, are most simply expressed in terms of the irreducible vertex parts, defined as follows. We start from the general correlation function of order $2n$,

$$G^{2n}(x_1, x_2, \ldots, x_{2n}) = \langle \phi(x_1)\phi(x_2) \ldots \phi(x_{2n}) \rangle. \quad (12.77)$$

This can be expanded by perturbation theory according to the same rules as $G^2 \equiv G$ except that diagrams for G^{2n} have $2n$ external lines each with an end at one of the points $x_1, x_2, \ldots, x_{2n}$. Diagrams representing contributions to G^4 are illustrated in Fig. 12.2. We then retain only the connected diagrams G_c^{2n}, namely those which cannot be separated into two parts without cutting a line. This diagram represents the true correlation function of order $2n$ after lower-order correlations have been subtracted out. For the present Hamiltonian the irreducible vertex part Γ^{2n} is then obtained from G_c^{2n} as follows. We ignore those connected diagrams which we can divide into two parts by cutting a single continuous line only. For example, the diagram shown in Fig. 12.3 which contributes to G_c^6 is ignored. Then Γ^{2n} comprises the remaining contributions to G_c^{2n} without the contributions from the $2n$ external legs. For example, the contribution of the first term in Fig. 12.2 to Γ^4 is simply $\bar{u}_0/3$. As a special case

$$\Gamma^2(p) = G(p)^{-1}. \quad (12.78)$$

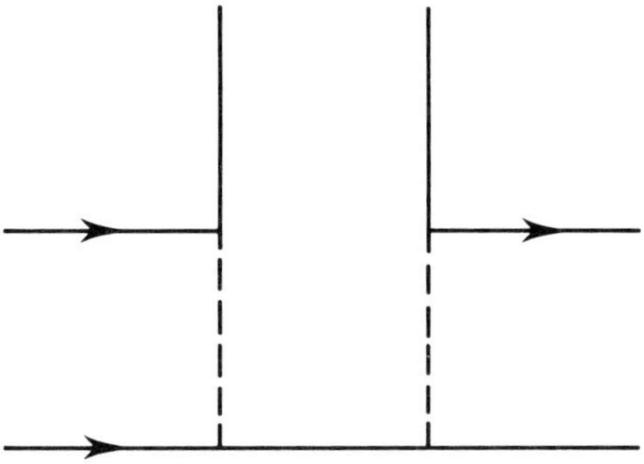

FIG. 12.3. An example of a reducible vertex function.

More formal definitions of the vertex functions can also be given (Amit 1978).

Now the renormalization group can be applied to critical phenomena in several different but equivalent ways. We describe just one of them which depends on the results (proofs of which are to be found in the references mentioned earlier) stated below.

1. Because the theory is renormalizable, the *bare* vertex functions satisfy certain linear equations in the limit of Λ very large. These equations can be written in several forms according to which variables are chosen as the independent ones. In particular, if the vertex functions are chosen to be functions of momenta p_i and of t, u_0 and Λ we find, asymptically as $\Lambda \to \infty$, that

$$\left\{\Lambda \frac{\partial}{\partial \Lambda} + \beta(u_0) \frac{\partial}{\partial u_0} - [\gamma_4(u_0) - \gamma_3(u_0)]\, t \frac{\partial}{\partial t} - \frac{N}{2} \gamma_3(u_0)\right\} \Gamma^N(p_i, t, u_0, \Lambda) = 0, \tag{12.79}$$

where the functions β, γ_i have to be determined but are independent of N. The equations (12.79) are the renormalization group equations for the bare functions.

2. The zeros u_0^* of the function $\beta(u_0)$ play a special role in the solution of equation (12.79) and are known as fixed points. If u_0^* is such that

$$\beta(u_0^*) = 0, \qquad d\beta/du_0^*(u_0^*) > 0, \tag{12.80}$$

then u_0^* is called an infrared stable fixed point. It has the property that, if u_0 is in the neighbourhood of u_0^*, then as $\Lambda \to \infty$, the solution of equation (12.79) tends to the solution of the same equation with $u_0 \equiv u_0^*$. In this asymptotic limit then, all theories with u_0 in the neighbourhood of u_0^* behave in the same way. This is the property of universality.

3. The limiting behaviour of the correlation function is, therefore, determined by the equations

$$\left[\Lambda\frac{\partial}{\partial\Lambda}-(\gamma_4^*-\gamma_3^*)t\frac{\partial}{\partial t}-\frac{N}{2}\gamma_3^*\right]\Gamma^N(p_i, t, u_0^*, \Lambda) = 0, \tag{12.81}$$

where

$$\gamma_i^* = \gamma_i(u_0^*). \tag{12.82}$$

We shall see that equation (12.81) leads to the scaling laws which depend on the two parameters γ_3^*, γ_4^*. In this way, one is led to a scaling theory with just two independent parameters.

4. Equation (12.81) can be transformed into a linear homogenous equation with constant coefficients by choosing ln t and ln Λ as the independent variables. Then one finds that

$$\Gamma^N(p_i, t, u_0^*, \Lambda) = \Lambda^{N\gamma_3^*/2}F^N[t/\Lambda^{\gamma_3^*-\gamma_4^*}, p_i] \tag{12.83}$$

where $F^N(x, p_i)$ is an arbitrary function of x (and p_i). However, dimensionally, we find that

$$[\Gamma^N] = [G_C^N]/[G^{(2)}]^N = [G^{(2)}p^{-d}]^{N/2}p^d[G^{(2)}]^{-N} = [\Lambda]^{-N(d/2-1)+d}, \tag{12.84}$$

and we also have, dimensionally,

$$[r] = [t] = [p^2] = [\Lambda^2]. \tag{12.85}$$

From (12.84) and (12.85) we must have

$$\Gamma^N(p_i, t, \Lambda) = \Lambda^{-N(d/2-1)+d}g^N(p_i/\Lambda, t/\Lambda^2). \tag{12.86}$$

For equation (12.83) to be consistent with this, the form of $F(x, p_i)$ is restricted.

In the limit $p_i = 0$, we must have

$$\Lambda^{N\gamma_3^*/2}F^N[t/\Lambda^{\gamma_3^*-\gamma_4^*}, 0] = \Lambda^{-N(d/2-1)+d}g^N(0, t/\Lambda^2). \tag{12.87}$$

This is possible if, and only if,

$$F^N(x, 0) = Ax^a, \qquad g^N(0, y) = A'y^b, \tag{12.88}$$

where equating powers of Λ and t in turn,

$$N\gamma_3^*/2 - a(\gamma_3^*-\gamma_4^*) = -N(d/2-1)+d-2b$$

and

$$a = b;$$

whence

$$\Gamma^N \sim A\Lambda^{-N(d/2-1)+d}(t/\Lambda^2)^b \tag{12.89}$$

with

$$b = \frac{N\gamma_3^*/2 + N(d/2-1) - d}{\gamma_3^* - \gamma_4^* - 2}. \tag{12.90}$$

This gives the limiting temperature dependence of each correlation function near T_c in the long-wavelength limit.

Similarly in the limit $t = 0$ we must have

$$\Lambda^{\gamma_3^*} F^2(0, p) = \Lambda^{-2(d/2-1)+d} g^2(p/\Lambda), \tag{12.91}$$

where we also use the fact that the second-order correlation function depends only on $|p|$. Again this is possible if and only if F and g have the forms given in (12.88). Now the equation which determine a and b are

$$\gamma_3^* = 2 - b, \qquad a = b.$$

Hence,

$$\Gamma^2(p, 0, u^*, \Lambda) \sim A\Lambda^2(p/\Lambda)^{2-\gamma_3^*}. \tag{12.92}$$

Thus γ_3^* is the conventional exponent η.

In a similar way other exponents are related to γ_3^*, γ_4^*.

5. This analysis provides a general basis for the results of scaling theory. To find the numerical values of all the exponents however, we need to determine the fixed point u_0^* and γ_3^*, γ_4^*.

At the present time the main field-theoretic method for finding these quantities is through an expansion in terms of $\epsilon(= 4 - d)$, known as the ϵ-expansion. It can be shown (see below) that as ϵ becomes small, u_0^* becomes small, and the vertex functions are required only for small values of their arguments. As the object is to calculate only three parameters, u_0^*, γ_3^* and γ_4^*, it is necessary to determine only three vertex functions which are often chosen to be $\Gamma^2(0, t, u_0^*, \Lambda)$, $\Gamma^4(0, t, u_0^*, \Lambda)$ and $\Gamma^2(p, 0, u_0^*, \Lambda)$. These functions are expanded in powers of u_0^* and ϵ. It is found that in order to determine the exponents to order ϵ^s it is necessary to include in the double expansion of Γ^2 and Γ^4/u_0^* all terms in $u_0^{*m}\,\epsilon^n$ with $m + n \leqslant s$.

In order to evaluate these quantities in $(4-\epsilon)$ dimensions we have to give a meaning to integrals over momentum in a non-integral number of dimensions. This can always be done by using mathematical devices in such a way that the momentum integrals are all of the form

$$I = (2\pi)^{-d} \int d\Omega_k \int_0^\infty k^{d-1}\,dk f(k), \tag{12.93}$$

where f depends on the magnitude of momentum only. The integral over solid angle is a number S_d which always occurs in combination with u_0 and $(2\pi)^{-d}$ and which therefore disappears from the final result. It is convenient to remove it from intermediate expressions by defining a new coupling constant u by

$$u = \tfrac{1}{2}u_0 S_d (2\pi)^{-d}. \tag{12.94}$$

The integral (12.93) is now reduced to a conventional one. However, it is often convenient to expand the integrand in powers of ϵ by using the Taylor series

$$\begin{aligned} k^{d-1} &= k^{3-\epsilon} = k^3 \exp(-\epsilon \ln k) \\ &= k^3[1 - \epsilon \ln k + \dots]. \end{aligned} \tag{12.95}$$

Since only the asymptotic forms of the vertex functions are required, terms which are of the order, t/Λ^2, r/Λ^2 or $|p|/\Lambda$, or smaller, relative to the leading terms can be neglected.

Since we shall compare the scaling relationships with the vertex functions and the latter are to be expanded in powers of ϵ, it is convenient to expand equations (12.89) and (12.92) appropriately. At the same time we shall express them in terms of the conventional coupling constants and in terms of r rather than t, since the vertex functions are calculated in terms of r.

From equation (12.92) we find that

$$\Gamma^2(p, 0, u^*, \Lambda) \propto \Lambda^2(p/\Lambda)^{2-\eta} \tag{12.96}$$

with

$$\eta = \gamma_3^*. \tag{12.97}$$

From equations (12.89) and (12.90) we have

$$\Gamma^2(0, t, u^*, \Lambda) = r \propto \Lambda^2(t/\Lambda^2)^{b_2} \tag{12.98}$$

with

$$b_2 = \frac{\gamma_3^* - 2}{\gamma_3^* - \gamma_4^* - 2}, \tag{12.99}$$

and

$$\Gamma^4(0, t, u^*, \Lambda) = \Lambda^\epsilon(t/\Lambda^2)^{b_4} \tag{12.100}$$

with

$$b_4 = \frac{2\gamma_3^* - \epsilon}{\gamma_3^* - \gamma_4^* - 2}. \tag{12.101}$$

However, according to Table 2,

$$r = \chi_T^{-1} \propto \Lambda^2(t/\Lambda^2)^\gamma. \tag{12.102}$$

Hence

$$b_2 = \gamma, \qquad b_4 = \gamma[(2\eta - \epsilon)/(\eta - 2)]$$

and

$$\Gamma^4(0, r, u^*, \Lambda) \propto \Lambda^\epsilon(r/\Lambda^2)^{(2\eta-\epsilon)/(\eta-2)}. \tag{12.103}$$

Equations (12.98), (12.102) and (12.103) express the three vertex functions in terms of the conventional exponents γ and η.

Now as $\epsilon \to 0$ and $r \to 0$, we find that $u^* \to 0, \gamma \to 1$ and $\eta \to 0$. We can therefore expand the vertex functions according to

$$\Gamma^2(p, 0, u^*, \Lambda) = p^2 \exp[-\eta \ln p/\Lambda] = p^2[1 - \eta \ln p/\Lambda + \ldots], \tag{12.104}$$

$$t/\Lambda^2 \propto (r/\Lambda^2)^{1/\gamma} = (r/\Lambda^2)[1 + (1/\gamma) - 1) \ln r/\Lambda^2 + \ldots] \tag{12.105}$$

and

$$\Gamma^4[0, r, u^*, \Lambda] \propto \Lambda^\epsilon\left[1 + \frac{2\eta - \epsilon}{\eta - 2} \ln(r/\Lambda^2) + \ldots\right]. \tag{12.106}$$

A comparison of these expansions with those obtained from perturbation theory determines u^*, γ and η.

As an example of how this programme is executed we consider the contributions to the exponents which are of first order in ϵ. This requires an expansion of Γ^2 to first order in u^* and ϵ and of Γ^4 to second order.

To first order, the self-energy is given by the diagrams of Fig. 12.1. Hence

$$\Gamma^2(p, r, u_0^*, \Lambda) = r + p^2, \tag{12.107}$$

with r related to t by equation (12.69). Evaluation of the integral according to the procedure outlined above yields

$$t = r\{1 - \tfrac{1}{6}u^*(n + 2) \ln(r/\Lambda^2) + \ldots\}. \tag{12.108}$$

Also

$$\Gamma^2(p, 0, u^*, \Lambda) = p^2. \tag{12.109}$$

Comparing equations (12.12), (12.11) with (12.07) and (12.105) we see that, to first order,

$$\eta = 0, \tag{12.110}$$

$$(1/\gamma) - 1 = -\tfrac{1}{6}u^*(n + 2). \tag{12.111}$$

The third relation is obtained from Γ^4. To second order in u^*, this is given by the diagrams of Fig. 12.4. These yield the expression

$$\Gamma^4_{ijkl}(0, t, u^*, \Lambda) = \tfrac{1}{3}(\delta_{ij}\delta_{kl} + \delta_{ik}\delta_{jl} + \delta_{il}\delta_{jk})$$
$$\times u^*\Lambda^\epsilon[2(2\pi)^d/S_d]\{1 + \tfrac{1}{6}(n + 8)u^* \ln(r/\Lambda^2) + \ldots\}. \tag{12.112}$$

Comparison with equation (12.101) shows that

$$\frac{2\eta - \epsilon}{\eta - 2} = \frac{n + 8}{6} u^*. \tag{12.113}$$

From equations (12.110), (12.111) and (12.113) we see that to first order in ϵ

$$u^* = 3/(n + 8), \tag{12.114}$$

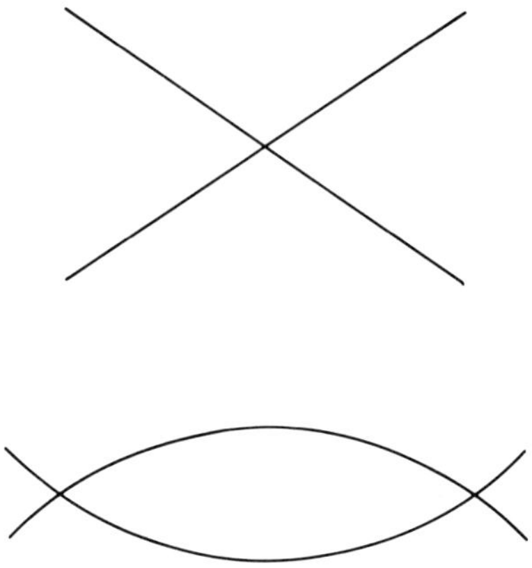

FIG. 12.4. First- and second-order contributions to Γ^4.

$$\eta = 0, \tag{12.115}$$

$$\gamma = 1 + \frac{(n+2)\epsilon}{2(n+8)}. \tag{12.116}$$

This completes the calculation to first order in ϵ.

The perturbation expansion has been carried out to the third order in ϵ for γ and to fourth order for η. The results are

$$\begin{aligned}\gamma = 1 &+ \frac{n+2}{2(n+8)}\epsilon + \frac{(n+2)}{4(n+8)^3}(n^2 + 22n + 52)\epsilon^2 \\ &+ \frac{(n+2)}{8(n+8)^5}\{n^4 + 44n^3 + 664n^2 + 2496n + 3104 \\ &- 48(5n+22)(n+8)\zeta(3)\}\epsilon^3 + O(\epsilon^4)\end{aligned} \tag{12.117}$$

and

$$\begin{aligned}\eta = \frac{(n+2)}{2(n+8)^2}\epsilon^2 \Bigg\{ 1 &+ \frac{(-n^2 + 56n + 272)}{4(n+8)^2}\epsilon + \frac{\epsilon^2}{16(n+8)^4} \\ &\times [-5n^4 - 230n^3 + 1124n^2 + 17920n \\ &+ 46144 - 384(5n+22)(n+8)\zeta(3)] \Bigg\} + O(\epsilon^5)\end{aligned} \tag{12.118}$$

where $\zeta(x)$ is the Riemann zeta function and

$$\zeta(3) \simeq 1.202.$$

The results for γ and η calculated from these expansions to second and third order are compared with the results for high-temperature

TABLE 12.4. Values of $\gamma(i)$ calculated to order ϵ^i from equation (12.117) compared with results from high-temperature series expansions for $\epsilon = 1$.

η	$\gamma(1)$	$\gamma(2)$	$\gamma(3)$	Series
1	1.167	1.244	1.195	1.250 ± 0.001
2	1.2	1.3	1.263	1.318 ± 0.01
3	1.227	1.346	1.324	1.405 ± 0.02

TABLE 12.5. Values of $\eta(i)$ calculated to order ϵ^i from equation (12.118) compared with results from high-temperature series expansions for $\epsilon = 1$.

η	$\eta(2)$	$\eta(3)$	$\eta(4)$	Series
1	0.019	0.038	0.029	0.047 ± 0.01
2	0.020	0.039	0.031	0.04 ± 0.01
3	0.021	0.039	0.031	0.040 ± 0.008

series expansions in Tables 12.4 and 12.5 for $n = 1, 2, 3$ and $d = 3$. For a value of ϵ as large as unity, agreement between $\gamma(2)$ and $\eta(3)$ and the numerical results from high-temperature expansions is surprisingly good. However, the agreement is rather poorer for $\gamma(3)$ and $\eta(4)$. This suggests that the expansion in terms of ϵ is at best an asymptotic expansion. Attempts have been made to re-sum the series to obtain improved approximations for the exponents. This is a matter of current research which goes beyond the scope of this book.

References

General

Amit, D. J. (1978). "Field Theory, the Renormalization Group, and Critical Phenomena". McGraw-Hill, New York.

Domb, C. and Green, M. S. (Eds.) (1976). "Phase Transitions and Critical Phenomena", Vol. 6. Academic Press, New York and London.

Ma, S. K. (1976). "Modern Theory of Critical Phenomena". Benjamin, Reading, Mass.

Pfeuty, P. and Toulouse, G. (1975). "Introduction au Groupe de Renormalisation et à ses applications". Press Universitaires, Grenoble. (Translation by G. Barton (1977). "Introduction to the Renormalization Group and Critical Phenomena". Wiley, New York.)

Stanley, H. E. (1971). "Introduction to Phase Transitions and Critical Phenomena". Clarendon Press, Oxford.

Wilson, K. G. and Kogut, J. (1974). *Physics Reports* **12C**, No. 2.

Special

Anderson, P. W. and Brinkman, W. F. (1975). "Helium Liquids" (Ed. J. G. M. Armitage and I. E. Farquhar), Chapter 8. Academic Press, New York and London.
Fisher, M. E. (1974). *Rev. Mod. Phys.* **46**, 587.
Kac, M., Uhlenbeck, G. E. and Hemmer, P. C. (1963). *J. Math. Phys.* **4**, 216.
Kadanoff, L. P., Götze, W., Hamblen, D., Hecht, R., Lewis, E. A. S., Palciauskas, V. V., Rayl, M., Swift, J., Aspnes, D. and Kane, J. (1967). *Rev. Mod. Phys.* **39**, 395.
Landau, L. D. (1965). "Collected Papers of L. D. Landau" (Ed. D. ter Haar). Gordon and Breach, New York.
Langer, J. S. (1964). *Phys. Rev.* **134**, 553.
Ma, S. K. (1974). *Rev. Mod. Phys.* **45**, 589.
Wilson, K. G. (1969). *Phys. Rev.* **179**, 1499.
Wilson, K. G. (1971a, b). *Phys. Rev.* **B4**, (a) 3174, (b) 3184.
Wilson, K. G. (1975). *Rev. Mod. Phys.* **47**, 773.

Problems

1. By expanding $\langle\phi_i(\boldsymbol{r})\rangle$, as given by equation (12.32), to first order in $h_i(\boldsymbol{r})$, show that the space-dependent susceptibility, $\chi_{ij}(\boldsymbol{r})$, is given by equation (12.34) where the averages are to be taken with $h_i(\boldsymbol{r})$ taken to be zero.
2. Use the method of steepest steepest descents to evaluate the free energy for a three-component order parameter. Hence show that equations (12.22) and (12.23) are satisfied.
3. Show that to second order in the interaction there is a single diagram which contributes to the self-energy and that its contribution is
$$\Sigma_2(k) = \left(\frac{\bar{u}_0}{3}\right)^2\left(\frac{n}{2}+1\right)\frac{1}{\mathscr{V}^2}\sum_{q,k'} G(k-q)G(k')G(k'+q).$$
4. Evaluate the integral in equation (12.69) in the limit of $d = 4$ and r/Λ very small to obtain equation (12.108).

Appendix A

Summary of the Results of Second Quantization

A collection of Bose or Fermi particles is described by a field operator $\psi(\boldsymbol{r})$, or $\psi_\sigma(\boldsymbol{r})$ if the particles possess spin, and its Hermitian conjugate $\psi^+(\boldsymbol{r})$ or $\psi_\sigma^+(\boldsymbol{r})$. These satisfy the commutation rules

$$\begin{aligned} \psi(\boldsymbol{r})\psi^+(\boldsymbol{r}') + \epsilon\psi^+(\boldsymbol{r}')\psi(\boldsymbol{r}) &= \delta(\boldsymbol{r}-\boldsymbol{r}'), \\ \psi(\boldsymbol{r})\psi(\boldsymbol{r}') + \epsilon\psi(\boldsymbol{r}')\psi(\boldsymbol{r}) &= 0, \end{aligned} \tag{A.1}$$

or

$$\begin{aligned} \psi_\sigma(\boldsymbol{r})\psi_{\sigma'}^+(\boldsymbol{r}') + \epsilon\psi_{\sigma'}^+(\boldsymbol{r}')\psi_\sigma(\boldsymbol{r}) &= \delta_{\sigma\sigma'}\delta(\boldsymbol{r}-\boldsymbol{r}'), \\ \psi_\sigma(\boldsymbol{r})\psi_{\sigma'}(\boldsymbol{r}') + \epsilon\psi_{\sigma'}(\boldsymbol{r}')\psi_\sigma(\boldsymbol{r}) &= 0, \end{aligned} \tag{A.2}$$

where

$$\epsilon = \begin{cases} +1, & \text{for fermions,} \\ -1, & \text{for bosons.} \end{cases}$$

In the case of spin-$\frac{1}{2}$ particles, such as electrons, it is convenient to write the two fields as components of a matrix. Thus, for such particles,

$$\psi(\boldsymbol{r}) = \begin{pmatrix} \psi_\uparrow(\boldsymbol{r}) \\ \psi_\downarrow(\boldsymbol{r}) \end{pmatrix}, \qquad \psi^+(\boldsymbol{r}) = (\psi_\uparrow^+(\boldsymbol{r}) \quad \psi_\downarrow^+(\boldsymbol{r})) \tag{A.3}$$

The field operators can be expanded in terms of a complete set of normalized classical single-particle wave functions $\phi_\alpha(\boldsymbol{r})$. Thus

$$\psi(\boldsymbol{r}) = \sum_{\alpha} c_\alpha \phi_\alpha(\boldsymbol{r}), \qquad \psi^+(\boldsymbol{r}) = \sum_{\alpha} c_\alpha^+ \phi_\alpha^+(\boldsymbol{r}), \tag{A.4}$$

where, in the case of spin, ϕ_α is a classical two-component matrix. The operators c_α^+ are creation operators for particles in the single-particle state $\phi_\alpha(\boldsymbol{r})$ while c_α are destruction operators for the same particles. The commutation relations of these operators follow from equations (A.1) and (A.2) and are

$$c_\alpha c_{\alpha'}^+ + \epsilon c_{\alpha'}^+ c_\alpha = \delta_{\alpha\alpha'}, \qquad c_\alpha c_{\alpha'} + \epsilon c_{\alpha'} c_\alpha = 0. \tag{A.5}$$

A complete set of orthonormal basis states for the many-particle system comprises the following:

$\|0\rangle$,	the vacuum state, which satisfies
$c_\alpha \|0\rangle = 0$,	for all α
$\|\alpha\rangle \equiv c_\alpha^+ \|0\rangle$,	all α, the state containing one particle in the single-particle state ϕ_α
$\|\alpha, \alpha'\rangle \equiv c_\alpha^+ c_{\alpha'}^+ \|0\rangle$,	the state containing two particles, one in ϕ_α and one in $\phi_{\alpha'}$,
$\|\alpha, \alpha', \alpha''\rangle = c_\alpha^+ c_{\alpha'}^+ c_{\alpha''}^+ \|0\rangle$,	the state containing three particles, one in each of the states with quantum numbers α, α', α''

and, in general, the state containing r particles in α, r' particles in α', r'' particles in α'' . . . , which is

$$[r!r'!r''! \ldots]^{-1/2}(c_\alpha^+)^r (c_{\alpha'}^+)^{r'} (c_{\alpha''}^+)^{r''} \ldots |0\rangle \equiv |r_\alpha, r'_{\alpha'}, r''_{\alpha''}, \ldots\rangle.$$

In the case of fermions,

$$(c_\alpha^+)^2 = 0, \tag{A.7}$$

and, at most, only one particle can be put in each single-particle state. From the commutation relations (A.5) and equation (A.6) it follows that

$$c_\alpha (c_\alpha^+)^r \ldots |0\rangle = r(c_\alpha^+)^{r-1} \ldots |0\rangle. \tag{A.8}$$

Hence, c_α operating on a basis state reduces the number of particles in the single-particle state α by 1. If there are no particles in α in the original state $|a\rangle$ then

$$c_\alpha |a\rangle = 0$$

and the state is destroyed. (For bosons c_α behaves formally like d/dc_α^+). Thus c_α^+ always creates an extra particle in the state α, if this is allowed, while c_α destroys one. Note, however, that for fermions,

$$c_\alpha^+|\alpha\rangle = c_\alpha^+ c_\alpha^+|0\rangle = (c_\alpha^+)^2|0\rangle = 0.$$

Thus c_α^+ destroys a fermion state which already contains a particle with quantum number α.

Any state containing exactly r particles with quantum number α is an eigenstate of $c_\alpha^+ c_\alpha$ with eigenvalue r. Thus $c_\alpha^+ c_\alpha$ is an operator which counts the number of particles with quantum number α in any state. This operator is, therefore, called a number operator. Number operators with different labels α commute. Since the states ϕ_α form a complete set, the operator which counts the total number of particles present is

$$N_\alpha = \sum_\alpha c_\alpha^+ c_\alpha = \int d^3r \psi^+(\mathbf{r})\psi(\mathbf{r}). \tag{A.9}$$

There is a one-to-one correspondence between operators in wave mechanics (or the coordinate representation), and operators in the present representation, the occupation number representation, such that the matrix elements of the conventional operators and the basic wave functions is the same as that of the corresponding operators and corresponding states given above. In wave mechanics, single-particle operators always appear in the form

$$\sum_i f(\mathbf{r}_i, \nabla_i),$$

where f can depend on both the position and momentum of the ith particle. The corresponding operator in the new representation is

$$\int d^3r \psi^+(\mathbf{r}) f(\mathbf{r}, \nabla)\psi(\mathbf{r}).$$

If $\Psi_{n_{\alpha_1} n_{\alpha_2}} \ldots (\mathbf{r}_1 \ldots \mathbf{r}_N)$ is the appropriately symmetrized and normalized wave function for N identical particles with n_{α_1} in the state α_1, n_{α_2} in the state α_2, etc., then

$$\int d^3r_1 \ldots d^3r_N \Psi^*_{n'_{\alpha_1} n'_{\alpha_2}} \ldots (\mathbf{r}_1 \ldots \mathbf{r}_N) \sum_{i=1}^{N} f(\mathbf{r}_i, \nabla_i) \Psi_{n_{\alpha_1} n_{\alpha_2}} \ldots (\mathbf{r}_1 \ldots \mathbf{r}_N)$$

$$= \left\langle n'_{\alpha_1}, n'_{\alpha_2} \ldots \middle| \int d^3r \psi^+(\mathbf{r}) f(\mathbf{r}, \nabla)\psi(\mathbf{r}) \middle| n_{\alpha_1}, n_{\alpha_2} \ldots \right\rangle. \tag{A.10}$$

This correspondence ensures that results obtained from the different representations are identical. We show in Table A.1 the correspondence of specific single-particle operators in the coordinate representation and the occupation number representation.

TABLE A.1. Correspondence of single-particle operators in the coordinate and occupation number representations.

Physical observable	Coordinate representation	Occupation number representation	Momentum basis§
Particle density at $\boldsymbol{r}$	$\sum_i \delta(\boldsymbol{r}_i - \boldsymbol{r})$	$\psi^+(\boldsymbol{r})\psi(\boldsymbol{r})$	$\sum_{\boldsymbol{k},\boldsymbol{q},\sigma} c^+_{\boldsymbol{k}\sigma} c_{\boldsymbol{k}+\boldsymbol{q}\sigma} \exp(i\boldsymbol{q}.\boldsymbol{r})$
Total number of particles	$\sum_i 1 = N$	$\int d^3r\,\psi^+(\boldsymbol{r})\psi(\boldsymbol{r})$	$\sum_{\boldsymbol{k},\sigma} c^+_{\boldsymbol{k}\sigma} c_{\boldsymbol{k}\sigma}$
Charge density at $\boldsymbol{r}$	$e\sum_i \delta(\boldsymbol{r}_i - \boldsymbol{r})$	$e\psi^+(\boldsymbol{r})\psi(\boldsymbol{r})$	$e\sum_{\boldsymbol{k},\boldsymbol{q},\sigma} c^+_{\boldsymbol{k}\sigma} c_{\boldsymbol{k}+\boldsymbol{q}\sigma} \exp(i\boldsymbol{q}.\boldsymbol{r})$
Current density at $\boldsymbol{r}$	$\frac{e}{2m}\sum_i [\boldsymbol{p}_i\delta(\boldsymbol{r}_i - \boldsymbol{r}) + \delta(\boldsymbol{r}_i - \boldsymbol{r})\boldsymbol{p}_i]$	$\frac{-ie}{2m}[\psi^+(\boldsymbol{r})\nabla\psi(\boldsymbol{r}) - \{\nabla\psi^+(\boldsymbol{r})\}\psi(\boldsymbol{r})]$	$\frac{e}{2m}\sum_{\boldsymbol{k},\boldsymbol{q},\sigma}(2\boldsymbol{k}+\boldsymbol{q}) c^+_{\boldsymbol{k}\sigma} c_{\boldsymbol{k}+\boldsymbol{q}\sigma} \exp(i\boldsymbol{q}.\boldsymbol{r})$
Kinetic energy	$\sum_i p_i^2/2m \equiv -\sum_i \nabla_i^2/2m$	$-\frac{1}{2m}\int d^3r\,\psi^+(\boldsymbol{r})\nabla^2\psi(\boldsymbol{r})$	$\sum_{\boldsymbol{k},\sigma} \frac{k^2}{2m} c^+_{\boldsymbol{k}\sigma} c_{\boldsymbol{k}\sigma}$
Potential energy in an external potential $V(\boldsymbol{r})$	$\sum_i V(\boldsymbol{r}_i)$	$\int d^3r\,\psi^+(\boldsymbol{r})V(\boldsymbol{r})\psi(\boldsymbol{r})$	$\mathscr{V}^{-1}\sum_{\boldsymbol{k},\boldsymbol{q},\sigma} c^+_{\boldsymbol{k}\sigma} c_{\boldsymbol{k}+\boldsymbol{q},r} \int d^3r\,V(\boldsymbol{r})\exp(i\boldsymbol{q}.\boldsymbol{r})$
Magnetic moment density at $\boldsymbol{r}$†	$(g/2)\sum_i \boldsymbol{\sigma}_i\delta(\boldsymbol{r}_i - \boldsymbol{r})$	$(g/2)\psi^+(\boldsymbol{r})\boldsymbol{\sigma}\psi(\boldsymbol{r})$	$(g/2)\sum_{\boldsymbol{k},\boldsymbol{q},\sigma} c^+_{\boldsymbol{k}\sigma} c_{\boldsymbol{k}+\boldsymbol{q}\sigma'} \exp(i\boldsymbol{q}.\boldsymbol{r}) u^+_\sigma \boldsymbol{\sigma} u_{\sigma'}$
Total magnetic moment†	$(g/2)\sum_i \boldsymbol{\sigma}_i$	$(g/2)\int d^3r\,\psi^+(\boldsymbol{r})\boldsymbol{\sigma}\psi(\boldsymbol{r})$	$(g/2)\sum_{\boldsymbol{k},\sigma,\sigma'} c^+_{\boldsymbol{k}\sigma} c_{\boldsymbol{k}\sigma'} u^+_\sigma \boldsymbol{\sigma} u_{\sigma'}$

† g is the gyromagnetic ratio of the particles and $\sigma_x, \sigma_y, \sigma_z$ are the 2×2 Pauli matrices.

§ The expressions are written for particles possessing spin. For particles which do not, the spin label should be ignored.

One can deal similarly with two-particle operators. There is, in fact, only one of importance in this book, namely the interaction energy of the particles. In the coordinate representation this is

$$\tfrac{1}{2}\sum_{i,j} V(r_i - r_j) \qquad \text{(A.11)}$$

and corresponds to

$$\tfrac{1}{2}\int d^3r\, d^3r'\, \psi^+(r)\psi(r)V(r-r')\psi^+(r')\psi(r'). \qquad \text{(A.12)}$$

This has a simple interpretation because, as can be seen from Table A.1, the last form can be written

$$\tfrac{1}{2}\int d^3r\, d^3r'\rho(r)V(r-r')\rho(r'). \qquad \text{(A.13)}$$

Thus we include the interaction between the charge density at r and r', integrating over all values of r and r'. Since the integration includes each pair of points twice we require the additional factor of $\frac{1}{2}$.

The forms (A.11) and (A.12) include the interaction energy of each particle with itself. Since this is a constant, it is often ignored. Then the interaction energy can be written

$$\tfrac{1}{2}\sum_{i\neq j} V(r_i - r_j) \qquad \text{(A.14)}$$

and this corresponds to

$$\tfrac{1}{2}\int d^3r\, d^3r'\, \psi^+(r)\psi^+(r')V(r-r')\psi(r')\psi(r). \qquad \text{(A.15)}$$

If the particles carry spin, the labels r, r' can conventionally be regarded as including spin indices. Then the integrals over r and r' imply sums over the spin indices as well. Explicitly, in this case, expression (A.15) stands for

$$\tfrac{1}{2}\sum_{\sigma\sigma'}\int d^3r\, d^3r'\, \psi^+_\sigma(r)\psi^+_{\sigma'}(r')V(r-r')\psi_{\sigma'}(r')\psi_\sigma(r). \qquad \text{(A.16)}$$

Many of the systems dealt with in this book are translationally invariant. In this case it is convenient to use as the basis single-particle wave functions

$$\phi_k(r) = \mathscr{V}^{-1/2}\exp(ik\cdot r) \qquad \text{(A.17)}$$

if the particles do not possess spin, and

$$\phi_{k\uparrow}(\boldsymbol{r}) = \phi_k(\boldsymbol{r})\begin{pmatrix}1\\0\end{pmatrix} = \phi_k(\boldsymbol{r})u_\uparrow, \qquad \phi_{k\downarrow}(\boldsymbol{r}) = \phi_k(\boldsymbol{r})\begin{pmatrix}0\\1\end{pmatrix} = \phi_k(\boldsymbol{r})u_\downarrow, \tag{A.18}$$

if they do. Then

$$\psi(\boldsymbol{r}) = \mathscr{V}^{-1/2}\sum_{k} c_k \exp(i\boldsymbol{k}\cdot\boldsymbol{r}) \tag{A.19}$$

or

$$\psi(\boldsymbol{r}) = \mathscr{V}^{-1/2}\sum_{k\sigma} c_{k\sigma} \exp(i\boldsymbol{k}\cdot\boldsymbol{r})u_\sigma$$

$$= \mathscr{V}^{-1/2}\sum_{k} \exp(i\boldsymbol{k}\cdot\boldsymbol{r})(c_{k\uparrow}u_\uparrow + c_{k\downarrow}u_\downarrow). \tag{A.20}$$

The corresponding forms of the physical single-particle operators are given in Table A.1. The two-particle interaction is then

$$\tfrac{1}{2}\sum_{\substack{k,k',q\\ \sigma,\sigma'}} c^+_{k\sigma}c^+_{k'+q\sigma'}\bar{V}(\boldsymbol{q})c_{k'\sigma'}c_{k+q\sigma}, \tag{A.21}$$

where

$$\bar{V}(\boldsymbol{q}) = \mathscr{V}^{-1}\int d^3r\, V(\boldsymbol{r}) \exp(i\boldsymbol{q}\cdot\boldsymbol{r}). \tag{A.22}$$

If the particles do not possess spin, the indices σ, σ' are simply dropped from the expression (A.21).

Although the field operators ψ, ψ^+ are introduced formally they have a clear physical interpretation. Consider the state

$$\psi^+(\boldsymbol{r})|0\rangle.$$

Because ψ^+ is the sum of creation operators (A.4) this state contains one particle. Now let us operate with the density operator $\rho(\boldsymbol{r}')$ on the state. Then

$$\begin{aligned}\rho(\boldsymbol{r}')\psi^+(\boldsymbol{r})|0\rangle &= \psi^+(\boldsymbol{r}')\psi(\boldsymbol{r}')\psi^+(\boldsymbol{r})|0\rangle\\ &= \psi^+(\boldsymbol{r}')[\delta(\boldsymbol{r}-\boldsymbol{r}') - \epsilon\psi^+(\boldsymbol{r})\psi(\boldsymbol{r}')]\,|0\rangle\\ &= \delta(\boldsymbol{r}-\boldsymbol{r}')\psi^+(\boldsymbol{r}')|0\rangle.\end{aligned} \tag{A.23}$$

Hence, this state contains one particle and definitely has the density $\delta(\boldsymbol{r}'-\boldsymbol{r})$. Thus $\psi^+(\boldsymbol{r})$ creates a particle at the position $\boldsymbol{r}$. Similarly $\psi(\boldsymbol{r})$ destroys a particle at $\boldsymbol{r}$.

We conclude this appendix by setting out some of the properties of independent particles which are used in the text. The Hamiltonian for independent particles moving in an external potential $V(\boldsymbol{r})$ is

$$H = \int \psi^+(r) \left[-\frac{\nabla^2}{2m} + V(r) \right] \psi(r)\, d^3 r. \tag{A.24}$$

Choose basic states $\phi_\alpha(r)$ which are eigenstates of the operator $[-(\nabla^2/2m) + V]$. Thus

$$\left[-\frac{\nabla^2}{2m} + V(r) \right] \phi_\alpha(r) = \epsilon_\alpha \phi_\alpha(r). \tag{A.25}$$

Then

$$H = \sum_\alpha \epsilon_\alpha c_\alpha^+ c_\alpha . \tag{A.26}$$

The eigenstates of H are then the states of the occupation number representation and the energies are

$$E = \sum_\alpha \epsilon_\alpha n_\alpha , \tag{A.27}$$

where each n_α can be any positive integer or zero for bosons or is 0 or 1 for fermions.

The Heisenberg equation of motion for a destruction operator is

$$\dot{c}_\alpha(t) = i[H, c_\alpha] = -i\epsilon_\alpha c_\alpha(t). \tag{A.28}$$

Hence,

$$c_\alpha(t) = c_\alpha(0) \exp(-i\epsilon_\alpha t) = \exp(-i\epsilon_\alpha t) c_\alpha , \tag{A.29}$$

and, similarly,

$$c_\alpha^+(t) = \exp(i\epsilon_\alpha t) c_\alpha^+ . \tag{A.30}$$

The sum-over-states in the grand canonical ensemble is

$$\begin{aligned} Z &= \operatorname{Tr} \exp[-\beta(H - \mu N)] \\ &= \operatorname{Tr} \exp\left[-\beta \sum_\alpha (\epsilon_\alpha - \mu) c_\alpha^+ c_\alpha^+ \right] \\ &= \sum_{n_{\alpha_1}, n_{\alpha_2}} \exp\left[-\beta \sum_\alpha (\epsilon_\alpha - \mu) n_\alpha \right] \\ &= \prod_\alpha \sum_{n_\alpha} \exp[-\beta(\epsilon_\alpha - \mu) n_\alpha] \end{aligned}$$

$$= \begin{cases} \prod_\alpha \{1 + \exp[-\beta(\epsilon_\alpha - \mu)]\}, & \text{for fermions,} \\ \prod_\alpha \{1 - \exp[-\beta(\epsilon_\alpha - \mu)]\}^{-1}, & \text{for bosons.} \end{cases} \tag{A.31}$$

These are of course the usual results and lead to the usual thermodynamic properties of independent particles.

The density matrix is

$$\rho_0 = Z^{-1} \exp[-\beta(H-\mu N)]. \tag{A.32}$$

Hence a corresponding one-particle correlation function

$$\langle c_\alpha(t) c_\gamma^+(t') \rangle = Z^{-1} \operatorname{Tr}\{\exp[-\beta(H-\mu N)]\, c_\alpha c_\gamma^+\} \exp(i\epsilon_\gamma t' - i\epsilon_\alpha t).$$

Since the basic states are diagonal in the numbers of particles and c_γ^+ creates a particle in state "γ" while c_α destroys one in α, the correlation function is zero unless $\gamma = \alpha$. Further, since the degrees of freedom belonging to different values of α are independent, the correlation function is

$$\langle c_\alpha(t) c_\gamma^+(t') \rangle = \delta_{\alpha,\gamma} \exp[i\epsilon_\alpha(t'-t)] \frac{\sum_{n_\alpha} \exp[-\beta(\epsilon_\alpha-\mu)n_\alpha](1-n_\alpha)}{\sum_{n_\alpha} \exp[-\beta(\epsilon_\alpha-\mu)n_\alpha]}$$

$$= \delta_{\alpha,\gamma} \exp[i\epsilon_\alpha(t'-t)](1-\bar{n}_\alpha),$$

where

$$\bar{n}_\alpha = \begin{cases} \{\exp[\beta(\epsilon_\alpha-\mu)]+1\}^{-1}, & \text{for fermions,} \\ \{\exp[\beta(\epsilon_\alpha-\mu)]-1\}^{-1}, & \text{for bosons.} \end{cases} \tag{A.33}$$

Appendix B

The Sums of Certain Series

In calculations involving temperature Green's functions one is very often faced with sums over the variable ω_ν. A number of tricks have been discovered for evaluating such sums and these are best explained in terms of examples.

Example 1

Evaluation of

$$S = \frac{1}{\beta} \sum_\nu F(i\omega_\nu), \tag{B.1}$$

where $F(\omega)$ is analytic except for simple poles in the complex plane. Take first the case that arises for Fermi particles where

$$\omega_\nu = (2\nu + 1)\pi/\beta. \tag{B.2}$$

Consider the function $g(\omega)$ of the complex variable ω given by

$$g(\omega) = F(\omega)/(\exp(\beta\omega) + 1). \tag{B.3}$$

This function is an analytic function of ω except where $F(\omega)$ has poles and where

$$\exp(\beta\omega) + 1 = 0. \tag{B.4}$$

The latter points are given by

$$\omega = i\omega_\nu,$$

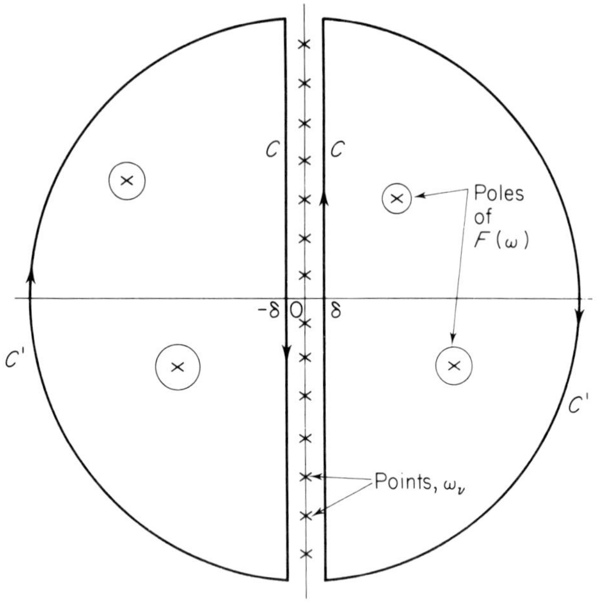

FIG. B.1. An illustration of the contours C and C' and poles of $F(\omega)$ in the complex ω-plane. These are used in evaluating the sum in equation (B.6).

and, provided that no poles of $F(\omega)$ happen to coincide with them, $g(\omega)$ has simple poles at these points with residues

$$-F(i\omega_\nu)/\beta. \tag{B.5}$$

The poles of $g(\omega)$ are illustrated in Fig. B.1. It follows that if C is a contour which encloses only the poles of $g(\omega)$ which are not poles of $F(\omega)$, then

$$\frac{i}{2\pi}\oint d\omega \frac{F(\omega)}{\exp(\beta\omega)+1} = \frac{1}{\beta}\sum_\nu F(i\omega_\nu) = S. \tag{B.6}$$

A suitable contour for C is shown in Fig. B.1; it comprises two lines at $\mathrm{Re}\,\omega = \delta$ and $\mathrm{Re}\,\omega = -\delta$. However, the integral can also be evaluated by forming closed contours in the half-planes $\mathrm{Re}\,\omega > 0$ and $\mathrm{Re}\,\omega < 0$ by adding large semi-circles to the two parts of C. These ae the contours C' and C'' illustrated in Fig. B.1. Provided that

$$\lim_{|\omega|\to\infty} |\omega g(\omega)| = 0, \tag{B.7}$$

the integrals along the semi-circles tend to zero as their radius tend to infinity, and, since we traverse C' and C'' in the clockwise direction,

$$S = \sum_a \frac{R(\omega_a)}{\exp(\beta\omega_a) + 1}, \tag{B.8}$$

where the sum is over all the poles ω_a of $F(\omega)$ and $R(\omega_a)$ is the residue of $F(\omega)$ at ω_a.

As a special case we evaluate

$$G(\tau) = \frac{1}{\beta} \sum_\nu \frac{\exp(i\omega_\nu \tau)}{i\omega_\nu - \epsilon}, \qquad \tau > 0. \tag{B.9}$$

In this case

$$g(\omega) = \frac{\exp(\omega\tau)}{(\omega - \epsilon)(\exp(\beta\omega) + 1)} \tag{B.10}$$

and satisfies the condition (A.7). $F(\omega)$ itself has a pole only at $\omega = \epsilon$ and

$$G(\tau) = \exp(\epsilon\tau)/(\exp(\beta\epsilon) + 1). \tag{B.11}$$

This is the one-particle temperature Green's function for independent particles and $\tau > 0$.

If $\tau < 0$, $g(\omega)$ defined by equation (B.10) does not satisfy condition (B.7). Indeed, as Im $\omega \to -\infty$, $|\omega g(\omega)|$ grows. However, the method can be used *mutatis mutandis* if one takes

$$g(\omega) = \frac{\exp(\omega\tau)}{(\exp(-\beta\omega) + 1)(\omega - \epsilon)}. \tag{B.12}$$

The result is

$$G(\tau) = -\frac{\exp(\epsilon\tau)}{\exp(-\beta\epsilon) + 1}, \qquad \tau < 0. \tag{B.13}$$

Thus the single-particle Green's function for independent fermions is

$$G(\tau) = \exp(\epsilon\tau)\{f(\epsilon)\theta(\tau) - [1 - f(\epsilon)]\theta(-\tau)\}. \tag{B.14}$$

Example 2

Another special case is the sum which occurs in equation (4.70) and is involved in the calculation of the conductivity of a metal, that is

$$I = \frac{1}{\beta} \sum_\lambda \{(i\omega'_\lambda - x)[i(\omega'_\lambda + \omega_\nu) - y]\}^{-1}, \tag{B.15}$$

with

$$\omega'_\lambda = (2\lambda + 1)\pi/\beta, \qquad \omega_\nu = 2\nu\pi/\beta. \tag{B.16}$$

In this case

$$F(\omega) = [(\omega - x)(\omega + i\omega_\nu - y)]^{-1}, \tag{B.17}$$

with simple poles at

$$\omega = x, y - i\omega_\nu \tag{B.18}$$

Hence,

$$I = [\exp(\beta x) + 1]^{-1} [x - y + i\omega_\nu]^{-1} + [\exp(\beta(y - i\omega_\nu)) + 1]^{-1} [y - x - i\omega_\nu]^{-1}.$$

But

$$\exp(-i\beta\omega_\nu) = \exp(-i2\nu\pi) = 1.$$

Hence,

$$I = \frac{f(x) - f(y)}{x - y + i\omega_\nu}, \tag{B.19}$$

where $f(x)$ is the Fermi–Dirac distribution function,

$$f(x) = 1/(\exp(\beta x) + 1).$$

Subject Index

T

U

V

W

Z